Free-Radical Substitution Reactions

FREE-RADICAL SUBSTITUTION REACTIONS

Bimolecular Homolytic Substitutions

(S_H2 Reactions) at Saturated Multivalent Atoms

K. U. INGOLD *Division of Chemistry, National Research Council of Canada, Ottawa*

B. P. ROBERTS *Department of Chemistry, University College, London*

Wiley-Interscience

A Division of John Wiley & Sons, Inc.

New York · London · Sydney . Toronto

Library of Congress Catalogue Card Number: 78-144332

ISBN 0-471-42770-5

Printed in the United States of America.

10 9 8 7 6 5 4 3 2 1

Preface

Bimolecular homolytic substitution (S_H2) reactions involve the attack on an atom in a molecule by an incoming radical to bring about replacement of a second radical originally bound to the atom. When the atom under attack is univalent, hence terminal, the substitution is more usually referred to as an atom abstraction, for example, hydrogen atom abstraction or chlorine atom abstraction. The majority of the chemical elements, however, are multivalent metals or metalloids, which show a diversity of behavior in S_H2 reactions that is only now beginning to be appreciated. The S_H2 process may be a component step in a reaction which is, overall, either a metathesis (e.g., $R'_3SnH + RSSR \rightarrow R'_3SnSR + RSH$), an addition (e.g., $R_3B + O_2 \rightarrow ROOBR_2$), or an elimination (e.g., $[RC(O)O]_2Hg \rightarrow RC(O)OHgR + CO_2$).

Although there have been numerous reviews dealing with atom abstractions, the somewhat scattered data concerning homolytic substitutions at multivalent atoms have been largely ignored. In this monograph we attempt to collect and correlate all the available information on this subject in the hope of stimulating both fundamental and applied research in the exciting new area of free-radical chemistry.

The material in this book is divided into chapters that correspond to the group of the Periodic Table to which the particular atom belongs. This simple arrangement obviates the need for a subject index, which would, after all, only list the compounds referred to in the text.

In a number of cases where the interest of the original investigators was centered more on the products of a reaction than on the mechanism of their formation we attempt to rationalize the experimental results in terms of the occurrence of S_H2 processes. Although this approach is not too rigorous and the conclusions may be biased, we hope it will provide the reader with a useful selection of reactions that require further mechanistic study.

We take great pleasure in expressing our gratitude to Professor Alwyn G. Davies who first raised our interest in the subject matter of this book and subsequently made many helpful suggestions during its preparation.

Ottawa, Canada
London, England
September 1970

KEITH U. INGOLD
BRIAN P. ROBERTS

Contents

Free-Radical Substitution Reactions

I Introduction

There is a large class of homolytic reactions which result in the net replacement of one radical or atom by another radical or atom.

$$R^{\cdot} + AB \rightarrow RA + B^{\cdot} \tag{1}$$

Such reactions can take place by a number of quite distinct mechanisms.

1. The reaction may involve the homolysis of AB followed by colligation of $R^{\cdot}$ and $A^{\cdot}$

$$AB \rightarrow A^{\cdot} + B^{\cdot} \tag{2}$$

$$R^{\cdot} + A^{\cdot} \rightarrow RA \tag{3}$$

Such a mechanism can be referred to[1] by the symbols S_H1 (substitution, homolytic, unimolecular) by analogy with unimolecular nucleophilic, S_N1, and electrophilic, S_E1, substitutions. That is, the process is regarded as a unimolecular substitution because the rate determining step is the unimolecular homolysis of AB. These S_H1 reactions are not the concern of the present review.

2. Reaction 1 may occur by a bimolecular process involving the homolytic attack of $R^{\cdot}$ on AB. This mechanism can be referred to by the symbols S_H2. The majority of bimolecular homolytic substitutions involve the attack of the radical $R^{\cdot}$ on a *univalent*, hence peripheral or terminal atom such as hydrogen or halogen. Homolytic substitutions at *multivalent* atoms are also known. Such reactions need not involve unreasonably high activation energies if A is coordinatively unsaturated as is the case, for example, if A forms a part of a multiply bonded system.

$$R\cdot + \underset{\displaystyle C}{\overset{}{\underset{\|}{AB}}} \longrightarrow \left(\underset{\displaystyle C\cdot}{\underset{|}{RAB}}\right)^{*} \longrightarrow \underset{\displaystyle C}{\underset{\|}{RA}} + B\cdot \tag{4}$$

[1] E. L. Eliel, in *Steric Effects in Organic Chemistry*, M. S. Newman (Ed.), Wiley, New York, 1956, Ch. 2.

Bimolecular homolytic substitution at coordinatively *unsaturated multivalent* atoms which do not form part of a multiply bonded system also need not involve high activation energies provided the atom under attack has low energy vacant orbitals available for increasing its co-ordination number (as does boron for example). Activation energies are normally high for S_H2 reaction at a coordinatively saturated multivalent atom (for example, sp^3-hybridized 4-coordinate carbon).

S_H2 reactions at multivalent atoms which do not form part of a multiply bonded system must, of necessity, involve radical attack on interior or nonterminal atoms. It is these reactions that form the subject of the present review.

Bimolecular homolytic substitutions at multivalent atoms can be conveniently subdivided into two types, synchronous and stepwise.

(a) S_H2 (*synchronous*). In this mechanism the loss of B is coordinated with the approach of R˙ and [RAB]˙ represents a *transition state*, not an intermediate.

$$R^{\cdot} + AB \rightarrow [R{-}{-}{-}A{-}{-}{-}B]^{\cdot\ddagger} \rightarrow RA + B^{\cdot} \tag{5}$$

(b) S_H2 (*stepwise*). This mechanism involves the attachment of R˙ to AB to give an intermediate (RAB)˙ which subsequently undergoes unimolecular decomposition to the products.

$$R^{\cdot} + AB \rightarrow (RAB)^{\cdot} \tag{6}$$

$$(RAB)^{\cdot} \rightarrow RA + B^{\cdot} \tag{7}$$

The distinction between a synchronous and a stepwise substitution would appear to lie in the behavior of the adduct radical (RAB)˙. If reaction 7 occurs within the time required for a single vibration ($\sim 10^{-13}$ sec) then (RAB)˙ is only a transition state and the substitution is synchronous. On the other hand, if the adduct radical has a finite lifetime the substitution is a stepwise process. In such a case it may even be possible to identify the adduct radical by electron spin resonance spectroscopy. For example,[2] the *t*-butoxytrimethylphosphoranyl radical has been identified in the reaction of *t*-butoxy radicals with trimethylphosphine.

$$(CH_3)_3CO^{\cdot} + (CH_3)_3P \rightarrow (C_3H)_3CO\dot{P}(CH_3)_3 \tag{8}$$

A signal due to the methyl radical was also identified[2] which implies that the phosphoranyl radical undergoes α-scission.

$$(CH_3)_3CO\dot{P}(CH_3)_3 \rightarrow (CH_3)_3COP(CH_3)_2 + CH_3^{\cdot} \tag{9}$$

With tri-*n*-butylphosphine, product studies by Buckler[3] have shown that the

[2] J. K. Kochi and P. J. Krusic, *J. Amer. Chem. Soc.*, **91**, 3944 (1969).
[3] S. A. Buckler, *ibid.*, **84**, 3093 (1962).

phosphoranyl radical can also decompose by β-scission with the elimination of a t-butyl radical, that is,

$$(CH_3)_3CO\dot{P}(Bu^n)_3 \rightarrow (CH_3)_3C^{\cdot} + OP(Bu^n)_3 \qquad (10)$$

3. If the adduct radical $(RAB)^{\cdot}$ is fairly stable, the homolytic substitution may require that $(RAB)^{\cdot}$ react with a second radical.

$$R^{\cdot} + AB \rightarrow (RAB)^{\cdot} \qquad (6)$$

$$(RAB)^{\cdot} + R'^{\cdot} \rightarrow RA + R'B \qquad (11)$$

The best known examples of such a mechanism are homolytic aromatic substitutions by, for example, phenyl or methyl radicals.[4–6]

· + ⟶ H $\xrightarrow{R'^{\cdot}}$ + R′H (12)

These reactions are not discussed further in this review.

4. The rate controlling step for the formation of the products of reaction 1 may involve the transfer of an electron between $R^{\cdot}$ and AB. Transfer from AB to $R^{\cdot}$ will yield the radical cation $\overset{+}{AB}{}^{\cdot}$ which may dissociate to A^+ and $B^{\cdot}$ or may undergo a nucleophilic attack by R^- to yield the same final products RA and $B^{\cdot}$, that is,

$$R^{\cdot} + AB \rightarrow R^- + \overset{+}{(AB)}{}^{\cdot} \rightarrow R^- + A^+ + B^{\cdot} \rightarrow RA + B^{\cdot} \qquad (13)$$

$$R^{\cdot} + AB \rightarrow R^- + \overset{+}{(AB)}{}^{\cdot} \rightarrow R^{-} \curvearrowright \overset{+}{(AB)}{}^{\cdot} \rightarrow RA + B^{\cdot} \qquad (14)$$

Similarly, with electron transfer from $R^{\cdot}$ to AB,

$$R + AB \rightarrow R^+ + \overset{-}{(AB)}{}^{\cdot} \rightarrow R^+ + A^- + B^{\cdot} \rightarrow RA + B^{\cdot} \qquad (15)$$

$$R^{\cdot} + AB \rightarrow R^+ + \overset{-}{(AB)}{}^{\cdot} \rightarrow R^{+} \curvearrowleft {}^{-}(AB)^{\cdot} \rightarrow RA + B^{\cdot} \qquad (16)$$

Prior electron transfer will be favored over direct substitution when the oxidation-reduction potentials of $R^{\cdot}$ and AB are suitably matched. Such redox processes are clearly established for many reactions, particularly those in which $R^{\cdot}$ or AB has a formal charge. A very well-known example in which the products correspond to a nonterminal S_H2 process but which proceeds by

[4] D. H. Hey, in *Vistas in Free Radical Chemistry*, Pergamon Press, London, 1959, p. 209.
[5] G. H. Williams, *Homolytic Aromatic Substitution*, Pergamon Press, London, 1960.
[6] R. O. C. Norman and G. K. Radda, *Adv. Heterocyclic Chem.*, **2**, 131 (1963).

electron transfer is the reaction of hydroperoxides with certain transition metal ions which happen to be free radicals, for example, Co^{II} and Mn^{II}.

$$Co^{II} + ROOH \rightarrow Co^{III} + OH^- + RO^{\cdot} \quad \text{or} \quad (Co^{II}OH) + RO^{\cdot} \quad (17)$$

Although such redox processes are beyond the scope of this review there are a host of reactions which might occur either by homolytic substitution or by electron transfer. Since there is probably some charge separation in the transition states of most homolytic substitutions, that is,

$$\overset{\delta+}{R}\cdots\cdot\overset{\delta-}{A}\cdots\cdot B \quad \text{and} \quad \overset{\delta-}{R}\cdots\cdot\overset{\delta+}{A}\cdots\cdot B$$

there really cannot be a sharp dividing line between S_H2 reactions and electron transfer processes. Moreover, if a dividing line were defined, an S_H2 reaction which had a fairly large degree of charge separation in its transition state in a nonpolar solvent might cross the line and become a redox reaction in a good ionizing solvent. In special cases, it may be possible to estimate the degree (or relative degree) of charge separation in the transition state from the polar effect of substituents remote from the reaction center as, for example, in systems to which the Hammett equation could be applied. More usually, however, the detailed mechanism of a considerable number of free-radical substitutions will remain uncertain until more widely applicable methods for determining the extent of electron transfer in transition states are developed.

5. The organic chemist ordinarily deals with reactions that are brought about by the introduction of the minimum amount of energy that will allow the reactions to proceed at a reasonable rate. The term hot atom or hot radical is used to refer to a species which has a large excess of electronic, translational, or vibrational energy over that it would have if it were in equilibrium with its surroundings. Of particular interest in connection with S_H2 reactions is the production of hot atoms by nuclear transformations. The atoms are formed with considerable excess kinetic energy and this enables them to initiate reactions which do not occur with thermalized atoms. Prominent among these hot-atom reactions are atom for atom and atom for radical substitutions. The atom for atom substitutions appear to be particularly efficient with recoil tritium substitutions for H or D but is rather less efficient for hot halogen atom substitutions[7] such as the recoil ^{128}I substitution for ^{127}I.

$$^{4}He(n, p)^{3}H \qquad ^{3}\overset{*}{H^{\cdot}} + C^{1}H_4 \rightarrow {}^{3}HC^{1}H_3 + {}^{1}H^{\cdot} \quad (18)$$

$$^{127}I(n, \gamma)^{128}I \qquad ^{128}\overset{*}{I^{\cdot}} + C_2H_5{}^{127}I \rightarrow C_2H_5{}^{128}I + {}^{127}I^{\cdot} \quad (19)$$

[7] R. J. Cross, Jr., and R. L. Wolfgang, *Radiochim. Acta*, **2**, 112 (1964).

Although these reactions are certainly bimolecular homolytic substitutions at a saturated multivalent atom they have not been included in this review. The reason for this is that hot-atom substitutions form a very distinct class of S_H2 reactions which have been adequately reviewed many times[8–13] and which, furthermore, bear little or no relation to substitutions by thermalized atoms.

The present review deals principally with reactions that appear to be straightforward S_H2 processes at covalently saturated multivalent atoms. The review is divided into sections corresponding to the Group of Periodic Table to which the atom under attack belongs. Reactions which probably proceed by a rate determining electron transfer and those which may proceed by radical addition to a π-electron system are not covered in detail.

Before discussing the individual reactions it is worthwhile considering the probable geometry of the transition state. For the $H^{\cdot} + H_2$ reaction, which is the simplest S_H2 process, the transition state is predicted on theoretical grounds to be linear and hence to involve backside attack by the $H^{\cdot}$ atom.[14–24]

$$H^{\cdot} + H_2 \rightarrow [H{-}{-}{-}H{-}{-}{-}H]^{\cdot} \rightarrow H_2 + H^{\cdot} \qquad (20)$$

Theoretical predictions regarding even slightly more complex S_H2 reactions are less certain. For instance, the transition state for chlorine atom attack on

[8] J. E. Willard, *Ann. Rev. Nuclear Sci.*, **3,** 193 (1953); *Ann. Rev. Phys. Chem.*, **6,** 141 (1955); *Nucleonics*, **19,** 61 (1961).
[9] A. P. Wolf, *Ann. Rev. Nuclear Sci.*, **10,** 259 (1960); *Adv. Phys. Org. Chem.*, **2,** 201 (1964).
[10] I. G. Campbell, *Adv. Inorg. Chem. Radiochem.*, **5,** 135 (1963).
[11] R. Wolfgang, *Ann. Rev. Phys. Chem.*, **16,** 15 (1965); *Progr. React. Kinet.*, **3,** 97 (1965); *Sci. Amer.*, **219,** 44 (1968).
[12] J. P. Adloff, *Radiochim. Acta*, **6,** 1 (1966).
[13] Y. N. Tsang and F. S. Rowland, *J. Phys. Chem.*, **72,** 707 (1968).
[14] H. Eyring and M. Polanyi, *Z. Physik. Chem.*, **12B,** 279 (1931).
[15] S. Glasstone, K. J. Laidler, and H. Eyring, *Theory of Rate Processes*, McGraw-Hill, New York, 1941, pp. 87–91, 107–115.
[16] R. E. Weston, *J. Chem. Phys.*, **31,** 892 (1959).
[17] R. E. Weston, *Science*, **158,** 332 (1967).
[18] I. Shavitt, *J. Chem. Phys.*, **31,** 1359 (1959); **49,** 4048 (1968).
[19] D. L. Bunker and N. C. Blais, *J. Chem. Phys.*, **41,** 2377 (1964).
[20] P. J. Kuntz, E. M. Nemeth, J. C. Polanyi, S. D. Rosner, and C. E. Young, *ibid.*, **44,** 1168 (1966).
[21] L. Pedersen and R. N. Porter, *ibid.*, **47,** 4751 (1967).
[22] D. J. LeRoy, B. A. Ridley, and K. A. Quickert, *Discuss. Faraday Soc.*, **44,** 92 (1967).
[23] I. Shavitt, R. M. Stevens, F. L. Stevens, and M. Karplus, *J. Chem. Phys.*, **48,** 2700 (1968).
[24] E. M. Mortensen, *ibid.*, **48,** 4029 (1968); **49,** 3526 (1968).

H_2 is probably linear but might possibly be triangular.[17,25–30] Although there have been a number of attempts to determine the stereochemistry of non-terminal S_H2 processes this is still not known with absolute certainty for any such reaction. The substitution might occur from the back, as in an S_N2 Walden inversion, or from the front or side. Unfortunately, S_H2 reactions at tetravalent atoms are relatively rare and attempts to substitute at an asymmetric optically active carbon have not led to any firm mechanistic conclusions (see Chapter V). However, it seems extremely probable that an S_H2 reaction at an sp^3 carbon atom will be a synchronous process and will proceed in a manner analogous to S_N2 reactions because in both cases there is an "excess" of electrons around the atom under attack. The split bond, R···· A···· B, holding the incoming and outgoing radicals will have an approximately planar surface (exactly planar if R = B) of zero electron density in which the three bonds to carbon which are not altered can lie. This arrangement minimizes the repulsive exchange energy between the altered and preserved bonds and between the unpaired electron and the electron pair in the bonds being altered.[31] This assumes that higher orbitals

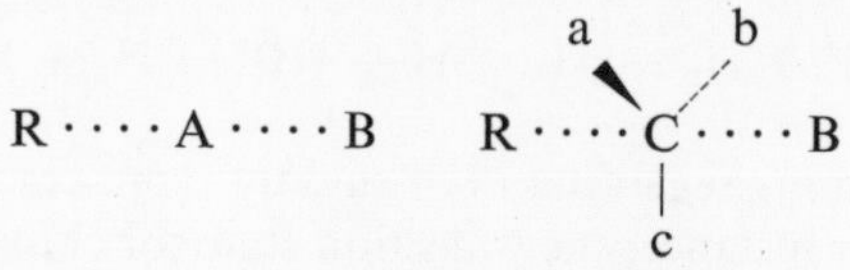

R····A····B   R····C(a)(b)(c)····B

are not available to take care of the incoming electron and might not, therefore, be true for homolytic substitution at elements heavier than carbon.

In contrast to Group IV atoms, homolytic substitution at boron, phosphorus, oxygen, and sulfur, is both common and well documented. Pryor and co-workers have made determined attempts to obtain the stereochemistry of substitution at divalent sulfur.[32–34] The order in which alkyl substituents reduce the rates of S_N2 reactions on disulfides is similar to their order for reduction of the rates of S_H2 reactions on disulfides. Since the S_N2 reactions probably occur by backside attack it was first concluded[33] that S_H2 reactions also involve backside attack and a linear transition state. However, Pryor and Smith[34] have now shown that all reactions for which

[25] Footnote 15, pp. 222–223.
[26] J. L. Magee, *J. Chem. Phys.*, **8,** 677 (1940).
[27] J. Bigeleisen and M. Wolfsberg, *ibid.*, **23,** 1535 (1955).
[28] K. S. Pitzer, *J. Amer. Chem. Soc.*, **79,** 1804 (1957).
[29] J. Bigeleisen, F. S. Klein, R. E. Weston, Jr., and M. Wolfsberg, *J. Chem. Phys.*, **30,** 1340 (1959).
[30] A. Persky and F. S. Klein, *ibid.*, **44,** 3617 (1966).
[31] C. K. Ingold, *Structure and Mechanism in Organic Chemistry*, 2nd ed., Cornell University Press, Ithaca, N.Y., 1969, p. 516.
[32] W. A. Pryor and T. L. Pickering, *J. Amer. Chem. Soc.*, **84,** 2705 (1962).
[33] W. A. Pryor and H. Guard, *ibid.*, **86,** 1150 (1964).
[34] W. A. Pryor and K. Smith, *ibid.*, **92,** 2731 (1970).

data were available could be correlated with the S_N2 reaction at the 1% level of significance. Consequently, a good correlation in a log-log graph between two such reactions is not diagnostic of mechanism. It was concluded that attack on sulfur by nucleophiles or radicals usually occurs from the back side. However, many of these reactions probably do not involve a synchronous process but rather proceed through an addition-elimination sequence involving a metastable intermediate in which sulfur has expanded its electronic octet (see Chapter VII for details).

The stereochemistry of substitution at boron, oxygen, and phosphorus has not been examined. If S_H2 reactions at boron are stepwise processes it seems reasonable to presume that the $(R_4B)^\cdot$ intermediate adduct radicals would have a tetrahedral structure. The e.s.r. spectrum of the *t*-butoxytrimethylphosphoranyl radical (see reaction 8) indicates that one methyl group is distinct from the other two.[2] A trigonal-bipyramidal configuration was proposed similar to that suggested for other phosphoranyl radicals.[35–37] An S_H2 reaction at phosphorus requires an α-scission of the phosphoranyl radical. No examples of the stereochemistry of this process have been reported but the stereochemistry of the alternate β-scission reaction has been examined by Bentrude and co-workers.[38,39] Thus, Bentrude, Hargis, and Rusek[38] have reported the stereospecific transfer of oxygen from *t*-butoxy radicals and of sulfur from *n*-butylthiyl radicals to give oxide or sulfide with retention of configuration at phosphorus in the geometrically isomeric (*cis-trans*) pair of 2-methoxy-5-*tert*-butyl-1,3,2-dioxaphosphorinans and also in optically active methylphenyl-*n*-propylphosphine.

(2-methoxy-5-*tert*-butyl-1,3,2-dioxaphosphorinan, Me–O–P:, with Bu^t) $\xrightarrow{Bu^tO^\cdot}$ (same ring, Me–O–P→O) (21)

(isomeric 2-methoxy-5-*tert*-butyl-1,3,2-dioxaphosphorinan, $\ddot{P}OMe$, with Bu^t) $\xrightarrow{Bu^tO^\cdot}$ (same ring, O←POMe) (22)

$:P(Me)(Pr^n)(Ph) \xrightarrow{Bu^tO^\cdot} Bu^{t\cdot} + O{\leftarrow}P(Me)(Pr^n)(Ph)$ (23)

[35] P. W. Atkins and M. C. R. Symons, *J. Chem. Soc.*, 4363 (1964).
[36] R. W. Fessenden and R. H. Schuler, *J. Chem. Phys.*, **45**, 1845 (1966).
[37] G. F. K. Koska and F. E. Brinckman, *Chem. Comm.*, 349 (1968).
[38] W. G. Bentrude, J. H. Hargis, and P. E. Rusek, Jr., *ibid.*, 296 (1969).
[39] W. G. Bentrude and R. A. Wielesek, *J. Amer. Chem. Soc.*, **91**, 2406 (1969).

In contrast, the thermolysis of ^{14}C-labeled di-*t*-butyl hyponitrite in the presence of tri-*t*-butyl phosphite gave almost quantitative yields of tri-*t*-butyl phosphate, approximately 75% of which was ^{14}C-labeled.[39] This result was interpreted in terms of an essentially irreversible formation of a phosphoranyl radical intermediate in a manner which allows statistical scrambling of the label. This could result either because the alkoxy groups are configurationally equivalent or because an unsymmetrical intermediate has a lifetime sufficient to allow equilibration of configurationally nonequivalent groups by, for example, pseudorotation. It was concluded[39] that the configuration, stability, and lifetime of phosphoranyl radicals probably depend on the nature of the substituents on the phosphorus.

Finally, although hot-atom chemistry is not meant to be a part of this review, it must be mentioned that the homolytic substitution of hydrogen by recoil tritium from an sp^3 hybridized carbon atom occurs with almost complete retention of configuration.[40–44] This was first conclusively demonstrated by Henchman and Wolfgang[43] who showed that when recoil tritium reacts with optically active 2-butanol in the gas phase, the hydrogen atom directly bonded to the asymmetric carbon is substituted with a $91 \pm 6\%$ retention of configuration. Since the experiment was performed in the gas phase the configurational retention was not due to any constraint imposed by a crystal or liquid cage. In similar experiments, Wai and Rowland[45] have shown that recoil ^{38}Cl substitutes for chlorine in either DL- or *meso*-2,3-dichlorobutane with almost complete retention of optical configuration. For both the recoil tritium and the recoil chlorine reactions attack must be from the front-side, not from the back-side. The results are in accord with a model which postulates that substitution involves a fast, localized interaction which occurs in about the characteristic period of a C–H or C–Cl vibration. There is no reason to believe that these hot-atom results have any direct bearing on S_H2 reactions between species in thermal equilibrium with their surroundings.

[40] F. S. Rowland, C. N. Turton, and R. Wolfgang, *ibid.*, **78**, 2354 (1956).
[41] H. Keller and F. S. Rowland, *J. Phys. Chem.*, **62**, 1373 (1958).
[42] J. G. Kay, R. P. Malsan, and F. S. Rowland, *J. Amer. Chem. Soc.*, **81**, 5050 (1959).
[43] M. Henchman and R. Wolfgang, *ibid.*, **83**, 2991 (1961).
[44] Y. -N. Tang, C. T. Ting, and F. S. Rowland, *J. Phys. Chem.*, **74**, 675 (1970).
[45] C. M. Wai and F. S. Rowland, *ibid.*, **71**, 2752 (1967); *ibid.*, **74**, 434 (1970).

II S_H2 Reactions at Group IA Elements

The alkaline earth elements are highly electropositive and generally form ionic compounds. Solutions of these compounds normally contain separately solvated anions and cations. Exceptions are provided by alkyllithium and aryllithium compounds in which the Li–C bond is considered to be predominantly covalent, and some reactions of organolithium compounds are thought to proceed by S_H2 attack at lithium.[1,2]

$$X^{\cdot} + LiR \rightarrow XLi + R^{\cdot} \tag{1}$$

However, lithium is formally a terminal atom[3] in an alkyllithium and these reactions will not be discussed further in this review.

[1] A. G. Davies and B. P. Roberts, *J. Chem. Soc.* (*B*), 1074 (1968).

[2] G. A. Russell and D. W. Lamson, *J. Amer. Chem. Soc.*, **91**, 3967 (1969).

[3] Association and solvation of organolithium compounds will make the lithium atom multivalent and nonterminal.

III S_H2 Reactions at Group II Elements

A. GROUP IIA

1. BERYLLIUM

There are no authenticated examples of bimolecular homolytic displacement reactions occurring at a beryllium atom. The Be–C bond is however subject to autoxidative cleavage[1] and, by analogy with the mechanism of autoxidation of other organometallic compounds, this may involve a free-radical chain process. The autoxidation of dimethylberyllium in diethyl ether gave beryllium methoxide which contained between 3 and 7% peroxidic oxygen, depending on the reaction conditions.[2] The suggested mechanism involved heterolytic insertion of oxygen into the Be–C bond to give methylberyllium methylperoxide. However a free-radical chain mechanism seems very likely and the product-forming propagation step of this would involve the S_H2 reaction of a methylperoxy radical at the beryllium atom in dimethylberyllium.

$$MeOO^{\cdot} + BeMe_2 \rightarrow MeOOBeMe + Me^{\cdot} \qquad (1)$$

2. MAGNESIUM

Although not common, there are several types of reaction of organomagnesium compounds which are recognized as having characteristics of

[1] T. G. Brilkina and V. A. Shushunov, *Reactions of Organometallic Compounds with Oxygen and Peroxides*, Iliffe, London, 1969.
[2] R. Masthoff, *Z. Anorg. Allgem. Chem.*, **336**, 252 (1965).

homolytic processes. Such radical reactions have been discussed by several authors.[3–5]

The highly electropositive character of magnesium results in the formation of ionic compounds especially with the more electronegative elements or groups. For an S_H2 reaction, as defined in this review, to occur at magnesium, attack of the incoming radical at the metal must be accompanied or followed by loss of the displaced group from magnesium.

$$Y^{\cdot} + MgX_2 \rightarrow [YMgX_2]^{\cdot} \rightarrow YMgX + X^{\cdot} \quad (2)$$

As the bond between Mg and Y becomes progressively more ionic, then the overall reaction is perhaps better represented as an electron transfer process

$$Y^{\cdot} + MgX_2 \rightarrow [\overset{-}{Y}\overset{+}{M}gX_2]^{\cdot} \rightarrow \overset{-}{Y}\overset{+}{M}gX + X^{\cdot} \quad (3)$$

Further, if the bond between Mg and X is also ionic, then the reaction may be independent of the metal entirely and be characteristic of $Y^{\cdot}$ and X^{-}.

$$Y^{\cdot} + \overset{-}{X}\overset{2+}{M}g\overset{-}{X} \rightarrow \overset{-}{Y}\overset{2+}{M}g\overset{-}{X} + X^{\cdot} \quad (4)$$

The bond which magnesium forms to carbon is usually regarded as being essentially covalent and radical attack on an alkylmagnesium or arylmagnesium compound to bring about displacement of a carbon radical has been suggested several times.

This discussion of the chemistry of organomagnesium compounds should be prefaced by the disclaimer that the molecular structure of the species involved is seldom as simple as indicated by the formulae used.[6] In general, the extent of our present knowledge of the homolytic reactions of organomagnesium compounds does not justify discussion of the results in terms of structures more complex than the two-coordinate representations used here.

Substitution by Oxygen-Centered Radicals

The absorption of oxygen by Grignard reagents to give alcohols after hydrolysis of the reaction mixture, has been known since the Grignard reagent was first prepared.

$$2RMgX + O_2 \rightarrow 2ROMgX \quad (5)$$

[3] M. S. Kharasch and O. Reinmuth, *Grignard Reactions of Non-Metallic Substances*, Prentice-Hall, New York, 1954.

[4] *Methods of Elemento-Organic Chemistry*, A. N. Nesmeyanov and K. A. Kocheshkov (Eds.), North-Holland Publishing Co., Amsterdam, 1967, Vol. 2.

[5] For recent accounts see: C. Blomberg, R. M. Salinger, and H. S. Mosher, *J. Org. Chem.*, **34**, 2385 (1969); H. M. Pelles, *ibid.*, **34**, 3687 (1969); J. J. Eisch and R. L. Harrell, Jr., *J. Organometal. Chem.*, **21**, 21 (1970).

[6] E. C. Ashby, *Quart. Rev. (London)*, **21**, 259 (1967).

As early as 1909, Wuyts[7] suggested the intermediate formation of a peroxide (of unspecified structure) during the autoxidation. In 1920 Porter and Steele[8] detected up to 6% of iodometric peroxide formed in the oxidation of ethylmagnesium bromide in ether-toluene at −70°, and they proposed that the reaction sequence was

$$RMgX + O_2 \rightarrow ROOMgX \tag{6}$$

$$ROOMgX + RMgX \rightarrow 2ROMgX \tag{7}$$

It was not until 1953, however, that Walling and Buckler[9,10] were able to prove the presence of the intermediate peroxide. By slow addition of the Grignard reagent in ether to ether kept saturated with oxygen at about −70°, followed by hydrolysis, they obtained alkyl hydroperoxides in 30–90% yields. Diphenylamine did not inhibit the autoxidation of a Grignard reagent, and butyraldehyde was not co-oxidized when present in the mixture, and on this evidence it was concluded that the autoxidation probably did not involve free radicals[10]; the following scheme was proposed:

$$RMgX + O_2 \longrightarrow \begin{matrix} R\diagdown \\ \bar{Mg}\text{—}X \\ :\ddot{O}\text{—}\ddot{O}:^{+} \end{matrix} \rightleftharpoons \begin{matrix} R\diagdown \\ \bar{Mg}\text{—}X \\ \ddot{O}\text{=}\ddot{O}^{+} \end{matrix} \updownarrow \begin{matrix} R\diagdown \\ \bar{Mg}\text{—}X \\ ^{+}\ddot{O}\text{—}\ddot{O}: \end{matrix} \longrightarrow ROOMgX \tag{8}$$

Walling never completely excluded the possibility that the autoxidation could proceed by a rapid free-radical chain process.[10,11] Subsequent work has tended to support the idea of involvement of free radicals in the autoxidation of Grignard reagents and of dialkylmagnesium or diarylmagnesium compounds[1,4,12]

Jensen and Nakamaye[13] suggested that the reactions of aliphatic Grignard reagents with electrophiles will normally involve retention of configuration at the carbon atom originally bound to magnesium. Retention was the stereochemical result of the cleavage of the Mg–C bond by carbon dioxide

[7] P. Wuyts, *Compt. Rend.*, **148,** 930 (1909); *Bull. Soc. Chim. France*, **36,** 222 (1927).
[8] C. W. Porter and C. Steele, *J. Amer. Chem. Soc.*, **42,** 2650 (1920).
[9] C. Walling and S. A. Buckler, *ibid.*, **75,** 4372 (1953).
[10] C. Walling and S. A. Buckler, *ibid.*, **77,** 6032 (1955).
[11] C. Walling, *Free Radicals in Solution*, Wiley, New York, 1957.
[12] H. Hock, H. Kropf, and F. Ernst, *Angew. Chem.*, **71,** 541 (1959).
[13] F. R. Jensen and K. L. Nakamaye, *J. Amer. Chem. Soc.*, **88,** 3437 (1966).

or mercuric bromide.[13] The Grignard reagent from hex-5-enyl bromide, which contains about 6% of cyclopentylmethylmagnesium bromide, reacts with carbon dioxide or dilute hydrochloric acid to give a product with structure corresponding to that of the Grignard reagent. However, the reaction with oxygen gave 74% of hex-5-en-1-ol and 26% of cyclopentylmethanol.[14]

$$CH_2{=}CHCH_2CH_2CH_2CH_2MgBr \xrightarrow[2.H_2O]{1.\ O_2} CH_2{=}CHCH_2CH_2CH_2CH_2OH\ (74\%) + \text{cyclo-}C_5H_9\text{—}CH_2OH\ (26\%) \quad (9)$$

This was interpreted[14] as evidence that the autoxidation occurred by way of a free-radical mechanism involving a free hex-5-enyl radical which is known to cyclize rapidly to give a cyclopentylmethyl radical.[15–18]

$$CH_2{=}CHCH_2CH_2CH_2\dot{C}H_2 \longrightarrow \cdot CH_2\text{—cyclo-}C_5H_9 \quad (10)$$

The most reasonable pathway was thought to involve one-electron transfer from the Grignard reagent.

$$RMgX + O_2 \rightarrow R^{\cdot} + XMgO_2^{\cdot} \quad (11)$$

$$R^{\cdot} + O_2 \rightarrow ROO^{\cdot} \quad (12)$$

$$ROO^{\cdot} + RMgX \rightarrow ROOMgX + R^{\cdot} \quad (13)$$

Reaction 11 might be the initiating step of a chain process or simply the first step of a nonchain mechanism. Cyclization of the hex-5-enyl radical would compete with reaction 12. Similar results were found and a similar interpretation was suggested for the autoxidation of hexa-3,5-dienylmagnesium bromide.[19] Studies of this type, however, suffer from the disadvantage that only the overall stereochemical result of the formation of the peroxide and its reduction by the Grignard reagent [reactions 6 and 7, respectively] is being observed and there is now some evidence that reaction 7 may involve intermediate alkyl radicals.[20–22a] For meaningful conclusions

[14] R. C. Lamb, P. W. Ayers, M. K. Toney, and J. F. Garst, *ibid.*, **88**, 4261 (1966).
[15] R. C. Lamb, P. W. Ayers, and M. K. Toney, *ibid.*, **85**, 3483 (1963).
[16] C. Walling and M. S. Pearson, *ibid.*, **86**, 2262 (1964).
[17] R. G. Garwood, C. J. Scott, and B. C. L. Weedon, *Chem. Comm.*, 14 (1965).
[18] D. J. Carlsson and K. U. Ingold, *J. Amer. Chem. Soc.*, **90**, 7047 (1968).
[19] M. E. Howden, A. Maerker, J. Burdon, and J. D. Roberts, *ibid.*, **88**, 1732 (1966).
[20] D. B. Bigley and D. W. Payling, *Chem. Comm.*, 938 (1968).
[21] A. G. Davies, J. Eley, and B. P. Roberts, unpublished results.
[22] D. Skelton, *Ph.D. thesis*, University of Aston, Birmingham, England, 1969.
[22a] J. M. Pabiot and R. Palland, *Compt. Rend. Acad. Sci. Paris, Sér. C*, **270**, 334 (1970).

to be drawn regarding the first step 6, the reactions would have to be carried out under conditions where the peroxides themselves could be studied.

Recently Walling and Cioffari[23] have reported that the amount of cyclization observed during the autoxidation of hex-5-enylmagnesium bromide increases markedly as the oxygen concentration in solution is decreased. Since no cyclization was observed in the reaction of *t*-butylperoxymagnesium bromide with hex-5-enylmagnesium bromide, the result is consistent with the radical chain mechanism for autoxidation in which cyclization of the hex-5-enyl radical competes with its reaction with oxygen.

The autoxidation of 1-methylheptylmagnesium bromide or 1-naphthylmagnesium bromide in dibutyl ether could not be inhibited by radical scavengers, probably because the organomagnesium compounds reacted rapidly with the inhibitors.[23a]

The autoxidation of norborn-2-ylmagnesium bromide (41% *exo*-59% *endo* and *ca.* 100% *endo*) and chloride (43% *exo*-57% *endo*), under conditions where the peroxide did not react further, gave in all cases a mixture of norborn-2-ylperoxy compounds similar to that (76% *exo*-24% *endo*) obtained in the autoxidation of *exo*-norbornylboranes and *endo*-norbornylboranes.[24] In the case of the boranes, inhibition of the autoxidation by radical scavengers afforded independent evidence for a free-radical chain process.[25] The proposed mechanism involved homolytic substitution by a norbornylperoxy radical at magnesium in a propagation step 15.

$$C_7H_{11}^{\cdot} + O_2 \rightarrow C_7H_{11}O_2^{\cdot} \qquad (14)$$

$$C_7H_{11}O_2^{\cdot} + C_7H_{11}MgX \rightarrow C_7H_{11}OOMgX + C_7H_{11}^{\cdot} \qquad (15)$$

$$C_7H_{11} = \text{norborn-2-yl};\ X = Cl, Br.$$

The epimeric constitution of the peroxide would be determined by reaction 14 and be independent of the nature of the organometallic compound. The observed ratio of **1**:**2** (*ca.* 76%:24%) is consistent with this proposal, reaction 14 being very rapid and involving little bond formation in the transition state.[26]

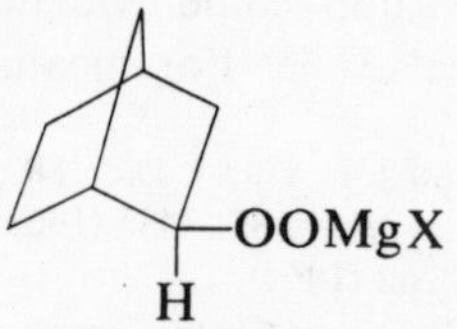

OOMgX
H
1

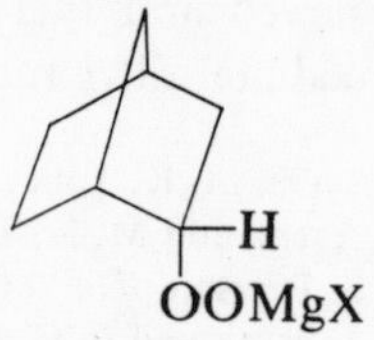

H
OOMgX
2

[23] C. Walling and A. Cioffari, *J. Amer. Chem. Soc.*, **92,** 6609 (1970).
[23a] See footnote 1, Chapter II.
[24] A. G. Davies and B. P. Roberts, *J. Chem. Soc.* (*B*), 317 (1969).
[25] A. G. Davies and B. P. Roberts, *ibid.*, 311 (1969).
[26] F. R. Jensen, L. H. Gale, and J. E. Rodgers, *J. Amer. Chem. Soc.*, **90,** 5793 (1968).

Autoxidation of the more ionic organomagnesium compounds may also be rationalized in terms of the oxidation of a carbanion (see above).[27,28] Triphenylmethylsodium reacts with oxygen in ether to form triphenylmethanol and bis(triphenylmethyl)peroxide in varying yields,[29–31] while triphenylmethylmagnesium bromide is reported to be converted to the peroxide by oxygen.[32] Russell[27,28] has interpreted these reactions in terms of a chain autoxidation of the triphenylmethyl carbanion involving electron transfer from the triphenylmethylperoxy radical without the involvement of S_H2 attack at the metal center.

$$Ph_3COO^{\cdot} + Ph_3C^- \rightarrow Ph_3COO^- + Ph_3C^{\cdot} \qquad (16)$$

The possibility of homolytic substitution at magnesium by an alkoxy radical does not appear to have been considered previously. The reactions of organomagnesium compounds with organic peroxides have been investigated a number of times[1,4] and suggested mechanisms have usually involved a heterolytic process with the organic group attached to magnesium acting as a nucleophile, often with a cyclic transition state.[33–36] However, it is reported that the reaction of di-*t*-butyl peroxide with Grignard reagents proceeds by competing heterolytic and homolytic pathways.[20,22,37,38] A complex is formed initially which may either rearrange to give an alkyl *t*-butyl ether or undergo homolysis to give *t*-butoxy and alkyl radicals.

$$\begin{array}{ccc} Bu^tO\overset{\overset{Bu^t}{|}}{O} + \underset{X}{MgR} & \rightleftharpoons & Bu^tO\overset{\overset{Bu^t}{|}}{O}\rightarrow \underset{X}{MgR} \\ & \swarrow \qquad \searrow & \\ Bu^tOR + Bu^tOMgX & & Bu^tO^{\cdot} + R^{\cdot} + Bu^tOMgX \end{array} \qquad (17)$$

The extent of the homolytic component increased in the order $R^p < R^s < R^t$ and in the order $X = Cl < Br < I$. The *t*-butoxy radicals produced in 17 were thought to abstract hydrogen from the solvent ether to give *t*-butyl alcohol which then reacted with the Grignard reagent to give Bu^tOMgX and

[27] G. A. Russell and A. Bemis, *ibid.*, **88**, 5491 (1966).
[28] G. A. Russell, *Pure Appl. Chem.*, **15**, 185 (1967).
[29] W. Schlenk and E. Marcus, *Chem. Ber.*, **47**, 1664 (1914).
[30] C. A. Kraus and R. Rosen, *J. Amer. Chem. Soc.*, **47**, 2739 (1925).
[31] W. Bachmann and F. Y. Wiselogle, *ibid.*, **58**, 1943 (1936).
[32] J. Schmidlin, *Chem. Ber.*, **39**, 628 (1906).
[33] T. W. Campbell, W. Burney, and T. L. Jacobs, *J. Amer. Chem. Soc.*, **72**, 2735 (1950).
[34] S.-O. Lawesson and N. C. Yang, *ibid.*, **81**, 4230 (1959).
[35] G. A. Baramki, H. S. Chang and J. T. Edwards, *Can. J. Chem.*, **40**, 441 (1962).
[36] D. T. Longone and A. H. Miller, *Tetrahedron Lett.*, 4941 (1967).
[37] L. Simet, *Ph.D. thesis*, University of New York, New York City (1957).
[38] S. Herbstman, *Ph.D. thesis*, University of New York, New York City (1963).

RH. Clearly homolytic displacement at magnesium is a more likely alternative in view of the comparable concentrations of Grignard reagent and solvent.

$$\mathrm{Bu}^t\mathrm{O}^{\cdot} + \mathrm{RMgX} \rightarrow \mathrm{Bu}^t\mathrm{OMgX} + \mathrm{R}^{\cdot} \tag{18}$$

$$\mathrm{R}^{\cdot} \xrightarrow{\text{solvent}} \mathrm{RH} \tag{19}$$

It is possible that the reaction of *t*-butyl hypochlorite with phenylmagnesium bromide to give chlorobenzene[39] might proceed by a free-radical chain process involving an S_H2 propagation step 18 (R = Ph, X = Br).

Although numerous preparative examples of 1,4-addition of a Grignard reagent to an α,β-unsaturated ketone are known, little mechanistic information is available.[3] The recent demonstration by H. C. Brown and his collabor-

$$\mathrm{RMgX} + \underset{1}{\mathrm{O}}{=}\underset{2}{\overset{|}{\mathrm{C}}}{-}\underset{3}{\overset{|}{\mathrm{C}}}{=}\underset{4}{\overset{|}{\mathrm{C}}}{-} \longrightarrow \mathrm{XMgO}{-}\overset{|}{\mathrm{C}}{=}\overset{|}{\mathrm{C}}{-}\underset{|}{\overset{|}{\mathrm{C}}}{-}\mathrm{R} \tag{20}$$

ators[40,41] that the corresponding 1,4-addition of a trialkylborane proceeds by a free-radical chain process, invites the extrapolation of this mechanism to the magnesium compounds. Addition of a Grignard reagent across the carbonyl group probably occurs by a heterolytic mechanism[42] and the ratio of 1,4- to 1,2-addition could be increased by adding cuprous chloride in catalytic amounts.[43,44] On the other hand the absolute yield of 1,4-addition product from the reaction of *s*-butyl crotonate with butylmagnesium bromide was decreased by the presence of trace amounts of iron and manganese as impurities in the magnesium used to prepare the Grignard reagent.[45] Ethylmagnesium bromide was found to react with β-chlorocrotonic esters to give 1,2- or 1,4-addition products depending on the stereochemistry of the ester and the nature of the catalyst employed.[46] It has been proposed that the copper-catalyzed 1,4-addition proceeds by electron transfer to the unsaturated ketone to give a conjugated radical anion system.[47] This is shown below for methylmagnesium halide.

[39] C. Walling and J. Kjellgren, *J. Org. Chem.*, **34,** 1487 (1969).
[40] G. W. Kabalka, H. C. Brown, A. Suzuki, S. Honma, A. Arase, and M. Itoh, *J. Amer. Chem. Soc.*, **92,** 710 (1970).
[41] H. C. Brown and G. W. Kabalka, *ibid.*, **92,** 712, 714 (1970).
[42] H. O. House and J. E. Oliver, *J. Org. Chem.*, **33,** 929 (1968).
[43] M. S. Kharasch and P. O. Tawney, *J. Amer. Chem. Soc.*, **63,** 2308 (1941).
[44] T. Luong, T. Ngoc, and R. Henriette, *Compt. Rend. Acad. Sci. Paris, Sér. C.*, **267,** 776 (1968).
[45] J. Hilden and J. Munch-Petersen, *Acta Chem. Scand.*, **21,** 1370 (1967).
[46] L. Decaux and R. Vessiere, *Compt. Rend. Acad. Sci. Paris, Sér. C.*, **267,** 738 (1968).
[47] H. O. House and W. F. Fischer, *J. Org. Chem.*, **33,** 949 (1968).

$$\text{ligand}-\dot{\text{Cu}}(\text{I}) + \text{MeMgX} \longrightarrow \begin{matrix}\text{Me} \\ \text{ligand}\end{matrix}\!\!>\!\overset{-}{\text{Cu}}(\text{I})\overset{+}{\text{Mg}}\text{X} \qquad (21)$$

$$\text{RCH}{=}\text{CH}{-}\underset{\underset{\text{O}}{\|}}{\text{C}}{-}\text{R} + \begin{matrix}\text{Me} \\ \text{ligand}\end{matrix}\!\!>\!\overset{-}{\dot{\text{Cu}}}(\text{I})\overset{+}{\text{Mg}}\text{X} \longrightarrow \dot{\text{RCH}}{-}\text{CH}{=}\underset{\underset{\text{O}^-}{|}}{\text{C}}{-}\text{R}\ \ \overset{+}{\text{Mg}}\text{X} + \begin{matrix}\text{Me} \\ \text{ligand}\end{matrix}\!\!>\!\text{Cu}(\text{II}) \qquad (22)$$

$$\downarrow$$

$$\text{ligand}-\dot{\text{Cu}}(\text{I}) + \underset{\underset{\text{Me}}{|}}{\text{RCH}}{-}\text{CH}{=}\underset{\underset{\text{O}^-}{|}}{\text{C}}{-}\text{R}\ \ \overset{+}{\text{Mg}}\text{X}$$

The alternative free-radical chain mechanism would be analogous to that suggested by Brown[40,41] for the conjugate addition of trialkylboranes, and would incorporate bimolecular homolytic displacement at magnesium.

$$\text{R}'^{\cdot} + \text{RCH}{=}\text{CH}{-}\underset{\underset{\text{O}}{\|}}{\text{C}}{-}\text{R} \longrightarrow \begin{matrix}\text{R}' \\ \text{R}\end{matrix}\!\!>\!\text{CH}{-}\text{CH}{=}\underset{\underset{\text{O}^{\cdot}}{|}}{\text{C}}{-}\text{R} \qquad (23)$$

$$\begin{matrix}\text{R}' \\ \text{R}\end{matrix}\!\!>\!\text{CH}{-}\text{CH}{=}\underset{\underset{\text{O}^{\cdot}}{|}}{\text{C}}{-}\text{R} + \text{R}'\text{MgX} \longrightarrow \begin{matrix}\text{R}' \\ \text{R}\end{matrix}\!\!>\!\text{CH}{-}\text{CH}{=}\underset{\underset{\text{OMgX}}{|}}{\text{C}}{-}\text{R} + \text{R}'^{\cdot} \qquad (24)$$

A synchronous process 25 involving a ketone-Grignard reagent complex is also possible.

$$\text{R}'^{\cdot}\ \underset{\underset{\text{R}}{|}}{\text{CH}}{=}\text{CH}{-}\underset{\underset{\text{R}}{|}}{\text{C}}{=}\text{O}\rightarrow\underset{\text{X}}{\text{Mg}}{-}\text{R}' \longrightarrow \qquad (25)$$

$$\text{R}'{-}\underset{\underset{\text{R}}{|}}{\text{CH}}{-}\text{CH}{=}\underset{\underset{\text{R}}{|}}{\text{C}}{-}\text{O}{-}\underset{\text{X}}{\text{Mg}} + \text{R}'^{\cdot}$$

Many other organometallic compounds react with α,β-unsaturated carbonyl compounds to give competing 1,2- and 1,4-addition.[47a,47b] The proportion of 1,2-addition was found to decrease as the polarity of the metal-carbon bond decreased, a result to be expected if 1,2-addition is a polar process whilst 1,4-addition occurs by a homolytic mechanism.

2,4,6-trisubstituted phenoxy radicals react with a Grignard reagent, RMgX, to give products derived from the alkyl radical, $R^{\cdot}$.[48–51]

$$ArO^{\cdot} + RMgX \rightarrow ArOMgX + R^{\cdot} \qquad ArO^{\cdot} = \text{2,4,6-trisubstituted phenoxy radical} \quad (26)$$

In the other extreme, this reaction can be written as an electron transfer process.[48]

$$ArO^{\cdot} + RMgX \rightarrow Ar\overset{-\;+}{O}MgX + R^{\cdot} \quad (27)$$

Diphenylnitroxide, $Ph_2NO^{\cdot}$, also reacts with Grignard reagents to give diamagnetic products[52] and probably this occurs by a similar radical displacement from magnesium. It is noteworthy that ethylmagnesium bromide adds to the carbonyl function in **3** faster than it reacts with the nitroxide group.[53]

3

Substitution by Carbon-Centered Radicals

In 1929 Gilman and Jones[54] reported that the exchange reaction between a Grignard reagent and an alkyl halide was catalyzed by cobaltous halides.

$$RMgX + R'X \xrightarrow{CoX_2} RX + R'MgX \quad (28)$$

[47a] H. Gilman and R. H. Kirby, *J. Amer. Chem. Soc.*, **63**, 2046 (1941).
[47b] J. J. Eisch, "*The Chemistry of Organometallic Compounds*" Macmillan, London, 1967.
[48] A. Rieker, E. Müller, and N. Beckert, *Z. Naturforsch.*, **17b**, 718 (1962).
[49] A. Rieker, *Angew. Chem. Int. Ed.*, **3**, 654 (1964).
[50] V. V. Erschov, A. A. Volod'kin, and M. V. Tarkhanova, *Izv. Akad. Nauk SSSR, Ser. Khim.*, 2470 (1967).
[51] A. A. Volod'kin, M. V. Tarkhanova, A. L. Buchachenko, and V. V. Erschov, *ibid.*, 63 (1968).
[52] K. Maruyama, *Bull. Chem. Soc. Japan*, **37**, 1013 (1964).
[53] E. G. Rozantsev, M. B. Neiman, and Yu. G. Mamedova, *Izv. Akad. Nauk SSSR, Otd. Khim. Nauk*, 1509 (1963).
[54] H. Gilman and H. L. Jones, *J. Amer. Chem. Soc.*, **51**, 2840 (1929).

This result was confirmed and the interaction of transition metal, particularly cobalt, halides with Grignard reagents was examined in detail by Kharasch.[3,55] For example, treatment with carbon dioxide of the product formed by the reaction of butylmagnesium bromide with bromobenzene in the presence of 1 mole % of cobaltous chloride, resulted in a yield of 7% of benzoic acid. Presumably the benzoic acid is derived from phenylmagnesium bromide. It was suggested[55] that phenyl radicals, generated in the presence of cobaltous chloride, could bring about homolytic displacement of butyl radicals from magnesium.

$$Ph^{\cdot} + BuMgBr \rightarrow PhMgBr + Bu^{\cdot} \quad (29)$$

It is now well established[56–58] that phenyl radicals would be formed in this system, and indeed that $R'^{\cdot}$ would be generated from R′X and RMgX in the presence of cobaltous chloride.

Zakharkin and co-workers[59–61] found that the uncatalyzed exchange reaction 28 occurred in more highly solvating ethers (e.g., dimethoxyethane) and this was taken as evidence that exchange, under these conditions, took place by a heterolytic mechanism.[59,60] However, heating of butylmagnesium iodide in the presence of alkyl or aryl halides in cumene at 80–140° gave appreciable yields of bicumyl.[61] A similar result had been observed by Kharasch and Urry much earlier.[62] When butylmagnesium iodide and methyl iodide were heated at 80° for 18 hr in heptane, without catalyst, methylmagnesium iodide was detected.[61] The reactions in nonsolvating solvents were interpreted in terms of a free-radical mechanism, alkyl radicals being generated by electron transfer from the Grignard reagent to the alkyl halide.[61]

$$RMgX + R'Y \rightarrow R^{\cdot} + \overset{+}{Mg}X + R'^{\cdot} + \overset{-}{Y} \quad (30)$$

It has also been reported that the exchange reaction 28 can be induced by irradiation with ultraviolet light.[63]

[55] M. S. Kharasch and C. F. Fuchs, *J. Org. Chem.*, **10**, 292 (1945).
[56] M. H. Abraham and M. J. Hogarth, *J. Organometal. Chem.*, **12**, 1, 497 (1968).
[57] D. I. Davies, D. H. Hey, and M. Tiecco, *J. Chem. Soc.*, 7062 (1965).
[58] D. I. Davies, J. N. Done, and D. H. Hey, *J. Chem. Soc.* (*C*), 1392, 2019, 2021 (1969).
[59] L. I. Zakharkin, K. A. Bilevitch, and O. Yu. Okhlobystin, *Dokl. Akad. Nauk SSSR*, **152**, 338 (1963).
[60] L. I. Zakharkin, O. Yu. Okhlobystin, and K. A. Bilevitch, *J. Organometal. Chem.*, **2**, 309 (1964); *Izv. Akad. Nauk SSSR, Otd. Khim. Nauk*, 1347 (1964).
[61] L. I. Zakharkin, O. Yu. Okhlobystin, and B. N. Strunin, *J. Organometal. Chem.*, **4**, 349 (1965).
[62] M. S. Kharasch and W. H. Urry, *J. Org. Chem.*, **13**, 101 (1948).
[63] I. A. Korshunov and A. P. Batalov, *Zh. Obshch. Khim.*, **29**, 4048 (1959).

There is now an ever increasing amount of chemical[64] and spectroscopic[65,66] evidence for the involvement of free radicals in the reaction between Grignard reagents and alkyl or aryl halides. When butyllithium and alkyl bromides (RBr) or iodides (RI) were mixed in a flow system before entering the cavity of an e.s.r. spectrometer, the spectrum due to R· was observed.[65] The results were rationalized in terms of reactions 31–34.

$$\mathrm{BuLi + RX \rightarrow Bu^{\cdot} + R^{\cdot} + LiX} \tag{31}$$

$$\mathrm{B\dot{u} + RX \rightarrow BuX + R^{\cdot}} \tag{32}$$

$$\mathrm{R^{\cdot} + BuLi \rightarrow RLi + Bu^{\cdot}} \tag{33}$$

$$\mathrm{R^{\cdot} + Bu^{\cdot} \rightarrow alkanes\ and\ alkenes} \tag{34}$$

It was suggested that this mechanism might be generalized to include other organometallic compounds and reaction 35 would occur in the case of a Grignard reagent.

$$\mathrm{R'^{\cdot} + RMgX \rightarrow R'MgX + R^{\cdot}} \tag{35}$$

Substitution by Nitrogen-Centered Radicals

The reactions of Grignard reagents with *N*-halo-amines and *N*-halo-imines reported before 1954 have been reviewed.[3] No attempts to elucidate the mechanism of the reaction have appeared. After hydrolysis of the product of the reaction of an *N*-chloroamine with a Grignard reagent, an amine was isolated in which the chlorine had been replaced either by hydrogen or by an alkyl group originally attached to magnesium.[67–71]

$$\mathrm{RR'NCl + R''MgX \rightarrow [RR'NMgX] + R''Cl} \tag{36a}$$

$$\mathrm{RR'NCl + R''MgX \rightarrow RR'NR'' + ClMgX} \tag{36b}$$

$$\mathrm{R,R' = alkyl\ or\ H.}$$

This dichotomy of behavior exactly parallels that shown by the trialkylboranes towards *N*-chloroamines,[72] where the reaction forming alkyl chloride

[64] R. G. Gough and J. A. Dixon, *J. Org. Chem.*, **33**, 2148 (1968).
[65] See footnote 2, Chapter II.
[66] H. R. Ward, R. G. Lawler, and R. A. Cooper, *J. Amer. Chem. Soc.*, **91**, 746 (1969).
[67] G. H. Coleman and C. R. Hauser, *ibid.*, **50**, 1193 (1928).
[68] G. H. Coleman and C. B. Yager, *ibid.*, **51**, 567 (1929).
[69] R. J. W. LeFévre, *J. Chem. Soc.*, 1745 (1932).
[70] G. H. Coleman, *J. Amer. Chem. Soc.*, **55**, 3001 (1933).
[71] G. H. Coleman and R. F. Blomquist, *ibid.*, **63**, 1692 (1941).
[72] A. G. Davies, S. C. W. Hook, and B. P. Roberts, *J. Organometal. Chem.*, **23**, C11 (1970).

is a free-radical chain process involving S_H2 attack of a dialkylamino radical at boron, while the analogue of 36b is a heterolytic process. If a similar mechanism were to obtain in the case of the Grignard reagent, reaction 36a would involve homolytic displacement of an alkyl radical from magnesium by the nitrogen-centered radical, $RR'N^{\cdot}$. A heterolytic process involving electrophilic attack of the relatively positive halogen in the *N*-chloroamine is however an attractive alternative in the case of a Grignard reagent.

$$\begin{matrix} R''\!-\!MgX \\ Cl\!-\!NRR' \end{matrix} \longrightarrow \begin{matrix} R'' \\ | \\ Cl \end{matrix} + \begin{matrix} MgX \\ | \\ NRR' \end{matrix} \qquad (37)$$

The reactions of Grignard reagents with nitric oxide[73,74] and nitrosyl chloride[73] invite further mechanistic study as potential sources of examples of homolytic substitution at magnesium.

Substitution by Sulfur-Centered Radicals

Grignard reagents react with dialkyl and diaryl disulfides to bring about cleavage of the S–S bond.[22,75,76] A similar reaction occurs with diselenides.[77]

$$RMgX + R'SSR' \rightarrow RSR' + R'SMgX \qquad (38)$$

Reaction 38 probably occurs by nucleophilic attack on sulfur,[22] and there is no evidence that free radicals are involved. The sterically hindered di-*t*-butyl disulfide does not react with Grignard reagents,[22] but when a mixture of *t*-butylmagnesium chloride and di-*t*-butyl disulfide in ether was irradiated with ultraviolet light while in the cavity of an e.s.r. spectrometer, a reaction occurred and the spectrum of the *t*-butyl radical was observed.[78] In the absence of the disulfide, photolysis of the Grignard reagent produced only a weak unidentified signal. Although not compelling, this result suggests that homolytic displacement by the *t*-butylthiyl radical at magnesium is a fairly rapid process.

$$Bu^tSSBu^t \xrightarrow{h\nu} 2Bu^tS^{\cdot} \qquad (39)$$

$$Bu^tS^{\cdot} + Bu^tMgCl \rightarrow Bu^tSMgCl + Bu^{t\cdot} \qquad (40)$$

[73] E. Müller and H. Metzger, *Chem. Ber.*, **89**, 396 (1956).
[74] M. H. Abraham, J. H. N. Garland, J. A. Hill, and L. F. Larkworthy, *Chem. Ind. (London)*, 1615 (1962).
[75] H. Wuyts, *Bull. Soc. Chim. France*, **35**, 166 (1906).
[76] H. Burton and W. A. Davy, *J. Chem. Soc.*, 528 (1948).
[77] T. W. Campbell and J. D. McCullough, *J. Amer. Chem. Soc.*, **67**, 1965 (1945).
[78] A. G. Davies and B. P. Roberts, unpublished results.

Substitution by Halogen Atoms

The mechanism of the cleavage by halogens of the C–Mg bond is not established. Further investigation might show that, in certain cases, the mechanism is homolytic, involving S_H2 attack by a halogen atom at magnesium.

3. CALCIUM, STRONTIUM, BARIUM, AND RADIUM

These elements generally form ionic compounds where an S_H2 reaction at the metal center would be unlikely.

A few organometallic compounds of these elements have been prepared, usually in highly solvating solvents.[4,79–81] As expected these organometallic compounds are subject to rapid autoxidation[1,4] which presumably occurs by a radical chain mechanism. However, these reactions are probably best considered as autoxidation of a carbanion, particularly in the case of organometallic compounds of the more electropositive elements in the group, rather than involving S_H2 attack at the metal atom.

B. GROUP IIB

1. ZINC AND CADMIUM

The metal–carbon bond in organozinc and organocadmium compounds is more polar than in organomercury compounds and so there is a predisposition towards heterolytic (electrophilic) mechanisms for the cleavage of the Zn–C and Cd–C bonds compared with Hg–C bonds. Therefore, although there are numerous examples of homolytic substitution at the metal atom in organomercurials there are very few examples of such reactions in the compounds of zinc or cadmium.

Substitution by Oxygen-Centered Radicals

The rapid autoxidation undergone by dialkylzinc and dialkylcadmium compounds has been known for many years.[1] Alkylmetal alkylperoxide and metal di(alkylperoxide) are the initial products of the autoxidation in solution, although subsequent reaction of these peroxides with unoxidized metal

[79] D. Bryce-Smith and A. C. Skinner, *J. Chem. Soc.*, 577 (1963).
[80] M. A. Coles and F. A. Hart, *Chem. Ind.* (*London*), 423 (1968).
[81] K. A. Kocheshkov, N. I. Sheverdina, and M. A. Zemlyanichenko, *Izv. Akad. Nauk SSSR Ser. Khim.*, 2090 (1969).

alkyl can lead to alkoxides which were thought at first to be the initial autoxidation products.[10,82–86]

$$R_2M + O_2 \rightarrow RMOOR \xrightarrow{O_2} (ROO)_2M \quad (41)$$

$$RMOOR + R_2M \rightarrow 2RMOR \quad (42)$$

$$M = Zn, Cd.$$

By analogy with the polar mechanism proposed for the autoxidation of the alkyls of boron and magnesium, it was suggested that the oxidation of organo-zinc and organocadmium compounds in solution proceeded by the heterolytic insertion of oxygen into the metal–carbon bond.[85,86]

$$R_2M + O_2 \longrightarrow \overset{R \curvearrowright O^+}{R\underset{-}{M}—O} \longrightarrow RMOOR \quad (43)$$

The results of studies of the vapor-phase autoxidation of dimethylzinc,[83,87] diethylzinc,[83,87] and dipropylzinc[87] were interpreted in terms of a chain mechanism, the chains originating on the walls of the reaction vessel. However, it was not known if free radicals were involved and no S_H2 processes were suggested.

The effect of small quantities of galvinoxyl, a free-radical scavenger, on the autoxidation of diethylzinc and of dimethylcadmium in solution has shown that these reactions proceed by free-radical chain mechanisms, in common with the autoxidative cleavage of most other carbon–metal bonds.[88] The autoxidation of dimethylcadmium to methylcadmium methylperoxide (MeCdOOMe) in anisole at room temperature was completely inhibited by the addition of 0.35 mole % of galvinoxyl to the solvent prior to addition of the organometallic compound. The inhibited autoxidation could be initiated at any time during the induction period by copper dibutyldithiocarbamate.[88] Galvinoxyl reacted rapidly with diethylzinc and the product of this reaction completely inhibited the autoxidation of the second Zn–C bond, although it had no effect upon the cleavage of the first bond by oxygen. Hence reaction 44 must occur by a free-radical chain process.

$$EtZnOOEt + O_2 \rightarrow (EtOO)_2Zn \quad (44)$$

The autoxidation of the first Zn–C bond in diethylzinc probably also occurs by a free-radical chain mechanism and the inability to observe inhibition of

[82] R. Demuth and V. Meyer, *Chem. Ber.*, **23**, 94 (1890).
[83] H. W. Thompson and N. S. Kelland, *J. Chem. Soc.* 746, 756 (1933).
[84] H. Hock and F. Ernst, *Chem. Ber.*, **92**, 2716, 2723 (1959).
[85] M. H. Abraham, *Chem. Ind. (London)*, 750 (1959); *J. Chem. Soc.*, 4130 (1960).
[86] A. G. Davies and J. E. Packer, *Chem. Ind. (London)*, 1177 (1958); *J. Chem. Soc.*, 3164 (1959).
[87] C. H. Bamford and D. M. Newitt, *J. Chem. Soc.*, 688 (1946).
[88] See footnote 1, Chapter II.

this stage was ascribed to the much higher rate of S_H2 reaction by $EtOO^\cdot$ at diethylzinc compared to that at ethylzinc ethylperoxide. Thus the autoxidation of the alkyls of zinc and cadmium in both solution and vapor phases probably involves homolytic substitution by an alkylperoxy radical at the metal center in one propagation step of the chain process, reaction 45 being faster than 46

$$R^\cdot + O_2 \rightarrow ROO^\cdot \qquad (12)$$

$$ROO^\cdot + R_2M \rightarrow ROOMR + R^\cdot \qquad (45)$$

$$ROO^\cdot + ROOMR \rightarrow (ROO)_2M + R^\cdot \qquad (46)$$

Davies and Packer[86] have noted that traces of pyridine strongly catalyzed the absorption of oxygen by solutions of dimethylcadmium in cyclohexane. Preliminary kinetic investigation of the pyridine-catalyzed autoxidation of dimethylcadmium in isooctane at 30° indicated that the pyridine exerted its effect by causing an increase in the rate of reaction 45 ($R = Me$) rather than by increasing the rate of chain initiation.[89] The rate of S_H2 reaction at cadmium increased steadily until one molar equivalent of pyridine had been added, and then as the concentration of pyridine was further increased, the rate of homolytic displacement at cadmium decreased. In general so long as the metal does not become *coordinatively saturated*, addition of a ligand may either increase or decrease the ease of homolytic substitution at the metal center depending on the specific reaction. The initial results of the autoxidation of dimethylcadmium indicate that 47 is faster than both 45 and 48.

$$MeOO^\cdot + Me_2Cd \leftarrow py \rightarrow MeOOCd(\leftarrow py)Me + Me^\cdot \qquad (47)$$

$$MeOO^\cdot + Me_2Cd \leftarrow py_2 \rightarrow MeOOCd(\leftarrow py_2)Me + Me^\cdot \qquad (48)$$

The autoxidation of diarylzinc and diarylcadmium compounds is more complex but is certainly homolytic[90,91] and probably involves S_H2 displacement by aryloxy radicals at the metal center as one step.

$$PhO^\cdot + Ph_2M \rightarrow PhOMPh + Ph^\cdot \qquad (49)$$

The e.s.r. signals due to methyl or ethyl radicals, respectively, were observed when dimethylcadmium or diethylzinc are caused to react with *t*-butoxy radicals in the cavity of the spectrometer.[92]

$$Bu^tO^\cdot + R_2M \rightarrow Bu^tOMR + R^\cdot \qquad (50)$$

$$M = Zn,\ R = Et;\ M = Cd,\ R = Me$$

[89] A. G. Davies, K. U. Ingold, B. P. Roberts, and R. Tudor, *J. Chem. Soc.* (*B*), to be published.

[90] V. N. Pankratova, U. N. Latyaeva, and G. A. Razuvaev, *Zh. Obshch. Khim.*, **35**, 900, (1965).

[91] R. F. Galiulina, O. N. Druzhkov, G. G. Petukhov, and G. A. Razuvaev, *ibid.*, **35**, 1164, (1965).

[92] A. G. Davies and B. P. Roberts, *J. Organometal. Chem.*, **19**, P17 (1969).

A similar homolytic substitution might account for the production of chlorobenzene in the reaction between *t*-butyl hypochlorite and diphenylzinc or diphenylcadmium.[39]

$$Bu^tO^{\cdot} + Ph_2M \rightarrow Bu^tOMPh + Ph^{\cdot} \quad (51)$$

$$Ph^{\cdot} + Bu^tOCl \rightarrow PhCl + Bu^tO^{\cdot} \quad (52)$$

However, electrophilic cleavage is also a possible mechanism and further work on this reaction is needed. The use of dialkylmetal instead of diphenylmetal compounds would favor a free-radical chain process.

Substitution by Carbon-Centered Radicals

There are no unambiguous examples of this class of S_H2 reaction reported in the literature for zinc and cadmium compounds. Possibly the small amount of ^{14}C found in the diphenylcadmium recovered from partial decomposition of the unlabeled organometallic in ^{14}C-labeled benzene[93] at 215–220° is due to the exchange reaction 53.

$$*Ph^{\cdot} + Ph_2Cd \rightarrow *PhCdPh + Ph^{\cdot} \quad (53)$$

Chain-transfer constants for diethylzinc and dibutylcadmium have been evaluated for the polymerization of styrene at 100°,[94] but the relevance of the results to the present discussion is not clear at present.

Substitution by Nitrogen-Centered Radicals

Once again no examples of this reaction are known for zinc and cadmium compounds. A likely candidate is afforded by the reaction of chlorine azide with dimethylcadmium in carbon tetrachloride.[95] Homolytic substitution by the azido radical might be involved in a free-radical chain mechanism.

$$N_3^{\cdot} + Me_2Cd \rightarrow MeCdN_3 + Me^{\cdot} \quad (54)$$

$$Me^{\cdot} + ClN_3 \rightarrow N_3^{\cdot} + MeCl \quad (55)$$

Substitution by Sulfur-Centered Radicals

The polarity of the Zn–C and Cd–C bonds in organozinc and organocadmium compounds makes it likely that their reaction with thiols[96–99] is a simple heterolytic acidolysis.

$$R_2M + R'SH \rightarrow RMSR' + RH \quad (56)$$

$$M = Zn, Cd$$

[93] G. A. Razuvaev and Y. N. Pankratova, *Zh. Obshch. Khim.*, **36,** 1702 (1966).
[94] T. Huff and E. Perry, *J. Polymer Sci., Part (A)*, **1,** 1553 (1963).
[95] K. Dehnicke, J. Strähle, D. Seybold, and J. Müller, *J. Organometal. Chem.*, **6,** 298 (1966).
[96] J. F. Nelson, *Iowa State Coll. J. Sci.*, **12,** 145 (1937); *Chem. Abstr.*, **32,** 3756 (1938).
[97] H. Gilman and J. F. Nelson, *J. Amer. Chem. Soc.*, **59,** 935 (1937).
[98] G. E. Coates and R. Ridley, *J. Chem. Soc.*, 1870 (1965).
[99] G. E. Coates and A. Lander, *J. Chem. Soc. (A)*, 264 (1966).

It is possible that some of these reactions might involve homolytic displacement by thiyl radicals at the metal center as a propagation step in a free radical chain mechanism for the cleavage of the metal–carbon bond. Similarly, the reaction of thiocyanogen with organozinc and organocadmium compounds to give metal thiocyanate[100,101] requires further investigation.

$$R_2M + (SCN)_2 \rightarrow RMSCN + RSCN \quad (57)$$

Substitution by Halogen Atoms

A considerable study has been made of the free-radical cleavage of organomercury compounds by halogens, but a comparable mechanistic examination has not been made of the corresponding reaction of organozinc and organocadmium compounds. A heterolytic (electrophilic) mechanism would be favored on going from mercury to cadmium and zinc, but the possible occurrence of radical processes involving homolytic substitution by halogen atoms at the metal center has not been satisfactorily eliminated for the halogen cleavage of organozinc and organocadmium compounds.

2. MERCURY

The weak and relatively nonpolar Hg–C bond[102] undergoes ready homolytic cleavage and, as expected, organomercury compounds take part in a variety of free-radical reactions, a large number of which involve bimolecular homolytic substitution at the mercury atom. Two coordinate mercury compounds, like two coordinate compounds of zinc and cadmium, can form donor–acceptor complexes in which the metal atom has increased its coordination number to three or four. It seems likely that a stepwise S_H2 mechanism will often operate, although the intermediate has never been detected.

$$Y^{\cdot} + HgX_2 \rightarrow Y\dot{H}gX_2 \rightarrow YHgX + X^{\cdot} \quad (58)$$

For historical reasons it would seem most appropriate to begin a survey of the S_H2 reactions of mercury compounds with substitutions accomplished by halogen atoms.

[100] F. Söderbäch, *Ann.*, **419**, 217 (1919).
[101] T. Wizemann, H. Müller, D. Seybold, and K. Dehnicke, *J. Organometal. Chem.*, **20**, 211 (1969).
[102] H. A. Skinner, *Adv. Organometal. Chem.*, **2**, 49 (1964) gives $D_{(RHg-R)} = 51.5$ (R = Me) 42.5 (Et), and $D_{(ClHg-R)} = 64.3$ (Me), 60.3 (Et), 66.4 (Ph) all in kcal/mole.

Substitution by Halogen Atoms

A great deal of work has been concerned with the reaction of halogens with organomercury compounds, principally in the study of electrophilic substitution processes.[103–105]

$$RHgX + Y_2 \rightarrow RY + YHgX \qquad (59)$$

$$Y_2 = \text{halogen}$$

The problem of differentiating between a polar (electrophilic) mechanism for 59 and a free-radical chain process, involving S_H2 reaction of a halogen atom at mercury, has been discussed by Jensen and Rickborn.[103] The interplay of radical and polar cleavages was first noted by Keller[106] in a study of the iodine cleavage of alkylmercuric iodides in dioxan. Keller showed that the cleavage was accelerated by light and peroxides and strongly retarded by oxygen. The rate of the retarded reaction was roughly second order in iodine, nearly independent of the alkylmercuric iodide concentration and approximately inversely proportional to the oxygen concentration. The reaction rate was insensitive to changes of structure in the alkyl group attached to mercury. These observations were consistent with a chain reaction involving attack of an iodine atom at mercury to displace an alkyl radical.

$$I^{\cdot} + RHgI \rightarrow R^{\cdot} + HgI_2 \qquad (60)$$

$$R^{\cdot} + I_2 \rightarrow RI + I^{\cdot} \qquad (61)$$

The retardation by oxygen would be due to trapping of the chain-carrying alkyl radicals to give the much less reactive alkylperoxy radical. The radical mechanism for the cleavage by iodine of organomercury compounds could be suppressed by the addition of iodide ions.[106]

Razuvaev and Savitskii[107] investigated the cleavage of a series of dialkylmercurials and diarylmercurials with iodine in carbon tetrachloride.

$$R_2Hg + I_2 \rightarrow RHgI + RI \qquad (62)$$

They reported that as the electronegativity of R was increased (from benzyl to *p*-anisyl) the mechanism changed from "crypto"-radical to "crypto"-ionic, accompanied by a decrease in activation energy (from 12.19 to 4.07 kcals/mole). However, the "crypto"-radical mechanism must be presumed to be a

[103] F. R. Jensen and B. Rickborn, *Electrophilic Substitution of Organomercurials*, McGraw-Hill, New York, 1968.
[104] O. A. Reutov and I. P. Beletskaya, *Reaction Mechanisms of Organometallic Compounds*, North-Holland Publishing Co., Amsterdam, 1968.
[105] O. A. Reutov, *Russ. Chem. Rev.*, **36**, 163 (1967).
[106] J. Keller, *Ph.D. thesis*, University of California, Los Angeles, 1948, as quoted by S. Winstein and T. G. Traylor, *J. Amer. Chem. Soc.*, **78**, 2596 (1956).
[107] G. A. Razuvaev and A. V. Savitskii, *Dokl. Akad. Nauk SSSR*, **85**, 575 (1952).

nonchain process, particularly in view of the first-order dependence on both iodine and mercurial exhibited by both dibenzyl- and di-*p*-anisyl-mercury.[107] Subsequently the homolytic cleavage by halogens of organomercury compounds has been studied on many occasions, primarily by Jensen[103] in the U.S.A. and by Reutov and Beletskaya[104] in the U.S.S.R.

Stereochemical Studies

Depending on the experimental conditions, the stereochemical result of the halogenodemercuration reaction of optically or geometrically isomeric organomercury compounds can range from complete loss of configuration to complete stereospecificity. It has been shown that the free-radical cleavage of the C–Hg bond by halogens results in loss of configuration at the carbon atom, whilst electrophilic cleavage by an S_E2 mechanism generally proceeds with retention of configuration at carbon.[103–105]

The cleavage by bromine of *cis*- and *trans*-4-methylcyclohexylmercuric bromide in carbon tetrachloride or carbon disulfide under nitrogen was investigated by Jensen and Gale.[108] Under these conditions free-radical cleavage predominated and the product distribution was 47.4% *cis*- and 52.6% *trans*-4-methylcyclohexyl bromide from either geometrical isomer of the mercurial. When more polar solvents (acetic acid, methanol), or hypobromous acid and other sources of more positive bromine were used to cleave the mercurial, each isomer gave predominantly the product formed with retention. Under these conditions, addition of air to the reaction vessel resulted in a marked increase in the degree of retention. With bromine in pyridine under air, the cleavages were quantitatively stereospecific, and even the *cis*-isomer of the mercurial, which was more susceptible to radical reaction, gave complete retention. The reason for the high degree of stereospecificity observed with bromine in pyridine is not completely clear.[103] The electrophile involved may be the (pyridine Br_2) complex, its dissociation product (pyridine Br^+), or even a small concentration of free bromine. The effect of the pyridine on the reaction mechanism may be to function as a free-radical scavenger (presumably for bromine atoms), to complex with the mercurial thereby increasing the carbanionic character of the organic group attached to mercury, or merely to serve as a very polar solvent, stabilizing the transition state for electrophilic substitution. It is probable that a combination of all these effects causes the predominance of the polar cleavage mechanism at the expense of the homolytic process.

Jensen[26,109] has further investigated the "radical" cleavage of 4-alkylcyclohexylmercuric halides by bromine or by sulfuryl chloride and has interpreted the distribution of geometrically isomeric halides produced in terms

[108] F. R. Jensen and L. H. Gale, *J. Amer. Chem. Soc.*, **81,** 1261 (1959); **82,** 148 (1960).
[109] F. R. Jensen and J. Miller, unpublished results reported in footnote 103, p. 79.

Table I Product Distribution from Radical Cleavages of *cis*- and *trans*-4-Alkylcyclohexylmercuric Halides[26]

Isomer	Alkylmercuric Halide	Conditions	Yield (%)	Product Distribution (%)	
				cis	*trans*
trans-4-*t*-Butyl	Br	Br_2, CCl_4, 0°	55	50.6	49.4
cis-4-*t*-Butyl	Br	Br_2, CCl_4, 0°	50	51.5	48.5
trans-4-Methyl	Br	Br_2, CCl_4, 0°	62	47.6	52.4
trans-4-Methyl	Br	Br_2, CS_2, 0°	—	47.2	52.8
cis-4-Methyl	Br	Br_2, CCl_4, 0°	67	47.4	52.6
trans-4-*t*-Butyl	Cl	SO_2Cl_2, CS_2, 46°	28	70.2	29.8
cis-4-*t*-Butyl	Cl	SO_2Cl_2, CS_2, 46°	38	69.2	30.8

of the torsional strain [110] developed in the transition state for the reaction of an intermediate 4-alkylcyclohexyl radical.[26] Some results are reproduced in Table I.

Propagation steps of a generalized free-radical chain mechanism which is consistent with the data are shown in equations 63–65.

$$X^{\cdot} + RHgX \longrightarrow \left[R\text{—}\dot{H}g(X)\text{—}X \leftrightarrow R\text{—}Hg(\dot{X})\text{—}X \leftrightarrow R^{\cdot}\,Hg(X)\text{—}X \text{ and so on} \right] \quad (63)$$

$$R\text{—}\dot{H}g(X)\text{—}X \longrightarrow R^{\cdot} + HgX_2 \quad (64)$$

$$R^{\cdot} + X_2 \longrightarrow RX + X^{\cdot} \quad (65)$$

$$X = Br$$

In the case of the reactions with sulfuryl chloride, it was considered that the mechanism involved 63 and 64 (X = Cl) followed by 66 and 67.

$$R^{\cdot} + SO_2Cl_2 \rightarrow RCl + \dot{S}O_2Cl \quad (66)$$

$$\dot{S}O_2Cl \rightleftharpoons SO_2 + Cl^{\cdot} \quad (67)$$

Because the reaction was conducted in refluxing solvent the equilibrium 67 would be displaced to the right, and the chlorine atom would be the species bringing about S_H2 displacement at mercury.[111] The step in which the alkyl halide is formed [equation 65] is very exothermic for the cleavage by bromine and little product structure will be developed in the transition state

[110] P. von R. Schleyer, *J. Amer. Chem. Soc.*, **89,** 699, 701 (1967).
[111] G. A. Russell and H. C. Brown, *ibid.*, **77,** 4031 (1955).

(Hammond's postulate). The activation energies for the formation of *cis*-halide or *trans*-halide will be both small and similar, and as axial or equatorial approach of the halogen to the 4-*t*-butylcyclohexyl radical are about equally probable, a 50:50 mixture of *cis*-products and *trans*-products should result.

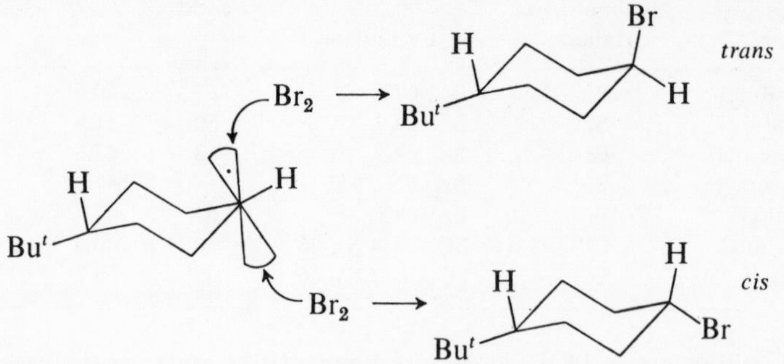

(68)

However, reaction 69 is less exothermic than 68 so that the transition state of 69 will take on more product character and torsional strain between the C_1–H_1 and C_2–$H_{2\ equatorial}$ (and C_6–$H_{6\ equatorial}$) bonds will be developed and contribute selectively to the activation energy for the formation of *trans*-product.[26] Thus the different product distributions resulting from the

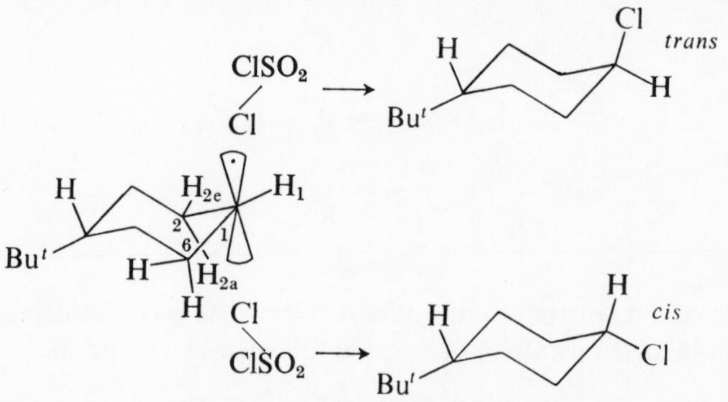

(69)

intermediacy of the same 4-alkylcyclohexyl radical are accounted for.

Jensen and coworkers have also studied the bromine cleavage of optically active *s*-butylmercuric bromide and depending on the conditions, each reaction path, radical or polar, could be made to predominate or to compete.[112,113] In nonpolar solvents complete racemization was observed.

$$(-)\mathrm{Bu^sHgBr} \xrightarrow[\mathrm{N_2}]{\mathrm{Br_2/CS_2}} (\pm)\mathrm{Bu^sBr} \qquad (70)$$

[112] F. R. Jensen, L. D. Whipple, D. K. Wedegaertner, and J. A. Landgrebe, *ibid.*, **81,** 1262 (1959).

[113] *Idem.*, *ibid.*, **82,** 2466 (1960).

Reutov and his collaborators investigated both the kinetics and stereochemistry of the cleavage by bromine of optically active *s*-butylmercuric bromide in carbon tetrachloride.[114,115] Although this light-catalyzed reaction was first order with respect to bromine and zero order with respect to mercurial, which was taken to be indicative of a completely radical cleavage, the rotation of the *s*-butyl bromide formed corresponded to 35–40% retention of configuration when the cleavage was carried out in the presence of air.[115] In the absence of oxygen, complete racemization was observed.[115,116] It seems unlikely that the optically active *s*-butyl bromide was formed by a homolytic mechanism (for example by S_H2 reaction at carbon[115]); the presence of oxygen probably results in inhibition of the radical process to such a degree that some electrophilic cleavage involving retention of configuration can compete.

A few other stereochemical studies have been carried out on the halogenodemercuration reaction. For example, *cis*-2-chlorovinylmercuric or *trans*-2-chlorovinylmercuric chloride reacts with iodine in nonpolar solvents (benzene or carbon tetrachloride) to give the same equimolar mixture of *cis*-chloroiodoethylene and *trans*-chloroiodoethylene, the rate of reaction depending on the degree of illumination.[117]

$$\begin{array}{c} \begin{matrix} H & & HgCl \\ & C{=}C & \\ Cl & & H \end{matrix} \\ OR \\ \begin{matrix} Cl & & HgCl \\ & C{=}C & \\ H & & H \end{matrix} \end{array} \xrightarrow[\text{benzene, } CCl_4]{I_2} \begin{array}{c} \begin{matrix} H & & I \\ & C{=}C & \\ Cl & & H \end{matrix} \quad 50\% \\ + \\ \begin{matrix} Cl & & I \\ & C{=}C & \\ H & & H \end{matrix} \quad 50\% \end{array} \qquad (71)$$

All the above stereochemical results are consistent with the products of the halogen cleavage of organomercurials being determined by the reaction of the carbon radical, produced by S_H2 cleavage of the C–Hg bond, by attack at mercury with the halogen.

[114] O. A. Reutóv, E. V. Uglova, I. P. Beletskaya, and T. B. Svetlanova, *Izv. Akad. Nauk SSSR, Ser. Khim.*, 1383 (1964).
[115] Footnote 104 (p. 199).
[116] E. V. Uglova, T. B. Svetlanova, I. P. Beletskaya, and O. A. Reutov, *Izv. Akad. Nauk SSSR, Ser. Khim.*, 1151 (1968).
[117] I. P. Beletskaya, V. I. Karpov, and O. A. Reutov, *Dokl. Akad. Nauk SSSR*, **161,** 586 (1965); *Izv. Akad. Nauk SSSR, Ser. Khim.*, 963 (1966).

Kinetic Studies

Keller[106] noted that the rate of cleavage by iodine of the C–Hg bond in alkylmercuric iodides, under conditions where the radical mechanism would predominate, was independent of the concentration of the mercurial. Subsequently this kinetic behavior has been confirmed on numerous occasions. For example, in the absence of added cadmium iodide, benzylmercuric bromide and ethyl α-bromomercuriphenylacetate in aqueous dioxan are readily cleaved by iodine, the former compound reacts instantaneously, the latter at a measurable rate accelerated by light and the reaction was independent of organomercury concentration and first order with respect to iodine.[118] In the presence of cadmium iodide, the radical cleavage was suppressed and the reaction became first order in both iodine and mercurial. These results seem fairly general, the radical reaction being independent of mercurial concentration and first order with respect to halogen,[119–123] whilst the electrophilic cleavage is first order with respect to both reactants.[103–105] In those cases where the rate of the radical reaction has been shown to be independent of mercurial concentration, initiation of the chain by homolytic scission of the halogen is indicated, although it appears that the reaction proceeds rapidly in the dark,[124] and some initiating radicals may arise from the molecular interaction of the reactants.[124]

$$RHgX + X_2 \rightarrow R^{\cdot} + HgX_2 + X^{\cdot} \tag{72}$$

The influence of illumination on the rate of the cleavage by bromine of *s*-butylmercuric bromide has been studied using light from an incandescent lamp to avoid dissociation of the C–Hg bond in the mercurial.[115] In the absence of air, the rate was proportional to the square root of the light intensity, while in air it was proportional to the first power of the intensity, the overall rate being much slower in the presence of oxygen. The quantum yield was greater than unity, confirming a chain reaction, and it was proposed that initiation occurred by photolytic cleavage of the bromine molecule into atoms. Alternative propagation steps were suggested involving homolytic substitution at either the carbon or the mercury atom of the C–Hg bond. An S_H2 reaction at the sp^3-hybridized carbon atom is thermodynamically less favorable[103] as well as being unlikely for kinetic reasons (see Chapter V).

[118] O. A. Reutov and I. P. Beletskaya, *Izv. Akad. Nauk SSSR, Otd. Khim. Nauk*, 1716 (1960).
[119] I. P. Beletskaya, O. A. Reutov, and V. I. Karpov, *ibid.*, 2125 (1961).
[120] I. P. Beletskaya, O. A. Reutov, and T. P. Gur'yanova, *ibid.*, 2178 (1961).
[121] I. P. Beletskaya, T. A. Azizyan, and O. A. Reutov, *ibid.*, 223 (1962).
[122] *Idem.*, *ibid.*, 1332 (1963).
[123] I. P. Beletskaya, A. V. Ermanson, and O. A. Reutov, *Izv. Akad. Nauk SSSR, Ser. Khim.*, 231 (1965).
[124] Footnote 103 (pp. 80–81).

The remaining chain mechanism, detailed in equations 73–78 is similar to that proposed previously.

Initiation: $$Br_2 \xrightarrow{h\nu} 2Br^{\cdot} \quad (73)$$

Propagation: $$Br^{\cdot} + Bu^{s}HgBr \rightarrow Bu^{s\cdot} + HgBr_2 \quad (74)$$

$$Bu^{s\cdot} + Br_2 \longrightarrow Bu^{s}Br + Br^{\cdot} \quad (75)$$

Termination:
$$\left.\begin{array}{l} Br^{\cdot} + Br^{\cdot} \quad (76) \\ Br^{\cdot} + Bu^{s\cdot} \quad (77) \\ Bu^{s\cdot} + Bu^{s\cdot} \quad (78) \end{array}\right\} \rightarrow \text{inactive products}$$

The dependence of the rate on the square root of the light intensity implies that termination is second order as shown in 76–78. In the presence of oxygen, it was proposed that termination occurred by trapping of the *s*-butyl radical by oxygen, giving rise to the observed dependence upon the first power of the light intensity.[115]

$$Bu^{s\cdot} + O_2 \rightarrow Bu^{s}OO^{\cdot} \quad (79)$$

The thermally initiated reaction of *s*-butylmercuric bromide with bromine (under nitrogen and in the dark) has also been studied.[125] The activation energy, 41.4 kcal/mole, is in fairly good agreement with the energy required for homolysis of molecular bromine (45.5 kcal/mole) (compare footnote 103). The lack of dependence of the rate of the cleavage on the concentration of the organomercury compound implies that reaction 74 is faster than 75, itself a very fast reaction, and termination would be mainly by 77 and 78. Although the mechanism shown in 73–78 probably does not represent the complete story, there seems no doubt that bimolecular homolytic substitution at mercury by halogen atoms is involved in the halogenodemercuration of organomercurials. Further kinetic study of this reaction in order to evaluate the absolute rate constants for S_H2 processes such as 74 should prove very fruitful.

Homolytic substitution by chlorine atoms at mercury has been suggested to account for the phenylmercuric chloride produced in the γ-radiolysis of diphenylmercury in chloroform,[126]

$$CHCl_3 \xrightarrow{\gamma} \dot{C}HCl_2 + Cl^{\cdot} \quad (80)$$

$$Cl^{\cdot} + Ph_2Hg \rightarrow PhHgCl + Ph^{\cdot} \quad (81)$$

however 82 and 83 would seem to be likely alternatives.

$$Ph_2Hg \xrightarrow{\gamma} PhHg^{\cdot} + Ph^{\cdot} \quad (82)$$

$$PhHg^{\cdot} + CHCl_3 \rightarrow PhHgCl + \dot{C}HCl_2 \quad (83)$$

[125] Footnote 104 (p. 201).

[126] C. Heitz and J.-P. Adloff, *Compt. Rend.*, **256,** 416 (1963); *J. Organometal. Chem.*, **2,** 59 (1964).

Substitution by Oxygen-Centered Radicals

In contrast to most metal–oxygen bonds, the Hg–O bond is relatively weak,[127] probably weaker than the Hg–C bond in dialkylmercurials and alkylmercuric compounds (40–60 kcal/mole). There are nevertheless a number of reported reactions which involve bimolecular homolytic substitution by oxygen-centered free radicals at mercury.

Until comparatively recently it was not appreciated that many organomercury compounds underwent autoxidation when exposed to air or oxygen. This autoxidation is especially facile when the mercury atom is linked to a secondary-alkyl or tertiary-alkyl group (or a benzylic or allylic group). The presence of aerial oxygen accelerated the thermal decomposition of diphenylmercury in benzene (at 260°) to mercury and biphenyl, as well as the exchange of phenyl groups between labeled benzene and diphenylmercury.[128]

When dry oxygen was bubbled through a solution of dicyclohexylmercury in isopropyl alcohol for 40 hr at 60°, mercury (64.4% based on mercurial), acetone (20%), cyclohexanol (44%), and cyclohexanone (43%) were produced. When the autoxidation was conducted in chloroform or carbon tetrachloride, high yields of cyclohexylmercuric chloride (about 70%) were obtained as well as other products typical of a free-radical reaction.[129,130] In spite of this, the oxidation in isopropanol was pictured as involving the formation of an unstable intermediate peroxide or oxygen complex the subsequent decomposition of which gave equimolar amounts of cyclohexanol and cyclohexanone.[129,130]

$$(C_6H_{11})_2Hg + O_2 \rightarrow [(C_6H_{11})_2HgO_2] \rightarrow Hg + C_6H_{11}OH + C_6H_{10}O \tag{84}$$

Kreevoy and Hansen[131] found that the reactions of isopropylmercuric and *t*-butylmercuric iodides with aqueous nonhalogen acids were profoundly influenced by the presence of oxygen. These authors, who did not know of Razuvaev's work,[129,130] speculated that the mechanism might involve the intermediate alkylperoxymercuric iodide, ROOHgI. A related effect of oxygen on the protolysis of dibenzylmercury has also been noted.[132]

[127] S. W. Benson, *J. Chem. Ed.*, **42,** 502 (1965) gives the bond energy in a hypothetical gas-phase molecule of mercuric oxide as, at most, 26 kcal/mole.
[128] G. A. Razuvaev, G. G. Petukhov, and Yu. A. Kaplin, *Dokl. Akad. Nauk SSSR*, **135,** 342 (1960).
[129] G. A. Razuvaev, G. G. Petukhov, S. F. Zhil'tsov, and L. F. Kudryavtsev, *ibid.*, **135,** 87 (1960).
[130] *Idem.*, *ibid.*, **144,** 810 (1962).
[131] M. M. Kreevoy and R. L. Hansen, *J. Phys. Chem.*, **65,** 1055 (1961).
[132] B. F. Hegarty, W. Kitching, and P. R. Wells, *J. Amer. Chem. Soc.*, **89,** 4816 (1967).

Razuvaev and co-workers investigated the autoxidation of diisopropylmercury in various solvents at 50° and again concluded that an intermediate oxygen complex was involved which rearranged to an alkylmercuric alkyl peroxide which could subsequently react with unchanged dialkylmercury or undergo Hg–O homolysis.[133]

$$R_2Hg + O_2 \longrightarrow [R_2Hg{:}O_2] \longrightarrow [ROOHgR] \quad \xrightarrow{R_2Hg} 2ROHgR \quad \text{or} \quad \longrightarrow RO_2^{\cdot} + {}^{\cdot}HgR \qquad (85)$$

$$RO_2^{\cdot} + R_2Hg \longrightarrow ROH + R_{-H}{=}O + RHg^{\cdot} \qquad (86)$$

The autoxidation of di-*t*-amylmercury and of dibenzylmercury was also reported.[133]

The mechanism of autoxidation of organomercury compounds was subsequently examined in greater detail by Razuvaev and collaborators.[134] The autoxidation of diisopropylmercury in nonane at 60–85° was found to be an autocatalytic process which could be inhibited by the addition of small quantities of *p*-hydroxydiphenylamine. Hence the oxidation occurred by a free-radical chain mechanism. The apparent activation energy for the reaction, obtained from the temperature dependence of the maximum rate, was 20 kcal/mole, and once the maximum rate had been attained it was independent of the pressure of oxygen above the solution in the range 200–500 mm Hg, implying chain termination by alkylperoxy radicals only. The products of the autoxidation, which was approximately three-halves order in mercurial up to a comparatively high degree of reaction, were mercury, isopropylmercuric isopropoxide, isopropylmercuric hydroxide, acetone, isopropyl alcohol, and small quantities of propylene and propane; some oxidation of the solvent also occurred.[134] At the beginning of the reaction, initiation was thought to occur by molecular interaction of oxygen with diisopropylmercury to give an intermediate peroxide which then decomposed homolytically.

$$R_2Hg + O_2 \longrightarrow RHgOOR \longrightarrow \begin{cases} \xrightarrow{R_2Hg} 2RHgOR \\ \longrightarrow RHg^{\cdot} + RO_2^{\cdot} \end{cases} \qquad (85)$$

R = isopropyl to reaction 96.

[133] G. A. Razuvaev, S. F. Zhil'tsov, O. N. Druzhkov, and G. G. Petukhov, *Dokl. Akad. Nauk SSSR*, **152**, 633 (1963).

[134] Yu. A. Aleksandrov, O. N. Druzhkov, S. F. Zhil'tsov, and G. A. Razuvaev, *ibid.*, **157**, 1395 (1964); *Zh. Obshch. Khim.*, **35**, 1440 (1965).

It should be noted that the reduction of the rate of oxygen uptake to virtually zero by the inhibitor precludes the formation of all but a minute amount of a peroxide by a molecular insertion mechanism. Alkylmercury radicals can decompose [$D_{(R-Hg\cdot)} \sim 6$ kcal/mole] or react with oxygen.

$$RHg^{\cdot} \begin{cases} \xrightarrow{O_2} RHgO_2^{\cdot} \\ \longrightarrow R^{\cdot} + Hg \end{cases} \tag{87}$$

$$R^{\cdot} + O_2 \longrightarrow ROO^{\cdot} \tag{12}$$

Formation of RHgOR was thought to occur by the interaction of $ROO^{\cdot}$ or $RHgO_2^{\cdot}$ with dialkylmercury, but the mechanism was not discussed, and S_H2 attack by the alkylperoxy radical to give the alkylmercuric alkylperoxide (RHgOOR) was apparently not considered.

$$ROO^{\cdot} + R_2Hg \rightarrow RHgOR + RO^{\cdot} \tag{88}$$

$$RHgO_2^{\cdot} + R_2Hg \rightarrow RHgOR + RHgO^{\cdot} \tag{89}$$

It was proposed that the alkoxy radical produced in 88 brought about homolytic displacement of an alkyl radical from mercury,

$$RO^{\cdot} + R_2Hg \rightarrow ROHgR + R^{\cdot} \tag{90}$$

but the radical $RHgO^{\cdot}$ was thought to displace an alkyl radical by attack at carbon, an unlikely reaction.

$$RHgO^{\cdot} + R_2Hg \rightarrow RHgOR + RHg^{\cdot} \tag{91}$$

After the initial stages of the autoxidation, the autocatalytic effect was shown to be due to homolytic decomposition of the product, RHgOR, addition of which, before the start of the reaction, caused the oxidation to begin at its maximum rate and show no further autocatalysis.

$$RHgOR \rightarrow RHg^{\cdot} + {}^{\cdot}OR \tag{92}$$

Mercuric isopropoxide, isopropylmercuric hydroxide, acetone, and isopropyl alcohol were thought to arise from further reactions of RHgOR.

$$RHgO_2^{\cdot} + RHgOR \rightarrow (RO)_2Hg + RHgO^{\cdot} \tag{93}$$

$$ROO^{\cdot} + RHgOR \rightarrow (RO)_2Hg + RO^{\cdot} \tag{94}$$

$$RHgO^{\cdot} + RHgOR \rightarrow RHgOH + Me_2CO + RHg^{\cdot} \tag{95}$$

$$RO^{\cdot} + RHgOR \rightarrow ROH + Me_2CO + RHg^{\cdot} \tag{96}$$

Reactions 95 and 96 involve α-hydrogen abstraction from the isopropylmercury group.

Jensen and Heyman[135] obtained similar products from the autoxidation of di-*s*-butylmercury. They also proposed a chain mechanism which differed from that of the Russian workers in that the formation of an alkylmercuric alkylperoxide was suggested to occur by a homolytic displacement reaction of an alkylperoxy radical at mercury.

$$ROO^{\cdot} + R_2Hg \rightarrow RHgOOR + R^{\cdot} \tag{97}$$

$$R^{\cdot} + O_2 \rightarrow ROO^{\cdot} \tag{12}$$

$$RHgOOR \rightarrow RHgO^{\cdot} + \dot{O}R \tag{98}$$

$$RO^{\cdot} + R_2Hg \rightarrow ROHgR + R^{\cdot} \qquad R = s\text{-butyl} \tag{90}$$

Homolytic substitution at mercury by an alkoxy radical was proposed by both groups of workers and some direct support for such a reaction has been obtained by Krusic and Kochi.[136] The e.s.r. spectrum of the *n*-butyl radical was observed when a solution containing di-*t*-butyl peroxide and di-*n*-butylmercury was irradiated with UV light while in the cavity of the spectrometer; presumably little or no signal was detected in the absence of peroxide.

$$Bu^tOOBu^t \xrightarrow{h\nu} 2Bu^tO^{\cdot} \tag{99}$$

$$Bu^tO^{\cdot} + Bu_2Hg \rightarrow Bu^tOHgBu + Bu^{\cdot} \tag{100}$$

Walling and Kjellgren[39] found that although there was little or no reaction between *t*-butyl hypochlorite and diphenylmercury in benzene in the dark, irradiation brought about reaction to give chlorobenzene, acetone, and *t*-butyl alcohol.

$$Bu^tOCl \xrightarrow{h\nu} Bu^tO^{\cdot} + Cl^{\cdot} \tag{101}$$

The incursion of a free-radical chain reaction of short length would seem probable, with propagation steps 52 and 102.

$$Ph^{\cdot} + Bu^tOCl \rightarrow PhCl + Bu^tO^{\cdot} \tag{52}$$

$$Bu^tO^{\cdot} + Ph_2Hg \rightarrow Bu^tOHgPh + Ph^{\cdot} \tag{102}$$

A better system for the observation of such a chain reaction would comprise dialkylmercury and *t*-butyl hypochlorite with mild thermal initiation from, for example, di-*t*-butyl hyponitrite. Reaction 102 would probably be involved in the thermal (100–130°) reaction of *t*-butyl hydroperoxide and di-*t*-butyl peroxide with diphenylmercury, which was thought to involve radical intermediates.[137]

[135] F. R. Jensen and D. Heyman, *J. Amer. Chem. Soc.*, **88**, 3438 (1966).

[136] P. J. Krusic and J. K. Kochi, *ibid.*, **91**, 3942 (1969).

[137] S. F. Zhil'tsov, L. F. Kudryavtsev, O. N. Druzhkov, M. A. Shubenko, and G. G. Petukhov, *Zh. Obshch. Khim.*, **37**, 2018 (1967).

In sharp contrast to the free-radical chain mechanism accepted for the autoxidation of dialkylmercurials, the autoxidation of bis(triethylgermyl)-mercury was thought to proceed by a molecular insertion mechanism.[138,139] This conclusion was based on the failure of 10 moles % of 2,6-di-*t*-butyl-4-methylphenol or of 2,4,6-tri-*t*-butylphenol to inhibit the autoxidation. At room temperature in octane the reaction is described by 103.

$$Et_3GeHgGeEt_3 + 0.5O_2 \rightarrow \underset{90\text{–}100\%}{Et_3GeOGeEt_3} + \underset{92\%}{Hg} \qquad (103)$$

At below −10°, less mercury is deposited and an unstable product $Et_3GeHgOGeEt_3$ is assumed to be formed. The peroxide $Et_3GeHgOOGeEt_3$ is a postulated intermediate in the autoxidation which was catalyzed by nucleophilic ligands, for example triethylamine and ammonia. Such a mechanistic change on going from a dialkylmercurial to the germanium analogue seems a little unlikely. It is conceivable that the products formed by the trapping with a phenol of a chain-carrying $Et_3GeOO\cdot$ radical undergo rapid radical-producing reactions under the experimental conditions (compare with the autoxidation of organoboron compounds). It is worth noting that the *lack* of inhibition by a known radical scavenger is poor evidence for the absence of a free-radical chain process.

Homolytic substitution by the benzoyloxy radical has been proposed to explain the high yield of isopropylmercuric benzoate obtained from the reaction of benzoyl peroxide with diisopropylmercury in benzene at 70°.[140]

$$PhCO_2^{\cdot} + Pr_2^iHg \rightarrow PhCO_2HgPr^i + Pr^{i\cdot} \qquad (104)$$

It is also possible that S_H2 reactions of oxy radicals at mercury are involved in the interaction of benzylmercuric nitrate with mercuric nitrate in aqueous or methanolic solution.[141]

Substitution by Carbon-Centered Radicals

Oswin, Rebbert, and Steacie[142] studied the photolysis of mixtures of hexadeuterioacetone and dimethylmercury in the gas phase and obtained CH_3CD_3

[138] G. A. Razuvaev, Yu. A. Aleksandrov, V. N. Glushakova, and G. N. Figurova, *J. Organometal. Chem.*, **14,** 339 (1968).
[139] Yu. A. Aleksandrov, G. A. Razuvaev, G. N. Figurova, and V. N. Glushakova, *Dokl. Akad. Nauk SSSR*, **185,** 1293 (1969).
[140] G. A. Razuvaev, S. F. Zhil'tsov, O. N. Druzhkov, and G. G. Petukhov, *Zh. Obshch. Khim.*, **36,** 258 (1966).
[141] J. H. Robson and G. F. Wright, *Can. J. Chem.*, **38,** 1 (1960).
[142] H. G. Oswin, R. Rebbert, and E. W. R. Steacie, *ibid.*, **33,** 472 (1955).

as one of the products, which they took as evidence for the homolytic displacement reaction 106.

$$CD_3COCD_3 \xrightarrow{h\nu} 2\dot{C}D_3 + CO \tag{105}$$

$$\dot{C}D_3 + (CH_3)_2Hg \longrightarrow CD_3HgCH_3 + \dot{C}H_3 \tag{106}$$

$$\dot{C}H_3 + \dot{C}D_3 \longrightarrow CH_3CD_3 \tag{107}$$

Repetition of this work confirmed the conclusion that the isotopic ethanes were produced mainly by methyl radical combination because the ratio

$$\frac{[CH_3CD_3]}{\{[C_2H_6][C_2D_6]\}^{1/2}}$$

was close to 2 and independent of pressure and light intensity.[143] Reaction 106 was again considered likely to account for the production of methyl radicals and an activation energy of 12.6 kcal/mole was assigned to it, a value which should be compared with that determined for hydrogen abstraction from the side chain to give CD_3H, namely, 10.0 ± 1.0 kcal/mole[142]

$$\dot{C}D_3 + (CH_3)_2Hg \rightarrow CD_3H + \dot{C}H_2HgCH_3 \tag{108}$$

The S_H2 reaction at mercury was also identified in the liquid phase by the detection of 12.5% of CD_3H in the methane fraction obtained after photolysis (>3600 Å) of azomethane and $(CD_3)_2Hg$.[143] However, the same authors cast some doubt on their interpretation of the results in terms of reaction 106 by determining the apparent deuterium isotope effect on this displacement to be about 3 at 453°K.[144] In order to resolve this point they followed the course of the photolysis of hexadeuterioacetone and dimethylmercury by the formation of CD_3HgCH_3.[144] This gave a lower rate for 106 than had been obtained by following the production of the isotopic methanes and ethanes, and it was clear that methyl radicals were being produced by reactions other than 106, for example 109 and 110.[144]

$$R^{\cdot} + (CH_3)_2Hg \rightarrow RHgCH_3 + \dot{C}H_3 \tag{109}$$

R = any radical in the system.

$$CH_3Hg\dot{C}H_2 \rightarrow \dot{C}H_3 + HgCH_2 \tag{110}$$

The general conclusion that homolytic displacement by methyl radicals at dimethylmercury could occur was still considered valid especially in the liquid phase experiments with azomethane.[144]

[143] R. E. Rebbert and P. Ausloos, *J. Amer. Chem. Soc.*, **85**, 3086 (1963).
[144] R. E. Rebbert and P. Ausloos, *ibid.*, **86**, 2068 (1964).

Jensen and Gale[145] have reported that the equilibration of *cis*-4-methylcyclohexylmercuric and *trans*-4-methylcyclohexylmercuric bromides can be brought about in dioxan at 98° in the presence of benzoyl peroxide as catalyst. Similarly, when *trans*-2-chlorovinylmercuric chloride was heated in various solvents in the presence of peroxide catalysts, the *cis*-isomer was isolated in 20–90% yield.[146] This latter isomerization was inhibited by hydroquinone. Both isomerizations probably proceed by way of a free-radical chain mechanism involving homolytic substitution at mercury.

$$X^{\cdot} + R^{*}HgHal \rightarrow XHgHal + R^{\cdot} \quad (111)$$

$$R^{\cdot} + R^{*}HgHal \rightleftharpoons RHgHal + R^{\cdot} \quad (112)$$

$$X = \text{initiating radical; Hal} = \text{Br, Cl,}$$

$$R = \text{2-chlorovinyl, 4-methylcyclohexyl.}$$

It is possible, though less likely, that the vinyl isomerization could proceed by reversible radical addition to the double bond. The substitution 112 may well proceed by way of an intermediate electron deficient three-coordinate mercury radical,[103] when the mechanism would be designated S_H2 (stepwise).

$$R^{\cdot} + R^{*}HgHal \rightleftharpoons [RR^{*}HgHal]^{\cdot} \rightleftharpoons RHgHal + R^{\cdot} \quad (113)$$

Similar *cis-trans* interconversions of unsaturated organomercury compounds can be brought about by means of initiation by UV light[147] or mild heating.[148] Such alkyl-exchange reactions could account for the racemization in solution observed for optically active 1-phenylethylmercuric chloride,[149] a compound which should be subject to facile free-radical autoxidation. A similar radical-induced racemization could account for the almost complete inactivity of the di-*s*-butylmercury produced by sodium stannite reduction of *s*-butylmercuric bromide,[112] a result which was originally thought to be inconsistent with the proposed mechanism.[150]

$$2RHgX \xrightarrow{\text{reducing agent}} RHgHgR \quad (114)$$

$$RHgHgR \rightarrow RHg^{\cdot} + Hg + R^{\cdot} \quad (115)$$

$$R^{\cdot} + \dot{H}gR \rightarrow R_2Hg \quad (116)$$

Borisov[151] reported that the liberation of alkylmercuric and arylmercuric chlorides from solutions of organomercury compounds in carbon tetrachloride was accelerated by the addition of acyl peroxides and thus was

[145] F. R. Jensen and L. H. Gale, *ibid.*, **81**, 6337 (1959).
[146] A. N. Nesmeyanov, A. E. Borisov, and V. D. Vil'chevskaya, *Izv. Akad. Nauk SSSR, Otd. Khim. Nauk*, 578 (1949).
[147] A. N. Nesmeyanov, A. E. Borisov, and A. N. Abramova, *ibid.*, 289 (1947).
[148] A. N. Nesmeyanov, A. E. Borisov, and N. A. Vol'kenau, *ibid.*, 992 (1954).
[149] D. S. Matteson and R. A. Bowie, *J. Amer. Chem. Soc.*, **87**, 2587 (1965).
[150] T. G. Traylor and S. Winstein, *J. Org. Chem.*, **23**, 1796 (1958).
[151] A. E. Borisov, *Izv. Akad. Nauk SSSR, Otd. Khim. Nauk*, 524 (1951).

probably a radical chain reaction. This reaction has been studied on many occasions since but radical displacements at mercury by the trichloromethyl radical do not appear to be involved.[152–154]

The use of alkylmercurials as a convenient source of alkyl radicals when heated with benzoyl peroxide has been proposed.[155] It was believed that this reaction involved stepwise S_H2 displacement of the alkyl radical from mercury by the phenyl radical (compare footnote 140, where the benzoyloxy radical was thought to be the displacing species). The ready displacement of alkyl radicals was ascribed to the increased strength of aryl carbon–mercury bonds.

$$PhC(O)OOC(O)Ph \rightarrow 2PhCO_2^{\cdot} \rightarrow 2Ph^{\cdot} + 2CO_2 \quad (117)$$

$$Ph^{\cdot} + RHgX \rightleftharpoons [Ph(R)HgX]^{\cdot} \quad (118)$$

$$[Ph(R)HgX]^{\cdot} \rightleftharpoons PhHgX + R^{\cdot} \quad (119)$$

It is possible that the small amount of exchange of phenyl groups between diphenylmercury and the labeled benzene solvent which takes place in the thermal decomposition of the mercurial[128,156] occurs by way of reaction 121.

$$Ph^{\cdot} + {}^{*}PhH \rightarrow PhH + {}^{*}Ph^{\cdot} \quad (120)$$

$${}^{*}Ph^{\cdot} + Ph_2Hg \rightarrow PhHgPh^{*} + Ph^{\cdot} \quad (121)$$

The diazo method of synthesizing arylmercury compounds, proposed by Nesmeyanov,[157] consists of the decomposition of the double salts of aryldiazonium halides and mercuric halides in solvents by means of copper powder or copper bronze.

$$(ArN_2X\ HgX_2) + 2Cu \rightarrow ArHgX + Cu_2X_2 + N_2 \quad (122)$$

The mechanism is thought to be homolytic: the metal reduces the diazonium cation to give the aryl radical and nitrogen and the former alkylates the mercuric salt.

Razuvaev and Ol'dekop have shown that alkylmercury and arylmercury compounds are formed in the acyl peroxide initiated decarboxylation of

[152] A. N. Nesmeyanov, A. E. Borisov, E. I. Golubeva, and A. I. Kovredov, *Izv. Akad. Nauk SSSR, Otd. Khim. Nauk*, 148 (1960); 1582 (1961); *Tetrahedron Lett.*, 25 (1960); *Tetrahedron* **18,** 683 (1962).

[153] A. N. Nesmeyanov, E. G. Perevalova, and O. A. Nesmeyanov, *Izv. Akad. Nauk SSSR, Otd. Khim. Nauk*, 47 (1962).

[154] See Chapter V of this review.

[155] F. R. Jensen and J. Rodgers, unpublished results reported in footnote 103 (p. 99).

[156] G. A. Razuvaev, G. G. Petukhov, and Yu. A. Kaplin, *Dokl. Akad. Nauk SSSR*, **152,** 1122 (1963).

[157] A. N. Nesmeyanov, *Zhur. russk. fiz.-khim. Obshch.*, **61,** 1393 (1929); *Chem. Ber.*, **62,** 1010 (1929).

mercuric salts of carboxylic acids without electronegative substituents.[158,159] The reaction has preparative potential[160] and in the decomposition of mercuric acetate with acetyl peroxide in glacial acetic acid the yield of methylmercuric acetate reaches 99% of the starting mercury salt and over 100% based on the peroxide, the latter indicating a chain process.

$$(\text{MeCO}_2)_2\text{Hg} \xrightarrow{(\text{MeCO}_2)_2} \text{MeHgO}_2\text{CMe} + \text{CO}_2 \quad (123)$$

The proposed mechanism involves S_H2 displacement of a carboxylate radical from mercury by an alkyl radical and it was suggested that the displacement was stepwise although no evidence was given.

$$(\text{RCO}_2)_2 \rightarrow 2\text{RCO}_2^{\cdot} \quad (124)$$

$$\text{RCO}_2^{\cdot} \rightarrow \text{R}^{\cdot} + \text{CO}_2 \quad (125)$$

$$\text{R}^{\cdot} + (\text{R}'\text{CO}_2)_2\text{Hg} \longrightarrow \left[\begin{array}{c}\text{R}'\text{CO}_2\dot{\text{H}}\text{gO}_2\text{CR}' \\ | \\ \text{R}\end{array}\right] \longrightarrow \text{RHgO}_2\text{CR}' + \text{R}'\text{CO}_2^{\cdot} \quad (126)$$

$$\text{R}'\text{CO}_2^{\cdot} \longrightarrow \text{R}'^{\cdot} + \text{CO}_2 \quad \text{and so on} \quad (127)$$

The occurrence of reaction 126 may reflect the low strength of the Hg–O bond, but loss of the carbon dioxide could possibly be concerted with the expulsion of R′·.

$$\text{R}^{\cdot} + \underset{\text{OC(O)R}'}{\underset{|}{\text{Hg}}}\text{—O—}\overset{\text{O}}{\overset{\|}{\text{C}}}\text{—R}' \longrightarrow \underset{\text{OC(O)R}'}{\text{R—}\underset{|}{\text{Hg}}} + \text{CO}_2 + \text{R}'^{\cdot} \quad (128)$$

The initiating radicals for the decarboxylation may be obtained from sources other than diacyl peroxides, for example from UV irradiation,[161,162] or from thermal decomposition of lead tetraacetate[163] or of phenyl iodosoacetate.[164]

[158] G. A. Razuvaev, Yu. A. Ol'dekop, and N. A. Maier, *Dokl. Akad. Nauk SSSR*, **98**, 613 (1954); *Zh. Obshch. Khim.*, **25**, 697 (1955).
[159] Yu. A. Ol'dekop and N. A. Maier, *ibid.*, **30**, 275, 299, 619, 3017, 3472 (1960).
[160] L. G. Markarova and A. N. Nesmeyanov, "The Organic Compounds of Mercury" in *Methods of Elemento Organic Chemistry*, A. N. Nesmeyanov and K. A. Kocheskov (Eds.), Vol. 4, North-Holland Publishing Co., 1967. Ch. 10.
[161] G. A. Razuvaev and Yu. A. Ol'dekop, *Dokl. Akad. Nauk SSSR*, **105**, 738 (1955).
[162] Yu. A. Ol'dekop and N. A. Maier, *Dokl. Akad. Nauk Belorussk. SSR*, **4**, 288 (1960); *Tr. po Khim. Khim. Tekhnol.*, **4**, 139 (1961), *Chem. Abstr.*, **55**, 27025b (1961); *Izv. Akad. Nauk SSSR, Ser. Khim.*, 1171 (1966).
[163] G. A. Razuvaev, Yu. A. Ol'dekop, Yu. A. Sorokin, and V. I. Tverdova, *Zh. Obshch. Khim.*, **26**, 1683 (1956).
[164] Yu. A. Ol'dekop, N. A. Maier, and A. V. Bystraya, *Dokl. Akad. Nauk Belorussk. SSR*, **6**, 503 (1962), *Chem. Abstr.* **58**, 1329g (1963).

Similar reactions could be achieved with mercurous carboxylates, but in this case the Hg–Hg bond was cleaved.[161,162,165–167] The key step involved is homolytic substitution by an alkyl radical at mercury.

$$R^{\cdot} + R'CO_2Hg—HgO_2CR' \rightarrow R'CO_2HgR + R'CO_2Hg^{\cdot} \qquad (129)$$

$$R', R = Me, Et, Pr^n, Ph$$

Very recently Ol'dekop and his coworkers have reported further synthetic applications of the radical-chain decarboxylation of mecurous[167a] and mercuric[167b–d] carboxylates.

Substitution by Sulfur-Centered Radicals

Diethylmercury reacts with thiophenol at 100° in dibutyl ether to evolve ethane.[96,97] Thiophenol also reacts with diphenylmercury during 3 hr at 130° to give mercury, diphenyl disulfide, mercuric thiophenoxide, and some benzene.[168] The C–Hg bond in divinylmercury, diallylmercury and bis(3-butenyl)mercury is also cleaved by thiophenol.[169] These reactions might be electrophilic acidolyses but evidence for the involvement of a free-radical chain process comes from the effect of initiators and inhibitors on the reaction of diphenylmercury with thiophenol in benzene at 35°. In the presence of 2 mole % of phenothiazine[170] the reaction ceases long before one molar equivalent of mercurial has been cleaved, whereas in the absence of inhibitor, 1 mole of thiophenol reacted in 1 hr. 1 Mole % of *t*-butyl hyponitrite, a radical initiator, caused a significant increase in the rate of uptake of

[165] Yu. A. Ol'dekop and M. M. Azanovskaya, *Zh. Obshch. Khim.*, **30**, 2291, 3472 (1960).
[166] Yu. A. Ol'dekop, N. A. Maier, and V. N. Pshenichnyi, *ibid.*, **34**, 317 (1964); **35**, 904 (1965); **36**, 1408 (1966); **38**, 1441 (1968); **39**, 536 (1969).
[167] N. A. Maier, A. A. Erdman, Yu. A. Dzhomidava, and Yu. A. Ol'dekop, *ibid.*, **37**, 128 (1967).
[167a] Yu. A. Ol'dekop, N. A. Maier, V. N. Pshenichnyi, and Z. F. Izmailova, *ibid.*, **40**, 308 (1970).
[167b] Yu. A. Ol'dekop, N. A. Maier, A. A. Erdman, and Yu. A. Dzhomidava, *ibid.*, **40**, 300, 637 (1970).
[167c] Yu. A. Ol'dekop, N. A. Maier, A. N. Erdman, and S. S. Stanovaya, *ibid.*, **40**, 305 (1970).
[167d] Yu. A. Ol'dekop, N. A. Maier, and Y. D. Butko, *ibid.*, **40**, 641 (1970).
[168] M. M. Koton, E. P. Moskvina, and F. S. Florinskii, *ibid.*, **20**, 2093 (1950).
[169] D. J. Foster and E. Tobler, *J. Amer. Chem. Soc.*, **83**, 851 (1961); *J. Org. Chem.*, **27**, 834 (1962).
[170] (phenothiazine structure: H, N, S) This compound is known to scavenge thiyl radicals; H. Low, *Ind. Eng. Chem., Research and Development*, **5**, 80 (1966).

thiophenol.[171] The cleavage under these conditions proceeds, at least in the later stages, by a free-radical chain mechanism with homolytic displacement at mercury as a propagation step.

$$PhS^{\cdot} + Ph_2Hg \rightarrow PhSHgPh + Ph^{\cdot} \tag{130}$$

Possibly a similar chain could account for the high rate of cleavage by thiophenol of phenylcarbomethoxymercury, which has been ascribed to a large degree of nucleophilic assistance by sulfur to electrophilic cleavage.[172]

$$PhS^{\cdot} + PhHg\underset{\substack{\|\\ O}}{C}OMe \longrightarrow PhHgSPh + \underset{\substack{\|\\ O}}{\dot{C}}OMe \tag{131}$$

$$\underset{\substack{\|\\ O}}{\dot{C}}OMe \longrightarrow CO_2 + Me^{\cdot} \tag{132}$$

The reaction of thiocyanogen with diphenylmercury[173] might also involve homolytic displacement at the metal.

$$S\dot{C}N + Ph_2Hg \rightarrow PhHgSCN + Ph^{\cdot} \tag{133}$$

$$Ph^{\cdot} + NCS{-}SCN \rightarrow PhSCN + S\dot{C}N \tag{134}$$

Substitution by Nitrogen-Centered Radicals

The reaction between organomercury compounds and *N*-bromoamides or *N*-bromoimides was studied some years ago by Razuvaev and co-workers who concluded that the mechanism involved a polar (electrophilic) cleavage of the C–Hg bond.[174–176]

For example,

$$MeHgNp^{\alpha} + \left[\begin{matrix}CO\\ \\ CO\end{matrix}\right\rangle NBr \longrightarrow \left[\begin{matrix}CO\\ \\ CO\end{matrix}\right\rangle N{-}HgMe + Np^{\alpha}Br \tag{135}$$

The interaction of benzylphenylmercury with *N*-bromosuccinimide gave bromobenzene without cleavage of the benzylmercury bond.[174] Studies of

[171] J. M. Smith, unpublished results.
[172] R. E. Dessy and F. E. Paulic, *J. Amer. Chem. Soc.*, **85,** 1812 (1963).
[173] F. Challenger, A. L. Smith, and F. J. Paton, *J. Chem. Soc.*, **123,** 1046 (1923).
[174] G. A. Razuvaev and N. S. Vasileiskaya, *Dokl. Akad. Nauk SSSR*, **67,** 851 (1949).
[175] *Idem.*, *ibid.*, **74,** 279 (1950).
[176] G. A. Razuvaev and Z. I. Bugaeva, *Uchenye Zapiski Gor'kov. Univ.* 143 (1953). *Chem. Abstr.*, **49,** 8230a (1955).

the stereochemistry of the reaction between *N*-halosuccinimide and *trans*-4-methylcyclohexylmercuric chloride indicated that complete retention occurred in the cleavage by the iodocompounds and bromocompounds, in agreement with an S_E2 mechanism.[177] However, when *N*-chlorosuccinimide

$$\text{Me-}C_6H_{10}\text{-HgCl} + \begin{bmatrix} CO \\ \quad \rangle NX \\ CO \end{bmatrix} \longrightarrow \text{Me-}C_6H_{10}\text{-X} + \begin{bmatrix} CO \\ \quad \rangle N\text{—HgCl} \\ CO \end{bmatrix}$$

$$X = I,\ Br \qquad (136)$$

was used, 7% of the *cis*-chloride was obtained and this was ascribed to the incursion of a homolytic cleavage. The halogen atom in the *N*-iodoimide and *N*-bromoimide would be highly polarized in the sense X^+, but the polarization would not be expected to be as great in the *N*-chloroimide, and this compound would be susceptible to free-radical reaction. Although it was not pointed out, this could be an example of homolytic substitution by the succinimidyl radical at mercury.

$$\begin{bmatrix} CO \\ \quad \rangle N^{\cdot} \\ CO \end{bmatrix} + R^{*}HgCl \longrightarrow \begin{bmatrix} CO \\ \quad \rangle N\text{—HgCl} \\ CO \end{bmatrix} + R^{\cdot} \qquad (137)$$

$$R^{\cdot} + \begin{bmatrix} CO \\ \quad \rangle NCl \\ CO \end{bmatrix} \longrightarrow RCl + \begin{bmatrix} CO \\ \quad \rangle N^{\cdot} \\ CO \end{bmatrix} \qquad (138)$$

Other related reactions whose mechanisms should be studied more closely are the formation of alkylmercuric azides by the action of chlorine azide on dialkylmercury compounds[95] and the reaction of *N*-halogenobis (trifluoromethyl) amine with dialkylmercurials to bring about C–Hg cleavage.[178]

$$R_2Hg + ClN_3 \xrightarrow{CCl_4} RHgN_3 + RCl \qquad (139)$$

$$R = Me,\ Et,\ Pr^i$$

$$Me_2Hg + (CF_3)_2NCl \rightarrow MeHgN(CF_3)_2 + MeCl \qquad (140)$$

[177] F. R. Jensen and J. Miller, unpublished results reported in footnote 103 (p. 89).
[178] R. C. Dobbie and H. J. Emeléus, *J. Chem. Soc.* (*A*), 367 (1966).

IV S_H2 Reactions at Group IIIA Elements

A. BORON

In contrast to the sp^3-hybridized carbon atom, the sp^2-hybridized boron atom is coordinatively unsaturated, and can accept one or two electrons from a donor species into its vacant *p*-orbital with consequent rehybridization. The ability to accept an electron pair results in the familiar behavior of 3-coordinate boron compounds as Lewis acids, forming 4-coordinate complexes, and providing an easy route for heterolytic (nucleophilic) substitution at boron, for example,

$$HO^- + BCl_3 \rightarrow [HOBCl_3]^- \rightarrow HOBCl_2 + Cl^- \quad (1)$$

It might be expected that colligation of a free radical to give a 4-coordinate boron species with seven bonding electrons might provide an easy route for homolytic substitution at boron, and this is increasingly being borne out by experiment.

$$X^{\cdot} + BY_3 \rightarrow [XBY_3]^{\cdot} \rightarrow XBY_2 + Y^{\cdot} \quad (2)$$

As already pointed out, the distinction between synchronous and stepwise mechanisms for homolytic displacement is a fine one, but it is possible to envisage two extreme cases. In the first, the bond to X is much stronger than the bonds to Y, the addition of X would be exothermic and the "adduct" $[XBY_3]^{\cdot}$ would probably represent the transition state for the concerted displacement of $Y^{\cdot}$ by $X^{\cdot}$. An example of such a process would be the displacement of an alkyl radical from a trialkylborane by a *t*-butoxy radical, since the B–O bond which is formed is some 30–40 kcal/mole stronger than the B–C bond broken.[1]

At the other end of the scale $X^{\cdot}$ could be a stable free radical and the bond to Y would be stronger than the full bond to X. The adduct could then be a

[1] D. S. Matteson, *J. Org. Chem.*, **29**, 3399 (1964).

true intermediate, which might itself be stable or undergo slow loss of $Y^\cdot$. In this connection it is perhaps worth mentioning that a complex $[Ph_3BCPh_3]^\cdot$ has been proposed to account for the red color produced when triphenylmethyl radicals are allowed to react with triphenylborane.[2] There are other possibilities in the choice of $X^\cdot$, for example an unhindered nitroxide $R_2NO^\cdot$.

ATOM FOR ATOM SUBSTITUTIONS

No homolytic substitution reactions of this type appear to have been reported for boron. Although halogen exchange reactions are common for boron halides, there is no evidence that these involve S_H2 displacement at the boron atom.

ATOM FOR RADICAL SUBSTITUTIONS

The halodeboronation reactions of organoboron compounds might be expected to offer possible examples of this class of reaction. In fact, although the first demonstration of halodeboronation was in 1930,[3] a radical mechanism involving an S_H2 reaction at boron was not proposed until recently.[4]

In polar solvents such as water or acetic acid the cleavage of B–C bonds by halogens appears to follow an electrophilic substitution mechanism and extensive studies of this type of reaction have been carried out by Kuivila and co-workers.[5] In the gas phase or in nonpolar solvents it seems likely that a radical mechanism may predominate. Although tributylborane is not attacked by iodine or bromine in carbon tetrachloride at room temperature, it does react with dry bromine in the absence of solvent to give butyl bromide and bromobutylboranes.[6] Similarly, tripropylborane reacted slowly with iodine above 140° leading to the replacement of alkyl groups by iodine giving iododipropylborane and propyl iodide.[7] There was no hydrogen iodide evolved and no other evidence for side chain halogenation as there was in the reaction of tributylborane with bromine.[6] A four-center mechanism was proposed, but the possible involvement of radicals was not excluded. A free-radical chain process with propagation steps 3 and 4 involving an S_H2 reaction of a halogen atom at the boron center would seem attractive.

$$X^\cdot + BR_3 \rightarrow XBR_2 + R^\cdot \tag{3}$$

$$R^\cdot + X_2 \rightarrow RX + X^\cdot \tag{4}$$

[2] G. E. Coates, K. Wade, and M. L. Green, *Organometallic Compounds*, Vol. 1, Methuen, 1968, p. 213.
[3] A. D. Ainley and F. Challenger, *J. Chem. Soc.*, 2171 (1930).
[4] J. Grotewold, E. A. Lissi, and J. C. Scaiano, *J. Organometal. Chem.*, **19,** 431 (1969).
[5] For example, H. G. Kuivila and E. J. Soboczenski, *J. Amer. Chem. Soc.*, **76,** 2675 (1954).
[6] J. R. Johnson, H. R. Snyder, and M. G. Van Campen, Jr., *ibid.*, **60,** 115 (1938).
[7] L. H. Long and D. Dolimore, *J. Chem. Soc.*, 3902 (1953).

Grotewold, Lissi, and Scaiano[4] found that although triethylborane and bromine did not react in the dark at room temperature in the gas phase, there was an immediate reaction on admission of light. The competitive bromination of triethylborane and cyclohexene was studied and the results rationalized in terms of reactions 5 and 6 occurring between bromine atoms and the borane.

$$Br^{\cdot} + CH_3CH_2BEt_2 \rightarrow BrBEt_2 + CH_3CH_2^{\cdot} \tag{5}$$

$$Br^{\cdot} + CH_3CH_2BEt_2 \rightarrow CH_3\dot{C}HBEt_2 + HBr \tag{6}$$

From runs using toluene in place of the borane it was deduced that reaction 6 was at least five times as fast as reaction 7 at room temperature.

$$Br^{\cdot} + PhCH_3 \rightarrow PhCH_2^{\cdot} + HBr \tag{7}$$

From this, the authors estimated the secondary CH bond energy in triethylborane as 80 ± 3 kcal/mole. The ease of hydrogen abstraction from C–H bonds adjacent to boron atoms has been noted previously and explained in terms of stabilization of the generated radical by the vacant *p*-orbital on boron.[4,8,9] Reaction 5 was shown to be slower than reaction 6 and when the boron atom is substituted with groups capable of $p\pi$–$p\pi$ back donation into its vacant orbital, S_H2 reaction at the boron atom is further suppressed in favor of side chain hydrogen abstraction. This is demonstrated by the fact that hydrogen chloride is evolved in the reaction of chlorine with tributylboroxine but that there is no B–C bond cleavage,[10] and that the free radical bromination of 2-(1-phenylethyl)-1,3,2-dioxaborolane gives no 1-phenylethyl bromide.[9] Only side chain chlorination products, Me_2BCH_2Cl and Me_2BCHCl_2, were isolated from the reaction of chlorine with trimethylborane at $-95°$.[11,12]

RADICAL FOR ATOM SUBSTITUTIONS

Examples of this type of reaction include the displacement of halogen atoms from boron by an incoming radical.

The reaction of aralkanes with tribromoborane initiated with azo-*bis*-isobutyronitrile has been used to prepare aralkyldibromoboranes and (by

[8] D. S. Matteson, *J. Amer. Chem. Soc.*, **82,** 4228 (1960).

[9] D. J. Pasto, J. Chow, and S. K. Arora, *Tetrahedron*, **25,** 1557 (1969).

[10] J. C. Perrine and R. C. Keller, *J. Amer. Chem. Soc.*, **80,** 1823 (1958).

[11] L. Zeldin and P. R. Giradot, *Abstracts of the 140th National Meeting of the American Chemical Society*, Chicago, Ill. (1961) p. 15N, *U.S. Patent* 3,083,230 (1963), *Chem. Abstr.*, **59,** 10116g (1963).

[12] R. Schaeffer and L. J. Todd, *J. Amer. Chem. Soc.*, **87,** 488 (1965).

hydrolysis of the latter) aralkyldihydroxyboranes.[13] For example, toluene and tribromoborane gave benzyldihydroxyborane and the corresponding dihydroxyboranes were obtained from *p*-xylene, 1-methylnaphthalene, diphenylmethane, and fluorene [with α-(carbamylazo) isobutyronitrile initiator]. Although the mechanism of this reaction has not received comment, it would seem that the only likely course involves a free-radical chain with propagation steps shown, for toluene, in equations 7 and 8.

$$PhCH_2^{\cdot} + BBr_3 \rightarrow PhCH_2BBr_2 + Br^{\cdot} \quad (8)$$

A similar displacement of halogen from boron by carbon-centered free radicals occurs in the formation of arylhaloboranes and alkylhaloboranes during the photolysis of mixtures of aryl and alkyl halides and trihaloboranes.[14] Triiodoborane and iodobenzene gave, after 48 hr UV irradiation at 110 ± 20°, a product which on hydrolysis yielded triphenylboroxine (40%) and 4-biphenyldihydroxyborane (0.4%). When trichloroborane was

$$PhI + BI_3 \xrightarrow[\text{2. hydrolysis and work-up}]{\text{1. } h\nu} (PhBO)_3 + C_6H_5\text{—}C_6H_4\text{—}B(OH)_2 \quad (9)$$

used only a 4.4% yield of triphenylboroxine was isolated. Photolysis of the iodo compounds with tribromoborane for 45–76 hr gave the following dihydroxyboranes or boroxines after work-up: *p*-tolyl (24%), *p*-chlorophenyl (10%), *m*-chlorophenyl (13.5%), 4-biphenylyl (44%), n-butyl (6.1%), *p*-iodophenyl (18.6%). Substitution at boron in dihalophenylboranes could also be affected.

$$PhI + PhBCl_2 \xrightarrow[40\text{ hr}]{h\nu} Ph_2BCl\ (23.4\%) \quad (10)$$

$$PhI + PhBBr_2 \xrightarrow[50\text{ hr}]{h\nu} Ph_2BBr\ (27\%) \quad (11)$$

In each case the main reaction appears to involve the photolysis of the aryl halide, the resulting aryl radicals bringing about S_H2 displacement of halogen atoms from the trihaloborane to form the aryldihaloborane. It has also been shown that triodoborane and iodobenzene react during 16 hours at 120° to give a 64% yield of diiodophenylborane but no comment was made on the mechanism.[15]

In the photochemical reaction of alkylbenzenes with tribromoborane the major products were those of dibromoboronation of the aromatic nucleus,

[13] A. K. Hoffmann and S. J. Groszos, U.S. Patent, 2,882,317 (1959); *Chem. Abstr.*, **53,** 16062i (1959).

[14] R. A. Bowie and O. L. Musgrave, *J. Chem. Soc.* (*C*), 566 (1966).

[15] M. Schmidt, W. Siebert, and F. Rittig, *Chem. Ber.*, **101,** 281 (1968).

which was considered to arise from excitation and subsequent rearrangement of a complex of tribromoborane and the hydrocarbon.[16] However dibromoboronation of the alkyl side chain occurred to a small extent and this was interpreted in terms of reactions similar to 7 and 8.

Bis(trifluoromethyl)nitroxide reacts with tribromoborane to quantitatively displace bromine.[17] This reaction may be an example of an S_H2 displacement at boron by the highly electronegative nitroxide. It would be interesting to extend the reaction of this radical to alkylboranes.

RADICAL FOR RADICAL SUBSTITUTIONS

The great majority of homolytic substitutions at boron that have been studied fall into this category.

Substitution by Oxygen-Centered Radicals

In an attempt to prepare dialkoxyhalomethylboranes, $XCH_2B(OR)_2$, Matteson[1] undertook a study of the halogenation of dihydroxymethylborane derivatives. Trimethylboroxine did not react with *N*-bromosuccinimide in refluxing carbon tetrachloride in the presence of free-radical initiators (with hindsight, an interesting reaction since homolytic displacement by the succinimidyl radical at boron is a possibility). However, trimethylboroxine reacted readily with *t*-butyl hypochlorite in benzene or without solvent at 0–15° under UV irradiation, but the reaction was very slow (30 hr at 25°) in in the dark. The methyl group was not chlorinated, instead the products were tri-*t*-butoxyboroxine and methyl chloride. A more or less concerted displacement [S_H2 (synchronous) mechanism] by the *t*-butoxy radical at boron was proposed to account for these results.

$$Bu^tOCl \xrightarrow{h\nu} Bu^tO^\cdot + Cl^\cdot \tag{12}$$

$$Bu^tO^\cdot + Me{-}B\begin{matrix}/\mathrm{O}{-}\\ \backslash \mathrm{O}{-}\end{matrix} \longrightarrow Bu^tOB\begin{matrix}/\mathrm{O}{-}\\ \backslash \mathrm{O}{-}\end{matrix} + Me^\cdot \tag{13}$$

$$Me^\cdot + Bu^tOCl \rightarrow MeCl + Bu^tO^\cdot \tag{14}$$

Reaction of *t*-butyl hypochlorite with dimethoxymethylborane, $MeB(OMe)_2$, gave mainly chlorination of the methoxy groups, and the desired *B*-methyl chlorination was accomplished using the hindered di-*t*-butoxymethylborane, $MeB(OBu^t)_2$, in which the S_H2 reaction at boron was inhibited. A similar reaction had been carried out much earlier by Johnson and co-workers[6]

[16] Y. Ogata, Y. Izawa, H. Tomioka, and T. Ukigai, *Tetrahedron*, **25**, 1817 (1969).
[17] H. J. Emeléus, P. M. Spaziante, and S. M. Williamson, *Chem. Comm.*, 768 (1969).

who showed that *t*-butyl hypochlorite when added slowly to tributylborane at $-80°$ produced a vigorous reaction of which *n*-butyl chloride was a major product, evidence also being found for attack on the side chains. Free-radical mechanisms for these reactions were not considered.

In a recent summary[17a] of his work on this problem, Matteson discusses the factors which affect the rates of S_H2 reaction at boron and side-chain hydrogen abstraction in methylboron compounds. He considers that the low reactivity of the β-methyl group towards hydrogen abstraction by the *t*-butoxy radical is due to the electrophilicity of the attacking radical. To whatever extent the boron atom can delocalise electrons towards itself, the energy gain must be nearly nullified by the reduction in availability of these electrons to the oxygen atom in the *t*-butoxy radical.

Further evidence for displacement of carbon radicals from boron by alkoxy radicals was obtained by the isolation of products derived from $R^\cdot$ after the thermal decomposition of di-*t*-butyl peroxide in refluxing chlorobenzene solutions of $(RBO)_3$ ($R = Ph(Me)CH$, Ph, *p*-MeC_6H_4).[18] When triphenylboroxine or tri-*p*-tolylboroxine was heated with dicyclohexyl peroxydicarbonate in chlorobenzene the appropriate arene and biaryl were detected after hydrolysis along with the *o*-chlorobiaryls and *m*-chlorobiaryls formed by homolytic substitution reactions of the displaced phenyl or *p*-tolyl radicals with the solvent.[19] Under the same conditions, tribenzylborane, with sufficient peroxide to attack one C–B bond per molecule of borane, gave toluene and bibenzyl.

The electron spin resonance spectra of the displaced alkyl radicals have been observed during UV irradiation of a solution containing di-*t*-butyl peroxide and a trialkylborane in the cavity of the spectrometer.[20–22]

$$Bu^tO^\cdot + BR_3 \rightarrow Bu^tOBR_2 + R^\cdot \tag{15}$$

The displacing butoxy radicals have also been generated by photolysis and thermolysis of di-*t*-butyl hyponitrite[20] and the displacement shown to occur from trialkylboroxines[20] and dibutylchloroborane.[23] The rate constant for reaction ($R = n$-Bu) has recently been determined by two independent methods.[24] Competitive experiments with cyclopentane were carried out in the cavity of an e.s.r. spectrometer. Comparison of the cyclopentyl and

[17a] D. S. Matteson, *Progress in Boron Chemistry*, **3**, 117 (1970).
[18] N. V. Kruglova and R. Kh. Freidlina, *Izv. Akad. Nauk. SSSR, Ser. Khim.*, 2044 (1965).
[19] N. V. Kruglova, *ibid.*, 1163 (1967).
[20] See footnote 92, Chapter III.
[21] See footnote 136, Chapter III.
[22] A. G. Davies and B. P. Roberts, *Chem. Comm.*, 699 (1969).
[23] See footnote 78, Chapter III.
[24] A. G. Davies, D. Griller, B. P. Roberts, and R. Tudor, *Chem. Comm.*, 640 (1970).

n-butyl radical signal intensities gave $k_{15} = 7.3 \times 10^6 \, M^{-1} \, \text{sec}^{-1}$ at 40° with an activation energy of 2 ± 3 kcal/mole. The alkoxy radical,

$$(CH_3)_2CHC(CH_3)_2O^{\cdot},$$

generated from the corresponding hypochlorite, underwent β-scission at a comparable rate to substitution. The relative yields of butyl chloride and acetone gave $k = 4.5 \times 10^7 \, M^{-1} \, \text{sec}^{-1}$ at 40° for the S_H2 reaction at Bu_3^nB.

Further results obtained by the e.s.r. technique have shown that the rate of S_H2 reaction falls with increasing branching of the alkyl group attached to boron in the tributylboranes.[24a] The relative reactivities at 30° are: Bu_3^nB (1), Bu_3^iB (0.034), Bu_3^sB (0.008). Attack at tri-*s*-butylborane is characterised by an activation energy 4.5 Kcals/mole greater than that for attack at tri-*n*-butylborane. It seems that steric congestion about boron is more important in determining the ease of *t*-butoxydealkylation of the tributylboranes than the stability of the alkyl radical which is displaced. For the isomeric tributylboroxines steric effects appear to be less important as can be seen from the relative reactivities at 30°: Bu_3^nB (1), $(Bu^nBO)_3$ (0.019), $(Bu^iBO)_3$ (0.034), $(Bu^sBO)_3$ (0.018), $(Bu^tBO)_3$ (0.088).

H. C. Brown and co-workers have recently shown that the 1,4-addition of a trialkylborane to an α,β-unsaturated carbonyl compound occurs by a free-radical chain mechanism.[25–26a] For example,[25] the reaction of triethylborane with methyl vinyl ketone was effectively inhibited by 5 mole % of galvinoxyl, a free radical scavenger.

$$CH_2{=}CHCOMe + Et_3B \longrightarrow EtCH_2{-}CH{=}\underset{\displaystyle Me}{\underset{|}{C}}{-}OBEt_2 \qquad (16)$$

The mechanism was thought to involve propagation steps 17 and 18.

$$Et^{\cdot} + CH_2{=}CHCOMe \longrightarrow EtCH_2{-}CH{=}\underset{\displaystyle Me}{\underset{|}{C}}{-}O^{\cdot} \qquad (17)$$

$$EtCH_2{-}CH{=}\underset{\displaystyle Me}{\underset{|}{C}}{-}O^{\cdot} + BEt_3 \longrightarrow EtCH_2CH{=}\underset{\displaystyle Me}{\underset{|}{C}}{-}OBEt_2 + Et^{\cdot} \qquad (18)$$

When the β-carbon of the unsaturated carbonyl compound was substituted, the addition usually failed in the absence of free radical initiation, probably

[24a] A. G. Davies, D. Griller, and B. P. Roberts, unpublished results.
[25] See footnote 40, Chapter III.
[26] See footnote 41, Chapter III.
[26a] A. Suzuki, S. Nozawa, M. Itoh, H. C. Brown, G. W. Kabalka, and G. W. Holland, *J. Amer. Chem. Soc.*, **92**, 3503 (1970).

because of short chain length. When oxygen or acyl peroxides were present as initiators,[26] 1,4-addition proceeded normally.

The reaction of alkaline hydrogen peroxide with organoboranes to give hydroxydeboronation is well known[27] and has been shown to proceed by a polar mechanism involving a 1,2-nucleophilic migration of the alkyl group

$$\text{R}\!-\!\underset{-}{\text{B}}\!<\!\!-\text{O}-\text{OH} \longrightarrow \;>\!\text{B}-\text{OR} + \bar{\text{O}}\text{H} \tag{19}$$

from boron to coordinated hydroperoxide anion. In *neutral* aqueous hydrogen peroxide however, radical reactions have recently been shown to predominate.[28] Trialkylboranes react with 30% aqueous hydrogen peroxide in tetrahydrofuran at 0° giving mainly hydrocarbon dimers derived from the alkyl groups originally attached to boron. For example, trihexylborane (from the hydroboration of hex-1-ene) after 3 hr reaction with 1.04 moles of hydrogen peroxide per mole of borane gave *n*-dodecane and 5-methylundecane and cleaved two of the three groups from boron. A 1:1 mixture of trihexylborane and tripentylborane gave an almost statistical distribution of dimers while if the reaction with trihexylborane was carried out in carbon tetrachloride or in tetrahydrofuran containing iodine, the yield of *n*-dodecane was drastically reduced and 1-chlorohexane and 1-iodohexane were formed respectively. Two possible mechanisms were proposed for the generation of free alkyl radicals leading to the observed products, both mechanisms involving an S_H2 reaction at boron. The first mechanism, shown in equations 20–22, involves production of alkyl radicals in a chain reaction the second propagation step [equation 22] being a homolytic substitution by the

$$>\!\text{B}-\text{R} + \text{HO}-\text{OB}\!< \;\longrightarrow\; >\!\text{BOH} + \text{R}^{\cdot} + \;>\!\text{BO}^{\cdot} \tag{20}$$

$$>\!\text{BO}^{\cdot} + \text{HOOH} \longrightarrow \;>\!\text{BOH} + \dot{\text{O}}\text{OH} \tag{21}$$

$$>\!\text{BR} + \dot{\text{O}}\text{OH} \longrightarrow \;>\!\text{BOOH} + \text{R}^{\cdot} \tag{22}$$

[27] H. C. Brown, *Hydroboration*, Benjamin, New York, 1962.
[28] See footnote 20, Chapter III.

hydroperoxy radical at boron. A nonchain process was also thought possible, the second step involving a homolytic substitution this time by the hydroxy radical.

$$\begin{matrix} R_2B & O\text{—}H \\ | & | \\ R & O\text{—}H \end{matrix} \longrightarrow R_2BOH + R^{\cdot} + {}^{\cdot}OH \qquad (23)$$

$$R_2BOH + {}^{\cdot}OH \longrightarrow RB(OH)_2 + R^{\cdot} \qquad (24)$$

This work has been extended to the reaction of the epimeric trinorbornylboranes and although the general conclusions are in agreement with those of the previous workers, it seems likely that the varying amounts of alcohol derived from the borane are produced in a competing polar process involving retention of stereochemistry in the norbornyl group and not by way of a homolytic displacement by the norbornyl radical at oxygen in a peroxy group.[29] Similar displacements at boron probably occur in the "abnormal" oxidation by alkaline hydrogen peroxide of phenylethyldihydroxyborane and styryldihydroxyborane derivatives and in their reaction with Fenton's reagent,[30] although these results were rationalized in terms of hydrogen abstraction from the side chain rather than S_H2 reaction at boron.

A species similar to the intermediate or transition state involved in S_H2 displacement by hydroxyl radicals from boron is probably involved in the anodic oxidation of butyldihydroxyborane in basic solution.[31]

$$[BuB(OH)_3]^- \xrightarrow{-e} [BuB(OH)_3]^{\cdot} \longrightarrow Bu^{\cdot} + B(OH)_3 \qquad (25)$$

Autoxidation

The autoxidation of organoboron compounds to give an organoperoxyborane is now a well-established reaction, which until recently was thought to proceed by a polar insertion of oxygen into the carbon–boron bond.[32]

$$R_3B + O_2 \longrightarrow \begin{matrix} R & O^+ \\ | & | \\ R_2\underline{B} & O \end{matrix} \longrightarrow R_2BOOR \qquad (26)$$

[29] A. G. Davies and R. Tudor, *J. Chem. Soc.*, to be published.
[30] D. J. Pasto, S. K. Arora, and J. Chow, *Tetrahedron*, **26**, 1571 (1969).
[31] A. A. Humffray and L. F. G. Williams, *Chem. Comm.*, 616 (1965).
[32] A. G. Davies, *Prog. Boron Chem.*, **1**, 265 (1964).

However, Davies and Roberts[33] showed that optically active 1-phenylethyl-dihydroxyborane yielded racemic peroxide in a reaction which could be inhibited by 1 mole % of galvinoxyl, an active free radical trap. A free-radical chain mechanism was proposed which involved an S_H2 reaction of the alkyl-peroxy radical at the boron center in the product-forming propagation step.

Initiation: $$\longrightarrow R^{\cdot} \text{ (rate, } R_i) \quad (27)$$

Propagation: $$R^{\cdot} + O_2 \longrightarrow ROO^{\cdot} \quad (28)$$

$$ROO^{\cdot} + \rangle B{-}R \xrightarrow{k_p} ROOB\langle + R^{\cdot} \quad (29)$$

Termination: $$2ROO^{\cdot} \xrightarrow{2k_t} \text{inactive products} \quad (30)$$

Subsequent work has confirmed the generality of this mechanism by studying the stereochemistry and effect of inhibitors on the autoxidation of a wide range of organoboranes.[34–37] A kinetic analysis of the autoxidation has yielded a series of absolute rate coefficients for homolytic substitution at boron by the alkylperoxy radical (equation 29).[38,39] In isooctane solution at 30° and at oxygen partial pressures between 150–760 mmHg, the rate of oxidation of a variety of organoboranes was found to follow equation 31, in which R_i is the rate of chain initiation.

$$\text{rate of oxidation} = \frac{k_p}{(2k_t)^{1/2}} [\text{organoborane}] R_i^{1/2} \quad (31)$$

This relation is similar to that commonly encountered in the autoxidation of hydrocarbons. The rate constants, k_p, were obtained by coupling the measured oxidation rates at known rates of chain initiation with the independently determined[40] rate constants for the self-reactions of the alkylperoxy radicals involved. The results are given in Table I.

Two major conclusions can be drawn from these results. Firstly, homolytic substitution by an *n*-butylperoxy radical at the boron atom in a tri-*n*-butylborane is a very fast process with a very small activation energy. The

[33] A. G. Davies and B. P. Roberts, *Chem. Comm.*, 298 (1966); *J. Chem. Soc.* (*B*), 17 (1967).
[34] See footnote 25, Chapter III.
[35] O. Grummitt and P. S. Korosec, *Abs. Amer. Chem. Soc. Meetings*, 151, K30 (1966).
[36] P. G. Allies and P. B. Brindley, *Chem. Ind.* (*London*), 319 (1967); 1439 (1968). *J. Chem. Soc.* (*B*), 1126 (1969).
[37] J. Grotewold, E. A. Lissi, and J. C. Scaiano, *ibid.*, 475 (1969).
[38] See footnote 89, Chapter III.
[39] K. U. Ingold, *Chem. Comm.*, 911 (1969).
[40] J. A. Howard and K. U. Ingold, *J. Amer. Chem. Soc.*, **90,** 1058 (1968); *Can. J. Chem.*, **46,** 2655 (1968).

Table I Absolute Rate Constants[a] for the Reaction

$$ROO^{\cdot} + \;>BR \xrightarrow{k_p} ROOB< \; + R^{\cdot}$$ **at 30° in Isooctane[38]**

Organoborane	Rate Constant (M^{-1} sec^{-1}) R = Bu^n	Bu^s	Bu^t
R_3B	2×10^6	8×10^4	—
R_2BOBR_2	3×10^5	7×10^4	—
R_2BOOR	3×10^4	1.5×10^4	—
R_2BOR'	5×10^3[b]	2×10^3[c]	—
$(RBO)_3$	1×10^3	5×10^4	3×10^4
$RB(OR')_2$	small[c]	4[b]	0.3[c]

a. Not statistically corrected for the number of displaceable alkyl groups per molecule.
b. R′ = Et.
c. R′ = Bu^n.

displacement rate coefficient is some 10^7 times greater than that for a similar reaction (usually considered an abstraction reaction) at a benzylic hydrogen in toluene.

$$ROO^{\cdot} + PhCH_3 \rightarrow ROOH + PhCH_2^{\cdot} \tag{32}$$

This difference accounts for the failure of early attempts to inhibit borane oxidations with the inhibitors then available.[32] Secondly, the rate of the S_H2 reaction decreases with increasing steric protection of the boron atom and with increasing oxygen substitution of the boron atom. The latter effect can be ascribed to the decreasing Lewis acidity of the boron atom, caused by $p\pi$–$p\pi$ interaction with oxygen lone pairs. This increases the activation energy for displacement, relative to a trialkylborane, since the incoming peroxy radical must disrupt the $\overset{-}{B}{=}\overset{+}{O}$ π-bond to some extent in forming the transition state (or intermediate) involved in displacement. It is noteworthy that although the oxidizability $[k_p/(2k_t)^{1/2}]$ of the substrate increases dramatically along the series $(Bu^nBO)_3 < (Bu^sBO)_3 < (Bu^tBO)_3$, because of variations in $2k_t$ the increase in k_p in going from $(Bu^nBO)_3$ to $(Bu^sBO)_3$ is quite small and k_p for $(Bu^tBO)_3$ is actually slightly less than that for the *s*-butyl analogue. For this reason and because rates of initiation can vary widely from compound to compound, previous semiquantitative estimates of the relative ease of oxidation of organoboron compounds[32] should not be taken as indicating relative ease of S_H2 reaction by alkylperoxy radicals at boron.

Some absolute rate coefficients for the S_H2 reaction of alkylperoxy radicals with a series of benzylic organoboron compounds are given in Table II.[40a]

[40a] S. Korcek and K. U. Ingold, unpublished results.

Table II Rate Constants k_p (M^{-1} sec^{-1}) for Autoxidation of Benzylic Organoboron Compounds at 30°.

R	$PhCH_2B(OR)_2$	$PhCH(Me)B(OR)_2$
Me	21	31
$-CH_2$ $-CH_2$	20	35
Bu^n	18	
$(PhCH_2BO)_3$	1×10^6	
$(PhCH(Me)BO)_3$		6×10^6

The high stability of the benzylic radicals displaced leads to a large increase in the rate of the S_H2 process compared with the butyl analogues (see Table I).

Armed with the knowledge of the rapidity of displacement by oxygen radicals at a trialkylborane, an attempt can be made to rationalize the large volume of confusing literature concerned with the initiation of free-radical vinyl polymerization by oxidizing organoborane systems. For example, one should not now be surprised by the lack of ethanol produced if an ethoxy radical were generated in the presence of triethylborane,[41] as the rate of displacement of an ethyl radical from triethylborane would be expected to be much greater than that of abstraction of hydrogen by ethoxy radicals to give ethanol. A chain transfer process involving displacement of an ethyl radical from triethylborane by an inhibitor radical has been suggested to account for the inefficiency of inhibition of methyl methacrylate polymerization in the presence of triethylborane.[42] The inhibitor radical is produced by reaction between a growing radical chain and phenolic or secondary aromatic amine inhibitors. More potent radical traps such as galvinoxyl, which is itself a stable free radical but reacts only slowly with a trialkylborane, can readily inhibit a similar polymerization.[43]

[41] R. L. Hansen and R. H. Hamann, *J. Phys. Chem.*, **67**, 2868 (1963).
[42] J. Grotewold, E. A. Lissi, and A. E. Villa, *J. Polymer Sci., Part A-1*, **6**, 3157 (1968).
[43] P. B. Brindley and R. G. Pearson, *J. Polymer Sci., Part B*, **6**, 831 (1968).

Substitution by Carbon-Centered Radicals

The interaction of methyl and acetyl radicals with triethylborane in the gas phase has been studied by Grotewold and Lissi.[44,45] When azomethane or acetone was photolyzed in the presence of triethylborane, ethane, propane, butane, and ethylene (the latter only in the case of acetone) were obtained as volatile products, the formation of which was taken as demonstrating the involvement of reaction 33.

$$Me^{\cdot} + BEt_3 \rightarrow MeBEt_2 + Et^{\cdot} \qquad (33)$$

The rate coefficient for 33 was estimated to be $5 \times 10^4\ M^{-1}\ sec^{-1}$ at room temperature, although this was only slightly larger than the rate coefficient for hydrogen abstraction from triethylborane ($\sim 10^4\ M^{-1}\ sec^{-1}$), again demonstrating the lability of the methylene hydrogens adjacent to boron. In a more complete study of the photolysis of biacetyl in the presence of triethylborane these authors[45] used the known rate of decarbonylation of acetyl radicals to estimate the rate of displacement of the ethyl radical from the borane by the acetyl radical.

$$MeC(O)C(O)Me \xrightarrow{h\nu} 2Me\dot{C}O \qquad (34)$$
$$Me\dot{C}O \longrightarrow Me^{\cdot} + CO \qquad (35)$$
$$Me\dot{C}O + Et_3B \longrightarrow MeCOBEt_2 + Et^{\cdot} \qquad (36)$$

Arrhenius parameters, found by varying the temperature from 316–413°K, were $\log_{10}(A_{35}/A_{36}) = 4.6 \pm 0.4$ (for A_{35}/A_{36} in mole l^{-1}.) and $E_{35} - E_{36} = 15.7 \pm 0.7$ kcal/mole. These results indicated an almost zero activation energy for reaction 36 and $\log_{10} A_{36} = 5.7$, which was considered a reasonable A factor for a reaction of this type. However, in the kinetic analysis a process involving the interaction of a biacetyl molecule in its excited triplet state with triethylborane producing an ethyl radical was considered unimportant.

$$^3\text{biacetyl} + Et_3B \rightarrow Et^{\cdot} \qquad (37)$$

The product of reaction 36, acetyldiethylborane, was not detected. It was later shown that triethylborane is a strong quencher of biacetyl phosphorescence in the gas phase,[46] but the uncertainty which this fact brings to the value of the rate coefficient for the displacement by the acetyl radical from triethylborane does not seem to have been pointed out. In view of the similar reactivity in hydrogen abstraction reactions of alkoxy radicals and carbonyl

[44] J. Grotewold and E. A. Lissi, *Chem. Comm.*, 21 (1965).
[45] J. Grotewold and E. A. Lissi, *J. Chem. Soc.* (*B*), 264 (1968).
[46] J. Grotewold and E. A. Lissi, *Chem. Comm.*, 1367 (1968).

triplet states, it would seem quite likely that the reaction shown in equation 38 would be very facile.

$$^3R'_2CO + BR_3 \rightarrow R'_2\dot{C}—OBR_2 + R^{\cdot} \quad (38)$$

A similar S_H2 reaction by carbon radicals has been carried out in solution using radicals generated by the thermolysis of phenylazotriphenylmethane to displace benzyl radicals from tribenzylborane, the final products being toluene and bibenzyl.[19] When a mixture of four parts methane and one part trichloroborane was passed through a high-voltage silent electrical discharge, the condensed gases contained $BHCl_2$, HCl, BCl_3, and $MeBCl_2$.[47] The dichloromethylborane may well be the product of homolytic substitution by the methyl radical at boron in trichloroborane.

The alkyl radical exchange reaction (equation 39) will be approximately thermoneutral and perhaps a true intermediate could be involved in some cases, the mechanism being of the S_H2 (stepwise) type.

$$R'^{\cdot} + BR_3 \rightarrow [R'BR_3]^{\cdot} \rightarrow R'BR_2 + R^{\cdot} \quad (39)$$

Although the possibility does not seem to have been considered previously, the oxidation of sodium tetraphenylborate[48] (anodic and with ceric ions) and of lithium tetrabutylborate[49] (by oxygen) may proceed *via* such intermediates, the relevant steps being shown in equations 40 and 41, respectively.

$$BPh_4^- - e \rightarrow [BPh_4]^{\cdot} \quad (40)$$

$$BBu_4^- + O_2 \rightarrow [BBu_4]^{\cdot} + O_2^{\dot{-}} \quad (41)$$

Recently, Bell and Platt[50] have proposed that the trifluoromethyl radical, produced by photolysis of hexafluoroacetone, will bring about displacement of a methyl radical from trimethylborane.

$$CF_3^{\cdot} + BMe_3 \rightarrow CF_3BMe_2 + Me^{\cdot} \quad (42)$$

The rate of S_H2 reaction decreased along the series $BMe_3 > SnMe_4 \gg GeMe_4 = SiMe_4$; the value of the rate coefficient for displacement at tetramethyltin was $1 \times 10^4\ M^{-1}\ sec^{-1}$ at 150°.

Substitution by Nitrogen-Centered Radicals

The first clear demonstration of the participation of an S_H2 reaction by nitrogen-centered radicals at a boron center appears to be the displacement of *n*-butyl radicals from tributylborane by dimethylamino radicals generated photolytically from tetramethyltetrazene.[51]

$$Me_2NN{=}NNMe_2 \xrightarrow{h\nu} 2Me_2N^{\cdot} + N_2 \quad (43)$$

$$Me_2N^{\cdot} + BBu_3 \longrightarrow Me_2NBBu_2 + Bu^{\cdot} \quad (44)$$

[47] J. L. Shephard, S. Witz, and E. I. Hormats, U.S. Patent 3,354,067 (1967).
[48] D. H. Geske, *J. Phys. Chem.*, **63,** 1062 (1959); **66,** 1743 (1962).
[49] R. Damico, *J. Org. Chem.*, **29,** 1971 (1964).
[50] T. N. Bell and A. E. Platt, *Chem. Comm.*, 325 (1970).
[51] A. G. Davies, S. C. W. Hook, and B. P. Roberts, *J. Organometal. Chem.*, **22,** C37 (1970).

When the reaction was conducted in the cavity of an e.s.r. spectrometer the spectrum of the displaced *n*-butyl radical is observed. The appropriate displaced alkyl radical has also been detected from tri-*s*-butylborane, oxybis-(dibutylborane) and tri-*s*-butylboroxine. Reaction 44 has also been identified as a propagation step in the reaction of *N*-chlorodimethylamine with tributylborane to give butyl chloride in a free-radical chain process.[52]

$$R'_2N^{\cdot} + BR_3 \rightarrow R'_2NBR_2 + R^{\cdot} \quad (45)$$

$$R^{\cdot} + R'_2NCl \rightarrow RCl + R'_2N^{\cdot} \quad (46)$$

A polar process giving butyldimethylamine competes with the radical cleavage of the first B–C bond in tributylborane, but both B–C bonds in dibutyl-(dimethylamino)borane are cleaved by the radical mechanism to give butyl chloride. In contrast, the reaction of chloramine itself with trialkylboranes gives alkylamine by a polar mechanism.[53]

Nitric oxide reacts with trialkylboranes and the reaction has been studied by various workers with a variety of results.[54–56] These results were obtained before it was established that oxygen reacted by a radical chain mechanism with organoboron compounds. In view of the similarities of the two molecules further work on the nitric oxide system might be expected to reveal that new homolytic displacements at boron are involved in the reaction. The reaction of nitrogen dioxide with trialkylboranes has also been mentioned.[57]

Substitution by Sulfur-Centered Radicals

Among neutral protic reagents, those most reactive towards trialkylboranes appear to be the alkanethiols; the reaction was mentioned by Gilman and Nelson[58,59] and was extensively studied by Mikhailov and co-workers.[60–64]

[52] See footnote 72, Chapter III.
[53] H. C. Brown, W. R. Heydkamp, E. Breuer, and W. S. Murphy, *J. Amer. Chem. Soc.*, **86,** 3665 (1964).
[54] See footnote 74, Chapter III.
[55] M. Inatome and L. P. Kuhn, *Adv. Chem. Ser.*, **42,** 183 (1964).
[56] S. J. Brois, *Tetrahedron Lett.*, 345 (1964).
[57] M. Inatome and L. P. Kuhn, *ibid.*, 73 (1965).
[58] See footnote 96, Chapter III.
[59] See footnote 97, Chapter III.
[60] B. M. Mikhailov, V. A. Vaver, and Yu. N. Bubnov, *Dokl. Akad. Nauk SSSR*, **126,** 575 (1959).
[61] B. M. Mikhailov and Yu. N. Bubnov, *Izv. Akad. Nauk SSSR, Otd. Khim. Nauk*, 1872 (1960).
[62] B. M. Mikhailov and Yu. N. Bubnov, *Zh. Obshch. Khim.* **31,** 160 (1961).
[63] B. M. Mikhailov and F. B. Tutorskaya, *ibid.*, **32,** 833 (1962).
[64] See also, K. Lang, German Patent 1,079,634 (1960).

One alkyl group is readily cleaved from the borane to give a dialkyl(alkylthio)borane and alkane.

$$R'SH + BR_3 \rightarrow R'SBR_2 + RH \qquad (47)$$

Mikhailov and Bubnov[65,65a] found that the reaction was initiated by peroxides and oxygen as well as UV light. Trialkylborane, freed from peroxides by heating at 100°, did not react with thiol below 150° but a trace of air, even that contained in broken porcelain, sufficed to start the reaction. Hydrogen peroxide or alkyl hydroperoxides were particularly effective initiators. Peroxides were also found to initiate the conversion of the ester

$$R_2BOCH_2CH_2SH$$

into alkane and 2-alkyl-1,3,2-oxathioborolane.

$$R_2BOCH_2CH_2SH \longrightarrow R{-}B\begin{array}{l} \diagup O{-}CH_2 \\ \qquad\quad | \\ \diagdown S{-}CH_2 \end{array} + RH \qquad R = Pr^n, Bu^n \qquad (48)$$

The reaction is thus a free-radical chain in which a propagation step involves an S_H2 displacement of an alkyl radical from boron by the alkylthiyl radical.

$$R^{\cdot} + R'SH \rightarrow RH + R'S^{\cdot} \qquad (49)$$

$$R'S^{\cdot} + BR_3 \rightarrow R'SBR_2 + R^{\cdot} \qquad (50)$$

Although the reaction of thiocyanogen with triphenylborane to yield diphenylthiocyanatoborane[66] could be a polar process,[5] it may alternatively involve a radical chain in which displacement of a phenyl radical by the thiocyano radical occurs.

$$Ph^{\cdot} + NCS{-}SCN \rightarrow PhSCN + S\dot{C}N \qquad (51)$$

$$S\dot{C}N + BPh_3 \rightarrow NCSBPh_2 + Ph^{\cdot} \qquad (52)$$

The addition of alkanethiols or dialkyl disulfides to trialkylboranes is reported to have an anti-oxidant effect on the latter which seems worthy of further investigation.[67]

It seems likely that radicals are involved in the reaction of elemental sulfur with trialkylboranes. In fact, mixtures of trialkylboranes with sulfur can

[65] B. M. Mikhailov and Yu. N. Bubnov, *Izv. Akad. Nauk SSSR, Ser. Khim.*, 2248 (1964).
[65a] B. M. Mikhailov, *Progress in Boron Chemistry*, **3**, 313 (1970).
[66] See footnote 101, Chapter III.
[67] A. Gross, German Patent 1,294,378 (1969), *Chem. Abstr.* **71**, 39142p (1969).

replace mixtures with oxygen as a catalyst in the polymerization of chloroprene.[68] Trialkylboranes reacted with sulfur at 140° to yield dialkylthioboranes.[69,69a]

$$nR_3B + S_n \rightarrow nR_2BSR \tag{53}$$

The mechanism of the reaction is unknown, but an S_H2 reaction at boron by sulfur-centered radicals may be involved.

B. ALUMINUM

There are few well-authenticated examples of homolytic displacement at a Group IIIA atom other than boron, and of those reported most reactions involve displacement of a carbon-centered radical from an organometallic compound. The lack of S_H2 reactions known for the remainder of the group is due both to the sparse attention that this type of process has received and to the increased tendency, compared to boron, of the organic derivatives of these metals to react by heterolytic mechanisms.

Substitution by Oxygen-Centered Radicals

The aluminum trialkyls are very readily autoxidized to alkoxides both in the gas phase and in solution where intermediate peroxides have been detected.[70,71] By analogy with the then presumed heterolytic mechanism for the autoxidation of organoboranes, it was proposed that in solution the aluminum trialkyls underwent a similar polar insertion of an oxygen molecule into the metal–carbon bond.[71] However, it was shown that the later

$$R_3Al + O_2 \longrightarrow R_2\underset{|}{\overset{R}{Al}}\!-\!\overset{O^+}{\underset{}{O}} \longrightarrow R_2AlOOR \begin{cases} \xrightarrow{R_3Al} 2R_2AlOR \\ \longrightarrow RAl(OR)_2 \end{cases} \tag{54}$$

stages of the autoxidation of trimethylaluminum in cyclohexane could be inhibited by the reaction product of galvinoxyl and trimethylaluminum[72]

[68] Farbenfabriken Bayer Akt. Ges., British Patent, 842,341 (1960).

[69] B. M. Mikhailov and Yu. N. Bubnov, *Izv. Akad. Nauk SSSR, Otd. Khim. Nauk*, 172 (1959); *Zh. Obshch. Khim.*, **29**, 1648 (1959).

[69a] Z. Yoshida, T. Okushi, and O. Manabe, *Tetrahedron Lett.*, 1641 (1970).

[70] A. M. Sladkov, V. A. Markevich, I. A. Yavich, L. K. Luneva, and V. N. Chernov, *Dokl. Akad. Nauk SSSR*, **119**, 1159 (1958).

[71] A. G. Davies and C. D. Hall, *J. Chem. Soc.*, 1192 (1963) and references cited.

[72] See footnote 1, Chapter II.

and that the autoxidation of a 50% *v/v* solution of tridodecylaluminum in benzene was retarded by addition of 0.25% of methylenebis-(2,6-di-*t*-butyl phenol), the reaction ceasing after about 50% of the theoretical oxygen uptake.[73] Although not compelling, this evidence may be taken to indicate that the autoxidation of aluminum alkyls and boron alkyls follow similar mechanisms involving a homolytic displacement by an alkylperoxy radical at the metal center in a propagation step of a chain reaction.[72]

$$ROO^{\cdot} + AlR_3 \text{ (or dimer)} \rightarrow ROOAlR_2 + R^{\cdot} \qquad (55)$$

That the reaction is not so simple as in the case of the organoboranes is shown by the detection of ethane, hydrogen and ethylene (the latter in lower concentration by an order of magnitude) as volatile products from the air oxidation of triethylaluminum in heptane solution.[74]

The vapor-phase oxidation of trimethylaluminum, which has been extensively studied by Cullis and co-workers,[75–77] also gave hydrogen as a product. These workers proposed a homolytic chain mechanism for the slow combustion of the alkyl involving $Me_2Al^{\cdot}$ and $Me_2AlOO^{\cdot}$ radicals as chain-carrying species.

Initiation:

$$Al_2Me_6 + O_2 \xrightarrow{\text{wall?}} Me_2AlOO^{\cdot} + Me_2Al^{\cdot} + 2Me^{\cdot} \qquad (56)$$

Propagation:

$$Me_2Al^{\cdot} + O_2 \rightarrow Me_2AlOO^{\cdot} \qquad (57)$$

$$Me_2AlOO^{\cdot} + Al_2Me_6 \rightarrow 2Me_2AlOMe + Me_2Al^{\cdot} \qquad (58)$$

Termination:

$$Me_2Al^{\cdot} \xrightarrow{\text{wall}} \text{inactive products} \qquad (59)$$

$$Me_2AlOO^{\cdot} \xrightarrow{\text{wall}} \text{inactive products} \qquad (60)$$

Chain branching:

$$Me_2AlOO^{\cdot} + Al_2Me_6 \rightarrow Me_2AlOOMe + 2Me_2Al^{\cdot} + Me^{\cdot} \qquad (61)$$

$$Me_2AlOOMe \rightarrow Me_2AlO^{\cdot} + {}^{\cdot}OMe \qquad (62)$$

$$Me_2AlO^{\cdot} + Al_2Me_6 \rightarrow Me_2AlOMe + 2Me_2Al^{\cdot} + Me^{\cdot} \qquad (63)$$

As the ignition limit is reached from the slow combustion region, the branching reactions 61–63 appear to become important and to lead to

[73] Personal communication from Dr. J. D. Johnson of the Ethyl Corporation, Baton Rouge, La., 1967.
[74] A. Gröbler, A. Simon, T. Kada, and L. Fazakas, *J. Organometal Chem.*, **7**, P3 (1967).
[75] C. F. Cullis, A. Fish, and R. T. Pollard, *Proc. Roy. Soc.* (*London*), **288A**, 123 (1965).
[76] *Idem, ibid.*, **289A**, 413 (1966).
[77] *Idem, ibid.*, **298A**, 64 (1967).

ignition.[76] It was supposed that the methyl radicals produced abstracted hydrogen from the trimethylaluminum faster than they reacted with oxygen under these conditions. In the proposed mechanism, step 58 seems rather improbable and steps 61 and 63, which involve S_H2 reactions at a saturated carbon atom, can probably be ruled out entirely (see Chapter V). It is difficult to see how the occurrence of reaction 55 (R = Me) can be excluded, although the rapid hydrogen abstraction from the methyl groups bound to aluminum by methyl radicals would be in line with the similarly rapid abstraction from alkylboranes.[44]

Autoxidation of triarylaluminums is less well defined than that of the aliphatic compounds. There is no doubt that the autoxidation of triphenylaluminum involves free radicals and has at least some chain character. For example, inactive triphenylaluminum oxidized in ^{14}C-labeled benzene for 1 hr at 80° gave, after removal of the solvent and hydrolysis of the solid residue, active biphenyl and phenol but inactive benzene.[78] The latter result indicates that reaction 64 does not occur under these conditions.

$$*Ph^{\cdot} + AlPh_3 \rightarrow *PhAlPh_2 + Ph^{\cdot} \qquad (64)$$

The thermally induced exchange of phenyl groups between triphenylaluminum and ^{14}C-labeled benzene solvent also does not appear to follow this mechanism.[79] The autoxidation of triphenylaluminum in benzene at room temperature is retarded by its reaction product with galvinoxyl.[72] No arylperoxide has ever been isolated, but the formation of phenylaluminum phenoxides can be envisaged as occurring by a homolytic displacement of a phenyl radical from aluminum by a phenoxy radical.

$$PhO^{\cdot} + AlPh_3 \rightarrow PhOAlPh_2 + Ph^{\cdot} \qquad (65)$$

A similar displacement was shown to occur[80,81] when the 2,4,6-tri-*t*-butylphenoxy radical (1 mol.) reacts with triphenylaluminum (1 mol.) in toluene at 20°.

$$ArO^{\cdot} + AlPh_3 \rightarrow ArOAlPh_2 + Ph^{\cdot} \qquad (66)$$

Ar = 2,4,6-tri-*t*-butylphenyl

The products were benzene (0.69 mol.), biphenyl (0.31 mol.) and 2,4,6-tri-*t*-butyl phenol (*ca.* 0.8 mol.). When the molar ratio of triphenylaluminum to aryloxy radical was 1:2, slightly more benzene (0.70 mol.) and biphenyl

[78] G. A. Razuvaev, E. V. Mitrofanova, G. G. Petukhov, and R. V. Kapline, *Zh. Obshch. Khim.*, **32,** 3454 (1962).

[79] G. A. Razuvaev, Yu. A. Kaplin and E. V. Mitrofanova, *Izv. Akad. Nauk SSSR, Ser. Khim.*, 1489 (1965).

[80] See footnote 49, chapter III.

[81] E. Müller, P. Ziemek, and A. Rieker, *Tetrahedron Lett.*, 207 (1964).

(0.35 mol.) were obtained and when the ratio was 1:2.01 the mixture exhibited an e.s.r. spectrum which was assigned to the aluminum containing radical **1**. It was suggested that this radical might be formed, perhaps reversibly, by electron transfer from the aryloxy radical to 2,4,6-tri-*t*-butylphenoxydiphenylaluminum and that a similar species might be an intermediate in the displacement of the first phenyl group from aluminum.

$$\left[Ph_2Al{-}O{-}C_6H_2(Bu^t)_3 \right]^{\dot{-}} \qquad {}^{+}O{-}C_6H_2(Bu^t)_3$$

1

Bimolecular homolytic substitution by alkoxy radicals has been invoked to explain the observation of the e.s.r. spectrum due to the ethyl radical when solutions of butoxydiethylaluminum[82] or triethylaluminum[83] and di-*t*-butyl peroxide were irradiated with UV light. The methyl radical was observed in a similar system containing trimethylaluminum and di-*t*-butyl peroxide.[82]

$$Bu^tO^{\cdot} + \;>Al{-}R \rightarrow Bu^tO{-}Al< \; + R^{\cdot} \qquad (67)$$

Relevant to these results is the report by Razuvaev and co-workers that the thermal reaction between triethylaluminum and di-*t*-butyl peroxide gives *t*-butoxydiethylaluminum, *t*-butyl ethyl ether, ethane, ethylene and butane.[84] It was proposed that ethyl radicals were generated by homolytic breakdown of a 2:1 complex of triethylaluminum and the peroxide.

$$\begin{array}{ccc} Et & & Et \\ \wr & & \wr \\ Et_2Al & & AlEt_2 \\ \uparrow & & \uparrow \\ O & \wr & O \\ / & & \backslash \\ Bu^t & & Bu^t \end{array} \longrightarrow 2Et_2AlOBu^t + 2Et^{\cdot} \qquad (68)$$

The similarity between the transition state for reaction 68 and that for S_H2 reaction of a *t*-butoxy radical at triethylaluminum is noteworthy, as is the

[82] See footnote 92, Chapter III.

[83] See footnote 136, Chapter III.

[84] G. A. Razuvaev, L. P. Stepovik, V. A. Dodonov, and G. V. Nesterov, *Zh. Obshch. Khim.*, **39**, 123 (1969).

fact that if free *t*-butoxy radicals were generated, by homolysis of a 1:1 complex, these would rapidly displace ethyl radicals from triethylaluminum to give the same products as reaction 68. A competing polar reaction was thought to give rise to the *t*-butyl ethyl ether.

Substitution by Sulfur-Centered Radicals

There appear to be no unambiguous examples of substitution by this class of radical. Elemental sulfur reacts with a trialkylaluminum to yield dialkylalkylthioaluminum[85,86] but there is no evidence for the involvement of alkylthiyl radicals.

$$nR_3Al + S_n \xrightarrow[70^\circ]{\text{hexane}} nR_2AlSR \tag{69}$$

Selenium[85,86] and tellurium[86] also react but give more complex mixtures of products. Sulfur chlorides react with trialkylaluminums[85] but the products do not appear to be consistent with a free-radical chain mechanism with S_H2 displacement of an alkyl radical from aluminum by a sulfur-centered radical.

Substitution by Carbon-Centered Radicals

It was mentioned previously that the exchange of phenyl groups with the solvent that occurs when triphenylaluminum is heated in ^{14}C-labeled benzene at 240–260° does not involve an S_H2 reaction of a phenyl radical at aluminum. This process is pictured as being an intramolecular reaction of a triphenylaluminum-solvent complex,[79] even though the exchange is accompanied by formation of biphenyl. Substitution by carbon-centered radicals may well be involved in the reaction of alkyl halides and polyhalides with alkylaluminum compounds. Although it has been reported that carbon tetrachloride does not react with triethylaluminum during 2 days at room temperature under nitrogen[87] other authors have reported a violent reaction.[88–90] When carbon tetrachloride was added dropwise to an equimolar amount of preheated triethylaluminum at 75°, an instantaneous reaction occurred in which chlorine atoms and ethyl and trichloromethyl radicals were thought to take part giving hydrogen chloride, butane, ethyl chloride, and chloroform as primary products.[89,90] The hydrogen chloride and chloroform then reacted further as protic reagents. However, it should be noted that the combination

[85] L. I. Zakharkin and V. V. Gavrilenko, *Izv. Akad. Nauk SSSR, Otd. Khim. Nauk*, 1391 (1960).
[86] H. Jenkner, West German Patent 1,031,306 (1958), *Chem. Abstr.*, **54,** 17269g (1960).
[87] C. Eden and H. Feilchenfeld, *J. Phys. Chem.*, **66,** 1354 (1962).
[88] J. W. Collette, *J. Org. Chem.*, **28,** 2489 (1963).
[89] H. Reinheckel, *Tetrahedron Lett.*, 1939 (1964).
[90] H. Reinheckel and R. Gensike, *J. Prakt. Chem.*, **37,** 214 (1968).

of reactions 70 and 71 gives the same products as reactions 72 and 73. Reaction 72 is an S_H2 reaction at an aluminum center.

$$Cl_3C^{\cdot} + AlEt_3 \rightarrow Cl_3CH + CH_3\dot{C}HAlEt_2 \quad (70)$$

$$Cl_3CH + AlEt_3 \rightarrow Cl_3CAlEt_2 + EtH \quad (71)$$

$$Cl_3C^{\cdot} + AlEt_3 \rightarrow Cl_3CAlEt_2 + Et^{\cdot} \quad (72)$$

$$Et^{\cdot} + AlEt_3 \rightarrow EtH + CH_3\dot{C}HAlEt_2 \quad (73)$$

Both homolytic and heterolytic mechanisms, perhaps occurring simultaneously, seem likely in the reactions of alkyl halides with organoaluminum compounds.[91,92] The observation of the e.s.r. spectrum due to $R^{\cdot}$ when solutions of the alkyl iodide, RI, and butyllithium were mixed in the cavity of the spectrometer was interpreted by Russell and Lamson as demonstrating that metal-halogen exchange reactions occur by a free-radical route.[93] Extension of this mechanism to the alkylaluminum system would predict the occurrence of reaction 75, as an as yet unreported alkyl-alkyl exchange.

$$R_3Al + R'X \rightarrow R^{\cdot} + R'^{\cdot} + R_2AlX \quad (74)$$

$$R'^{\cdot} + AlR_3 \rightarrow R'AlR_2 + R^{\cdot} \quad (75)$$

Substitution by Nitrogen-Centered Radicals

Homolytic substitution at aluminum by the dimethylamino radical provides an explanation of the results obtained in the thermal reaction between tetramethyltetrazene and triethylaluminum or trimethylaluminum or the trimethylamine complexes of these compounds and of aluminum hydride.[94] Homolysis of a complex of the two reactants might yield alkyl radicals or hydrogen atoms in a reaction which could also involve free dimethylamino radicals (compare the reaction of di-*t*-butyl peroxide with triethylaluminum).

$$R_3Al + Me_2NN{=}NNMe_2 \rightarrow [R_3AlMe_2NNN{=}NMe_2] \quad (76)$$

$$[R_3AlMe_2NN{=}NNMe_2] \rightarrow R^{\cdot} + R_2AlNMe_2 + N_2 + \dot{N}Me_2 \quad (77)$$

$$Me_2N^{\cdot} + AlR_3 \rightarrow Me_2NAlR_2 + R^{\cdot} \quad (78)$$

A 1:1 complex of triethylaluminum and tetramethyltetrazene was isolated at temperatures below those required for reaction (85°). Irradiation with UV light of a mixture of triethylaluminum and the tetrazene in toluene at low temperatures in the cavity of the spectrometer afforded the e.s.r. spectrum of

[91] H. Kaar and G. Schwindlerman, *Izv. Akad. Nauk Eston. SSR, Ser. Tekh. i Fiz.-Mat. Nauk*, **13,** 148 (1964); *Chem. Abstr.*, **61,** 10695e (1964).

[92] D. B. Miller, *J. Org. Chem.*, **31,** 908 (1966).

[93] See footnote 2, Chapter II.

[94] N. R. Fetter and B. Bartocha, *Can. J. Chem.*, **40,** 342 (1962).

the ethyl radical in addition to broader lines due to an unidentified species.[51] In this case reaction 77 is being brought about photolytically.

Nitric oxide reacts slowly with the ether complex of triethylaluminum to give products which after hydrolysis, yielded $Cu[ON(NO)Et]_2$ on treatment with hydrated cupric oxide.[95] Chlorodiethylaluminum in cyclohexane absorbed 2.1 mol. of nitric oxide in about 1 hr at room temperature.[96] A polar mechanism was suggested [reaction 79], but, as with the nitrozation reaction of trialkylboranes, homolytic substitution by nitric oxide is an attractive alternative.

$$R_2\overset{\overset{\displaystyle R}{|}}{Al} + \dot{O}N \longrightarrow R_2\overset{\overset{\displaystyle R}{|}}{\underset{-}{Al}}—O—\overset{+}{N}^{\cdot} \longrightarrow R_2AlO\overset{\overset{\displaystyle R}{|}}{N}^{\cdot} \xrightarrow{N\dot{O}} R_2AlO\overset{\overset{\displaystyle R}{|}}{N}—NO \quad (79)$$

The reaction of trialkylaluminums with chlorine azide[97,98] provides an example of a process which might possibly follow a free-radical chain mechanism involving homolytic substitution by azido radicals.

$$Et_3Al + ClN_3 \xrightarrow{0°} Et_2AlN_3 + EtCl \quad (80)$$

Substitution by Halogen Atoms

The reaction between trialkylaluminum and halogen proceeds very vigorously even at low temperatures, for example spontaneous ignition accompanies the reaction of triethylaluminum and chlorine at $-60°$. All three alkyl groups are readily cleaved except by iodine with which it is difficult to remove the third group. Mixtures of alkylaluminum compounds with chlorine or bromine have been shown to be catalysts for the polymerization of vinyl chloride at low temperatures.[99] It is possible that the reaction with halogens may involve a free-radical chain, at least in part.

$$R^{\cdot} + X_2 \rightarrow RX + X^{\cdot} \quad (81)$$

$$X^{\cdot} + AlR_3 \rightarrow XAlR_2 + R^{\cdot} \quad (82)$$

C. GALLIUM, INDIUM, AND THALLIUM

These elements will be considered together as very little is known about the homolytic reactions of their compounds.

[95] E. B. Baker and H. H. Sisler, *J. Amer. Chem. Soc.*, **75**, 5193 (1953).
[96] See footnote 74, Chapter III.
[97] See footnote 95, Chapter III.
[98] J. Müller and K. Dehnicke, *J. Organometal. Chem.* **12,** 37 (1968).
[99] G. A. Razuvaev, K. S. Minsker, and Yu. A. Sangalov, *Dokl. Akad. Nauk SSSR*, **159,** 158 (1964).

Substitution by Oxygen-Centered Radicals

The best example of this class is the reaction between photolytically generated *t*-butoxy radicals and triethylgallium which yields ethyl radicals observed by e.s.r.[100]

$$Bu^tO^\cdot + GaEt_3 \rightarrow Bu^tOGaEt_2 + Et^\cdot \tag{83}$$

Di-*t*-butyl peroxide reacts with triethylindium and triethylthallium at 50–100° to give ethane, small amounts of ethylene, traces of *t*-butanol and much Et_2MOBu^t (M = In, Tl).[101] Thermal decomposition of a 1:1 complex between the organometallic and peroxide could give ethyl and *t*-butoxy radicals, the latter reacting with metal trialkyl to displace more ethyl radicals and form more Et_2MOBu^t. It seems that the reaction of benzoyl peroxide with the trialkyl derivatives of the Group IIIA elements can occur by both polar and radical mechanisms (compare the reaction of hydrogen peroxide with trialkylboranes).[29] Thus, while triethylaluminum and benzoyl peroxide react to give almost entirely benzoyloxydiethylaluminum and ethyl benzoate with only a small yield of products derived from ethyl radicals, triethylthallium in benzene[102] reacts quantitatively at room temperature according to equations 84 and 85.

$$2Et_3Tl + (PhCOO)_2 \rightarrow 2Et_2TlOC(O)Ph + 2Et^\cdot \tag{84}$$

$$2Et^\cdot \rightarrow C_2H_6 + C_2H_4 \quad \text{or} \quad C_4H_{10} \tag{85}$$

Reaction 84 might occur in two stages, breakdown of a 1:1 complex between triethylthallium and peroxide giving a benzoyloxy radical which subsequently displaces a second ethyl radical from triethylthallium.

$$PhC(O)O^\cdot + TlEt_3 \rightarrow PhC(O)OTlEt_2 + Et^\cdot \tag{86}$$

Dicyclohexyl peroxydicarbonate reacts similarly with triethylthallium.[103]

The trialkyls of gallium and indium, like those of boron and aluminum, undergo vigorous autoxidation,[104] and a similar mechanism, involving displacement of alkyl radicals from the metals by incoming alkylperoxy radicals, seems plausible to account for at least part of the reaction. Triethylthallium is reported to be stable to dry air[105] but it seems possible that the trialkyls of thallium, like the dialkyls of mercury, will turn out to be slowly

[100] See footnote 136, Chapter III.

[101] S. F. Zhil'tsov, V. I. Shcherbakov, and O. N. Druzhkov, *Zh. Obshch. Khim.* **39,** 1327 (1969).

[102] G. A. Razuvaev and E. V. Mitrofanova, *ibid.*, **38,** 249 (1968).

[103] G. A. Razuvaev, V. A. Dodonov, and E. V. Mitrofanova, *ibid.*, **39,** 690 (1969).

[104] See footnote 1, Chapter III.

[105] E. G. Rochow and L. M. Dennis, *J. Amer. Chem. Soc.*, **57,** 486 (1935).

autoxidized. Triphenylthallium in benzene reacts slowly with oxygen, phenol being isolated after hydrolysis,[106] implying that phenoxy radicals may be capable of carrying out a displacement of phenyl radicals from thallium.

The autoxidation of triethylindium in the gas phase at 40–100° has been studied by Cullis and co-workers.[107] It was concluded that the oxidation takes place by a free-radical chain mechanism, giving an intermediate peroxide, the subsequent decomposition and rearrangement of which produces acetaldehyde which acts as a chain-branching agent. In a vessel with a clean pyrex or carbon surface, the total pressure change on autoxidation was small and ethane, ethylene, ethanol, and diethyl ether were the only products; the build-up of oxidation products on the walls was accompanied by acetaldehyde formation and further products. The proposed mechanism involved neither ethyl nor ethylperoxy radicals in the propagation steps.

Initiation:

$$Et_3In + O_2 \xrightarrow{\text{walls}} Et_2In^{\cdot} + EtOO^{\cdot} \qquad (87)$$

Propagation:

$$Et_2In^{\cdot} + O_2 \rightarrow Et_2InOO^{\cdot} \qquad (88)$$

$$Et_2InOO^{\cdot} + Et_3In \rightarrow Et_2InOOEt + Et_2In^{\cdot} \qquad (89)$$

Termination:

$$\left.\begin{matrix} Et_2In^{\cdot} \\ Et_2InOO^{\cdot} \end{matrix}\right\} \xrightarrow{\text{walls}} \text{inactive products} \qquad (90)$$

Volatile products from oxidations in clean vessels were accounted for by reactions 91–95.

$$Et_2InOOEt \rightarrow Et_2InO^{\cdot} + {}^{\cdot}OEt \qquad (91)$$

$$EtO^{\cdot} + Et_3In \rightarrow Et_2O + Et_2In^{\cdot} \qquad (92)$$

$$Et^{\cdot} + Et_3In \rightarrow EtH + Et_2In\dot{C}HCH_3 \qquad (93)$$

$$EtO^{\cdot} + Et_3In \rightarrow EtOH + Et_2In\dot{C}HCH_3 \qquad (94)$$

$$Et_2In\dot{C}HCH_3 \rightarrow C_2H_4 + Et_2In^{\cdot} \qquad (95)$$

Reactions 89 and 92 involve homolytic displacement at an sp^3-hybridized carbon atom. No unambiguous example of such a displacement is known (see Chapter V). Propagation steps involving ethyl and ethylperoxy radicals are more likely.

$$Et^{\cdot} + O_2 \rightarrow EtOO^{\cdot} \qquad (96)$$

$$EtOO^{\cdot} + InEt_3 \rightarrow EtOOInEt_2 + Et^{\cdot} \qquad (97)$$

[106] H. Gilman and R. G. Jones, *ibid.*, **61,** 1513 (1939).
[107] C. F. Cullis, A. Fish, and R. T. Pollard, *Trans. Faraday Soc.*, **60,** 2224 (1964).

By analogy with the known displacement of alkyl radicals from the trialkyls of boron, aluminum, and gallium, reaction 98 should also be included in the overall scheme for the oxidation of triethylindium.

$$EtO^{\cdot} + InEt_3 \rightarrow EtOInEt_2 + Et^{\cdot} \quad (98)$$

Substitution by Other Radicals

There are no instances of homolytic displacement from the metal atom in gallium, indium and thallium compounds except by oxygen-centered radicals. For the sake of completeness some reactions which may involve S_H2 processes will be mentioned.

The mechanism of the cleavage of the M–C bond (M = Ga, In, Tl) by halogens[108–110] has not been investigated, but may involve homolytic displacement of carbon radicals by halogen atoms. Homolytic displacement by hydrogen atoms is also conceivable in the reaction between trimethylgallium and hydrogen induced by an electrical discharge.[111]

$$2Me_3Ga + H_2 \xrightarrow[\text{discharge}]{\text{electrical}} (R_2GaH)_2 \quad (99)$$

The reaction of alkanethiols with trialkylboranes occurs by a free-radical chain mechanism but there is no evidence that the reaction of methanethiol with trimethylgallium,[112] trimethylindium,[113] or trimethylthallium[114] is not a simple heterolytic acidolysis.

The trialkyls of gallium, indium, and thallium, like those of aluminum, react with chlorine azide to produce the dialkyl metal azides.[98]

$$3Et_3M + 3ClN_3 \rightarrow (Et_2MN_3)_3 + 3EtCl \quad (100)$$

$$M = Ga, In$$

$$Et_3Tl + ClN_3 \rightarrow Et_2Tl^+N_3^- + EtCl \quad (101)$$

Again, these reactions may involve bimolecular homolytic substitution by the azido radical.

[108] W. C. Schumb and H. J. Crane, *J. Amer. Chem. Soc.*, **60,** 306 (1938).
[109] E. Wiberg, T. Johannsen, and O. Stecher, *Z. Anorg. Chem.*, **251,** 114 (1943).
[110] P. A. Fowell and C. T. Mortimer, *J. Chem. Soc.*, 3734 (1958).
[111] E. Wiberg and T. Johannsen, *Angew. Chem.*, **55,** 38 (1942).
[112] G. E. Coates and R. G. Hayter, *J. Chem. Soc.*, 2519 (1953).
[113] R. Didchenko, J. E. Alix, and R. H. Toeniskoetter, *J. Inorg. Nucl. Chem.*, **14,** 35 (1960).
[114] A. G. Lee and G. M. Sheldrick, *J. Organometal. Chem.*, **17,** 487 (1969).

V S_H2 Reactions at Group IVA Elements

A. CARBON

Probably more effort has been expended on the search for homolytic substitutions at sp^3 hybridized carbon than on any other multivalent element. Quite apart from the ubiquity of compounds containing carbon, interest in this reaction arises from the fact that substitution at an asymmetric carbon would answer the question as to whether the radical attacks from the backside with inversion of configuration (S_N2 like[1])

$$R^{\cdot} + abcCB \longrightarrow R\cdots\underset{b\quad c}{\overset{a}{C}}\cdots B \longrightarrow RCabc + B^{\cdot} \qquad (1)$$

or from the front side with retention of configuration (S_E2 like[1]).

$$R^{\cdot} + abcCB \longrightarrow \underset{B\quad b\quad c}{\overset{R\quad a}{C}} \longrightarrow RCabc + B^{\cdot} \qquad (2)$$

In spite of the many attempts to discover a homolytic substitution at an unstrained sp^3 hybridized carbon there would appear to be no unequivocal example of such a reaction. This does not necessarily mean that such reactions do not or cannot occur. Rather, most experimental attempts to detect an S_H2 reaction at carbon seem to have been rather poorly designed. A large number of the anticipated substitutions are nearly or exactly thermoneutral, for example, the iodine for iodine exchange that has received so much

[1] See footnote 31, Chapter I.

attention (see below). A thermoneutral process is obviously a poor choice if one wishes to observe a reaction which does not proceed readily.

In this connection, it is worth noting that the activation energy for the simple, almost thermoneutral, S_H2 reaction at carbon of deuterium with methane, namely,

$$D^{\cdot} + CH_4 \rightarrow CH_3D + H^{\cdot} \quad (3)$$

has been estimated to be 40 kcal/mole[2] and 37 kcal/mole,[3] it having been assumed in the calculations that the transition state is S_N2 like. Since the activation energy for the S_H2 reaction at hydrogen,

$$D^{\cdot} + CH_4 \rightarrow CH_3^{\cdot} + HD \quad (4)$$

is only ~8–10 kcal/mole[4] it is easy to understand why the substitution at carbon is indescribably slow compared with the hydrogen atom abstraction process.

Chou and Rowland[5] have generated hot tritium atoms of various energies by the photolysis of TBr at various wavelengths and have observed the experimental onset of the substitution reaction,

$$^{*}T^{\cdot} + CD_4 \rightarrow CD_3T + D^{\cdot} \quad (5)$$

by monitoring the yield of CD_3T. This yield is negligible at energies below about 35 kcal/mole. The agreement of this energy with the calculated activation energy for reaction 3 is perhaps fortuitous since the calculations relate to an inversion process whereas, in molecules in which it has been possible to test the stereochemical outcome, substitution (with more energetic tritium atoms) has involved retention of configuration (see Chapter I). It is, of course, not known whether reaction 5 proceeds with retention or inversion. The onset of tritium substitution in solid cyclohexane has been studied by Menzinger and Wolfgang[6] using mono-energetic tritium atoms from a molecular beam.

$$T^{\cdot} + C_6H_{12} \rightarrow C_6H_{11}T + H^{\cdot} \quad (6)$$

This reaction becomes significant at energies somewhat above 23 kcal/mole.

It appears to the present authors that the positive identification of an S_H2 reaction at sp^3 hybridized carbon will probably have to await studies on more strongly exothermic reactions than most of those which have been examined so far. That is, reactions 1 or 2 will require that the R to carbon bond

[2] F. O. Rice and E. Teller, *J. Chem. Phys.*, **6,** 489 (1938).
[3] E. Gorin, W. Kauzmann, J. Walter, and H. Eyring, *ibid.*, **7,** 633 (1939).
[4] A. F. Trotman-Dickenson, *Adv. Free Radical Chem.*, **1,** 1 (1965).
[5] C. C. Chou and F. S. Rowland, *J. Chem. Phys.*, **50,** 2763 (1969).
[6] M. Menzinger and R. Wolfgang, *J. Amer. Chem. Soc.*, **89,** 5992 (1967).

strength is large but the carbon to B bond strength is low because of resonance stabilization of B. Further, the carbon should be relatively open to attack by R˙. Various possible candidates for homolytic substitutions at carbon will no doubt suggest themselves to the reader. However, with regard to the stereochemical outcome of such a reaction it is worth reiterating that backside attack, reaction 1, is to be expected because it minimizes the internal energy of the transition state by minimizing the positive exchange energy between the electron pairs in the preserved bonds (C–a, C–b, and C–c) and the electrons in the changing partial bonds R···C and C···B.[1]

ATOM FOR ATOM SUBSTITUTIONS

The first attempt to characterize an S_H2 reaction was made in 1935 by Ogg and Polanyi.[7] These workers studied the gas phase racemization of optically active 2-iodobutane which is caused by the iodine produced by its own decomposition at 240–280°. The kinetics indicated that racemization involved reaction of 2-iodobutane with an iodine atom and this was interpreted as a substitution with Walden inversion, that is,

$$d\text{-Bu}^s\text{I} \rightarrow \text{Bu}^{s\cdot} + \text{I}^\cdot \tag{7}$$

$$\text{I}^\cdot + d\text{-Bu}^s\text{I} \rightarrow l\text{-Bu}^s\text{I} + \text{I}^\cdot \tag{8}$$

However, the possibility that racemization occurred by way of the *sec*-butyl radical was not satisfactorily eliminated.

$$\text{I}^\cdot + d\text{-Bu}^s\text{I} \rightleftarrows \text{Bu}^{s\cdot} + \text{I}_2 \rightleftarrows \text{I}^\cdot + l\text{-Bu}^s\text{I} \tag{9}$$

Subsequent work has revealed the difficulties that are involved in an unambiguous choice between the two mechanisms as well as the complexities that may be involved in these apparently simple reactions.[8–17] Hermann and Noyes[9] studied the racemization of 2-iodobutane in 1,3-hexachlorobutadiene at 130–170° in the presence of isotopically labeled elementary iodine. If iodine atoms caused exchange only by direct substitution involving inversion at carbon, the ratio of the rates of racemization to exchange would be 2.0; if

[7] R. A. Ogg, Jr., and M. Polanyi, *Trans. Faraday Soc.*, **31,** 482 (1935).
[8] D. Clark, H. O. Pritchard, and A. F. Trotman-Dickenson, *J. Chem. Soc.*, 2633 (1954).
[9] R. A. Hermann and R. M. Noyes, *J. Amer. Chem. Soc.*, **78,** 5764 (1956).
[10] J. E. Bujake, Jr., M. W. T. Pratt, and R. M. Noyes, *ibid.*, **83,** 1547 (1961).
[11] M. Nakashima and R. M. Noyes, *Int. J. Chem. Kinetics*, **1,** 391 (1969).
[12] J. H. Sullivan, *J. Phys. Chem.*, **65,** 722 (1961).
[13] S. W. Benson and H. E. O'Neal, *J. Chem. Phys.*, **34,** 514 (1961).
[14] S. W. Benson, *ibid.*, **38,** 1945 (1963).
[15] D. M. Golden and S. W. Benson, *Chem. Rev.*, **69,** 125 (1969).
[16] H. R. Ward, R. G. Lawler, and R. A. Cooper, *Tetrahedron Lett.*, 527 (1969).
[17] A. J. Kassman and D. S. Martin, Jr., *J. Amer. Chem. Soc.*, **91,** 6237 (1969).

they reacted only by abstracting iodine the ratio would be 1.0. The observed ratio was 1.54 ± 0.06, independent of temperature and of the concentration of either reactant. Although this result is consistent with the simultaneous occurrence of both exchange mechanisms it could also be accounted for if a large fraction of the butyl radicals formed by iodine atom abstraction reacted with the *same* iodine molecule that was produced in the act of radical formation. (If a butyl radical always reacted with the identical molecule of iodine formed at its own genesis the ratio of racemization to exchange would be 2.0, as for the direct substitution.) Bujake, Pratt, and Noyes[10] subsequently showed that the rates of exchange of several primary and secondary alkyl iodides with iodine-131 were reduced by small amounts of oxygen to limiting values that were not further changed by increasing the oxygen pressure. The inhibited exchange was ascribed to trapping of the butyl radicals by oxygen, this preventing their reaction with molecular iodine.

$$Bu^{s\cdot} + O_2 \rightarrow Bu^sOO^\cdot \tag{10}$$

The residual exchange was ascribed to either the direct substitution reaction or to the inability of oxygen to scavenge the butyl radicals while they are still present in the cage with the iodine produced in the abstraction step, reaction 9. Yet a third possibility has recently been ruled out by Nakashima and Noyes.[11] This involved a β-hydrogen abstraction in a step having the same kinetics as the iodine atom abstraction, namely,

$$R'R''CHCH_2I + I^{*\cdot} \rightarrow R'R''C{=}CH_2 + HI^* + I^\cdot \tag{11}$$

This step could lead to isotopically labeled organically-bound iodine if the hydrogen iodide exchanged with alkyl iodide by a four center mechanism or if it added to the olefin formed in reaction 11.

Benson and co-workers[13–15] have very carefully investigated the kinetics of the gas-phase pyrolysis reactions of a large number of alkyl iodides and the kinetics of their reactions with hydrogen iodide. This work has provided a reliable source of thermodynamic data on the free radicals and molecules involved in these reactions. Although the kinetics can be quite complicated relatively few elementary reactions are involved. One of the important processes is the equilibrium,

$$I^\cdot + RI \rightleftarrows R^\cdot + I_2 \tag{12}$$

The forward reaction has a normal pre-exponential factor for an atom abstraction reaction and an activation energy which is not significantly in excess of its estimated endothermicity (~18 kcal/mole for primary iodides[13]). The reverse reaction occurs about once in a hundred collisions. The direct substitution of iodine for iodine has not been observed in any of these gas-phase studies and it seems unlikely, therefore, that it can play any more than a minor role in solution. At all events, the iodine atom-alkyl iodide reaction is clearly a poor candidate for the identification of an S_H2 reaction at carbon in view of the importance of the S_H2 reaction at iodine.

There have been few other attempts to investigate homolytic halogen for halogen substitution at carbon. One which might be mentioned is Davidson and Sullivan's study of the exchange of labeled bromine with bromotrichloromethane in the vapor phase.[18]

$$Br^{*\cdot} + CCl_3Br \overset{a}{\rightleftarrows} CCl_3^{\cdot} + Br_2^{*} \underset{a}{\rightleftarrows} Br^{\cdot} + CCl_3Br^{*} \quad (13)$$

The presence of $CCl_3^{\cdot}$ radicals in the reaction was demonstrated. A subsequent study[19] of the thermal bromination of chloroform yielded Arrhenius parameters for reaction 13*a* which were identical, within the limits of experimental error, to the values found from the exchange of the radioactive bromine. Any nonterminal S_H2 process was too slow to detect.

A number of other early attempts to demonstrate halogen for hydrogen homolytic substitutions at carbon served merely to confirm that these reactions proceed via a terminal S_H2 process, that is, by hydrogen atom abstraction.

ATOM FOR RADICAL SUBSTITUTIONS

The fluorination of organic compounds with molecular fluorine in the gas or liquid phase normally results in large yields of fission products, particularly tetrafluoromethane.[20–22] It has frequently been suggested or implied that the fission products arise by a direct S_H2 reaction in which a fluorine atom, or even a fluorinated carbon radical attacks a saturated carbon atom.[20–24]

$$F^{\cdot} + CF_3B \rightarrow CF_4 + B^{\cdot} \quad (14)$$

However, as Tedder in particular has pointed out,[22] there is normally incomplete disposal of the heat formed in these vigorous and highly exothermic reactions. The yields of both scission products and polymeric products can be

[18] N. Davidson and J. H. Sullivan, *J. Chem. Phys.*, **17,** 176 (1947).
[19] J. H. Sullivan and N. Davidson, *ibid.*, **19,** 143 (1951).
[20] L. A. Bigelow, *Chem. Rev.*, **40,** 51 (1947).
[21] L. A. Bigelow, in *Fluorine Chemistry*, Vol. I, J. H. Simons (Ed.), Academic Press, New York, 1950, p. 373.
[22] J. M. Tedder, *Adv. Fluorine Chem.*, **2,** 104 (1961).
[23] F. P. Avonda, J. A. Gervasi, and L. A. Bigelow, *J. Amer. Chem. Soc.*, **78,** 2798 (1956).
[24] B. C. Bishop, J. B. Hynes, and L. A. Bigelow, *ibid.*, **85,** 1606 (1963).

greatly reduced[25,26] or even entirely eliminated[27–31] if stringent precautions are taken to carry out the reaction under isothermal conditions at reasonable temperatures. At the present time there does not appear to be any real evidence in favor of reaction 14.

The attack of hydrogen atoms on toluene at high temperatures (680–850°) yields both hydrogen and methane.[32] The methane is not formed by the substitution of hydrogen for phenyl, but rather via the substitution of methyl for hydrogen,[33] presumably by the reaction sequence,

$$H^{\cdot} + CH_3C_6H_5 \rightarrow (H)(CH_3)C_6H_5^{\cdot} \rightarrow C_6H_6 + CH_3^{\cdot} \quad (15)$$

$$CH_3^{\cdot} + CH_3C_6H_5 \longrightarrow CH_4 + {}^{\cdot}CH_2C_6H_5 \quad (16)$$

The opening of the cyclopropane ring by chlorine atoms,[34–38] bromine atoms,[39] and iodine atoms[40,41] can perhaps be considered as examples of S_H2 processes at saturated carbon, for example,

$$I^{\cdot} + \Delta \rightarrow ICH_2CH_2CH_2^{\cdot} \quad (17)$$

$$(\log k_{17} = 9.62 - 17{,}500/2.3RT\ M^{-1}\ sec^{-1})$$

However, in addition to ~28 kcal/mole of strain energy[42] the cyclopropane ring has considerable "unsaturated" character as was first recognized by Walsh.[43] There are a number of physical and chemical properties which

25 E. A. Tyczkowski and L. A. Bigelow, *ibid.*, **77**, 3007 (1955).
26 A. F. Maxwell, F. E. Detoro, and L. A. Bigelow, *ibid.*, **82**, 5827 (1960).
27 J. M. Tedder, *Chem. Ind. (London)*, 508 (1955).
28 P. C. Anson and J. M. Tedder, *J. Chem. Soc.*, 4390 (1957).
29 P. C. Anson, P. S. Fredricks, and J. M. Tedder, *ibid.*, 918 (1959).
30 P. S. Fredricks and J. M. Tedder, *ibid.*, 144 (1960).
31 G. C. Fettis, J. H. Knox, and A. F. Trotman-Dickenson, *ibid.*, 1064 (1960).
32 M. Szwarc, *Chem. Rev.*, **47**, 75 (1950).
33 A. T. Blades and E. W. R. Steacie, *Can. J. Chem.*, **32**, 1142 (1954).
34 J. D. Roberts and P. H. Dirstine, *J. Amer. Chem. Soc.*, **67**, 1281 (1945).
35 P. G. Stevens, *ibid.*, **68**, 620 (1946).
36 D. E. Applequist, G. P. Fanta, and B. Henrikson, *ibid.*, **82**, 2368 (1960).
37 C. Walling and P. S. Fredricks, *ibid.*, **84**, 3326 (1962).
38 M. L. Poutsma, *ibid.*, **87**, 4293 (1965).
39 M. S. Kharasch, M. Z. Fineman, and F. R. Mayo, *ibid.*, **61**, 2139 (1939).
40 R. A. Ogg, Jr., and W. J. Priest, *J. Chem. Phys.*, **7**, 736 (1939).
41 S. W. Benson, *ibid.*, **34**, 521 (1961).
42 S. W. Benson, *Thermochemical Kinetics*, Wiley, New York, 1968, p. 48.
43 A. D. Walsh, *Trans. Faraday Soc.*, **45**, 179 (1949).

illustrate the close relationships between cyclopropyl bonds and olefinic double bonds and which arise from a basic similarity in their bonding.[44] The π-like cyclopropyl orbitals (formed by the overlap of two sp^5 hybrid bonds[45]) are subject to radical and electrophilic attack just like the π electrons of olefins. The electronic structure of the transition state is therefore likely to be similar to that involved in halogen atom additions to olefins.[37] That is, the reaction could be interpreted as an addition of the halogen to the cyclopropane ring followed by ring opening rather than as a synchronous S_H2 process.

Reaction 17 has an activation energy of 17.5 kcal/mole in the gas phase[41] which indicates that these atom induced ring openings do not occur very readily since this reaction is approximately thermoneutral. The extent of ring cleavage by chlorine atoms in both the gas and liquid phases increases as the temperature is lowered.[36,37] This implies that the ring opening reaction 18 has a lower activation energy than the hydrogen abstraction reaction 19.

$$Cl^{\cdot} + \Delta \rightarrow ClCH_2CH_2CH_2^{\cdot} \tag{18}$$

$$Cl^{\cdot} + \Delta \rightarrow HCl + \Delta^{\cdot} \tag{19}$$

The relatively high activation energy of reaction 19 is presumably due to the high C–H bond strength in cyclopropane (~101 kcal/mole[46]).

9,10-Dehydrodianthracene, a strained Dewar form of anthracene, reacts with molecular bromine by a free-radical chain process.[47] Presumably, a bromine atom attacks the hydrocarbon and breaks the 9,10 transannular

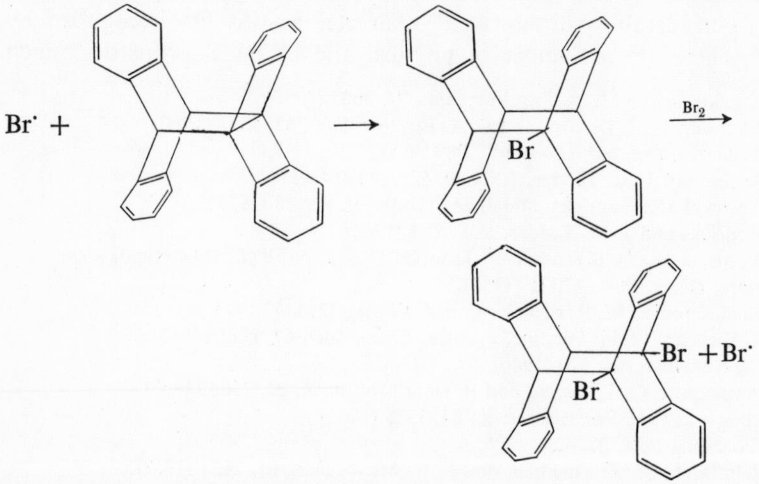

Br· + [structure] → [structure with Br] $\xrightarrow{Br_2}$ [structure with Br, Br] + Br· (20)

44 W. A. Bernett, *J. Chem. Educ.*, **44**, 17 (1967).
45 F. J. Weigert and J. D. Roberts, *J. Amer. Chem. Soc.*, **89**, 5962 (1967).
46 J. A. Kerr, *Chem. Rev.*, **66**, 465 (1966).
47 D. E. Applequist and R. Searle, *J. Amer. Chem. Soc.*, **86**, 1389 (1964).

bond. The fact that bromination occurs at all establishes the stereochemical point that radical displacement on carbon may occur with inversion of configuration. However, it does not determine the preferred stereochemistry of S_H2 reactions at carbon.

Similarly, the highly strained 1,3-dehydroadamantane is readily cleaved by bromine or iodine to yield strain-free 1,3-dihalo-adamantanes.[48] If, as seems likely, these reactions involve free radicals, the ease of the reaction may be due less to the presence of a cyclopropyl ring than to the fact that all four bonds at carbon atoms 1 and 3 are extended almost from one side of these atoms.

It is clear that homolytic substitutions by atoms at strained, but formally saturated, carbon atoms can occur reasonably readily. However, there is very little evidence that such substitutions can occur at unstrained, sp^3 hybridized, carbon.

RADICAL FOR RADICAL SUBSTITUTION

Holroyd and Noyes[49] investigating the photolysis of mercury dimethyl at 2600 Å and 175° reported that more than two methyl radicals appeared to be formed (as methane and ethane) per quantum absorbed. This was offered as possible evidence for the occurrence of reaction 21.

$$CH_3^{\cdot} + CH_3HgCH_3 \rightarrow C_2H_6 + Hg + CH_3^{\cdot} \qquad (21)$$

Some support for this process came from a study of the reaction of $CD_3^{\cdot}$ radicals with mercury dimethyl.[50] The $CD_3^{\cdot}$ radicals were produced by the photolysis of perdeuteroacetone at wavelengths above 2800 Å where mercury dimethyl does not absorb. Some CH_3CD_3 was produced but it was pointed out that this did not necessarily arise entirely from reaction 21 because there might be a transfer of energy from an excited acetone molecule or acetyl radical to the mercury dimethyl with subsequent decomposition of the latter. It was also concluded[50] from an analysis of the data that if reaction 21 occurred it had a low activation energy. This is inconsistent with our present views regarding the difficulty of S_H2 reactions at sp^3 hybridized carbon. Subsequent work has shown that some $CH_3^{\cdot}$ radicals are produced by the homolytic substitution at mercury,

$$CD_3^{\cdot} + CH_3HgCH_3 \rightarrow CD_3HgCH_3 + CH_3^{\cdot} \qquad (22)$$

The CH_3CD_3 is formed principally by the combination of methyl and

[48] R. E. Pincock and E. J. Torupka, *ibid.*, **91,** 4593 (1969).
[49] R. A. Holroyd and W. A. Noyes, Jr., *ibid.*, **76,** 1583 (1954).
[50] See footnote 142, Chapter III.

deuteromethyl radicals. That is, reaction 22 but not reaction 21 plays a role in the decomposition of dimethyl mercury (see Chapter III).[51,52]

Reactions analogous to 21 have been proposed by Nesmeyanov and co-workers[53] who reported that the benzoyl peroxide induced reaction between carbon tetrachloride and dialkylmercury compounds gave alkylmercury chlorides and a trichloromethyl compound (or the corresponding carboxylic acid after alkaline hydrolysis).

$$CCl_4 + RHgR' \rightarrow RHgCl + R'CCl_3 \quad (23)$$

It was suggested that the chain reaction involved the step,

$$CCl_3^{\cdot} + RHgR' \rightarrow R'CCl_3 + RHg^{\cdot} \quad (24)$$

As Jensen and Guard have pointed out,[54] it is difficult to conceive of a mechanism which would yield the reported products that does not proceed through an S_H2 reaction. However, in their reinvestigation of these processes Jensen and Guard were unable to repeat the reactions reported by Nesmeyanov et al.[53] For example, the major products from dibutylmercury were 1,1,1,3-tetrachloropentane, 1-chlorobutane, chloroform, butylmercuric chloride, and mercury. None of the expected 1,1,1-trichloropentane was found. The tetrachloropentane arose from the addition of carbon tetrachloride to 1-butene (which is a major product at $-78°$). The products were accounted for by proposing an elimination (possibly a concerted elimination) of olefin from the dibutyl mercury under the action of a $CCl_3^{\cdot}$ radical.

$$CCl_3^{\cdot} + Bu^nHgBu^n \rightarrow CCl_3H + Bu^nHg\cdot + CH_3CH_2CH{=}CH_2 \quad (25)$$

$$CH_3CH_2CH{=}CH_2 \xrightarrow{CCl_4} CH_3CH_2CHClCH_2CCl_3 \quad (26)$$

That is, the $CCl_3^{\cdot}$ radical enters into an S_H2 reaction at hydrogen, not at carbon or mercury (see also Chapter III).

The results of Jensen and Guard are in basic agreement with the earlier work of Razuvaev et al.[55] who reported that the reaction of diisopropyl mercury with $CDCl_3$ and CCl_4 at 130° yielded isopropylmercury chloride, mercury, propane, propylene, isopropyl chloride, and chloroform. The first solvent also yielded 2-deuteriopropane and monodeuteriodichloromethane, while hexachloroethane and a small amount of mercurous chloride were also

[51] C. E. Waring and R. Pellin, *J. Phys. Chem.*, **71,** 2044 (1967).

[52] C. J. Wolf and J. Q. Walker, *ibid.*, **72,** 3457 (1968).

[53] See footnote 152, Chapter III.

[54] F. R. Jensen and H. E. Guard, *J. Amer. Chem. Soc.*, **90,** 3250 (1968).

[55] G. A. Razuvaev, S. F. Zhil'tsov, O. N. Druzhkov, and G. G. Petukhov, *Dokl. Akad. Nauk SSSR*, **156,** 393 (1964).

formed in CCl_4. At these high temperatures the chlorinated solvent does not add to the double bond of the liberated olefin.

A radical–radical disproportionation which would seem to require an S_H2 reaction at carbon has been proposed in the thermal decomposition of trimethylantimony.[56]

$$2Me_2Sb^{\cdot} \rightarrow Me_3Sb + MeSb \tag{27}$$

However, the kinetic peculiarities which led to this reaction being proposed could well arise from some other cause since the decomposition was carried out at high temperatures (475–664°) and in the presence of excess toluene.

Probably the most clear cut evidence that homolytic substitutions at carbon by carbon radicals do not occur under normal conditions has been obtained by Slaugh[57] and by Trahanovsky and Doyle.[58] Slaugh[57] showed that the 3-phenoxypropyl radical does not yield cyclopropane.

$$PhOCH_2CH_2\dot{C}H_2 \not\rightarrow PhO^{\cdot} + \Delta \tag{28}$$

That is, an S_Hi reaction (substitution, homolytic, intramolecular) at a saturated carbon does not occur even when the displaced radical is resonance stabilized.

Trahanovsky and Doyle[58] measured the rates and identified the products from the thermal decomposition of *t*-butylperoxy 6-bromohexanoate and *t*-butylperoxy hexanoate in various solvents at 100° and, for the former compound, in the gas phase at 250°. It was anticipated that cyclopentane formation by an intramolecular homolytic substitution of the initial radical was quite likely since the energy required to break the C–Br bond should be more than returned by the formation of the C–C bond. Moreover, the

$$\dot{C}H_2\text{–}(CH_2)_4\text{–Br (ring)} \rightarrow (\text{cyclopentane}\cdots Br)^{\cdot} \rightarrow \text{cyclopentane} + Br^{\cdot} \tag{29}$$

five-membered ring in the transition state should not be unduly strained and substitution at this position should be favored by the fact that it would occur at an unhindered primary carbon. In contrast to their reasonable expectations, Trahanovsky and Doyle found that cyclopentane was not produced under any circumstances. The intramolecular homolytic substitution shown in reaction 29 must be very unfavorable since it failed to occur even when the radical was generated in relatively inert solvents such as benzene, chlorobenzene and hexafluorobenzene or in the gas phase. Thus, the energy requirements for this type of radical displacement reaction must be very high.

[56] S. J. W. Price and A. F. Trotman-Dickenson, *Trans. Faraday Soc.*, **54,** 1630 (1958).
[57] L. H. Slaugh, *J. Amer. Chem. Soc.*, **83,** 2734 (1961).
[58] W. S. Trahanovsky and M. P. Doyle, *J. Org. Chem.*, **32,** 146 (1967).

Trahanovsky and Doyle's failure to find any evidence for radical cyclization by intramolecular substitution at sp^3 hybridized carbon under their ideal conditions makes other less detailed reports of this type of radical reaction highly suspect. For example, it has been reported that the heptyl radical generated by treating heptyl bromide with sodium vapor at 300° cyclizes to form cyclohexane with the elimination of a methyl radical.[59] The cyclohexane was identified by g.l.c. retention times and mass spectrometry. However, it may have been produced by some path other than an intramolecular substitution. Similarly, numerous reports attest to the fact that $\alpha\omega$-dihalo (particularly diiodo) propanes, butanes and pentanes can yield the corresponding cyclic hydrocarbons under conditions where the first halogen at least is removed in such a way as to leave behind a free radical.[60–63] Suggestions that cyclization is due to an intramolecular homolytic substitution,

$$\mathrm{YCH_2(CH_2)_{\mathit{n}}CH_2X} \xrightarrow{\mathrm{R\cdot}} \mathrm{YCH_2(CH_2)_{\mathit{n}}\dot{C}H_2} \longrightarrow \overbrace{\mathrm{CH_2(CH_2)_{\mathit{n}}CH_2}} + \mathrm{Y^{\cdot}} \tag{30}$$

seem unlikely to be correct in view of Trahanovsky and Doyle's results[58,64] and have been positively disproved for the 4-iodobutyl radical by Garst and Barbas.[62] In addition, it is worth noting that both Trahanovsky and Doyle, and Garst and Barbas are agreed that there is no evidence that ω-halo radicals are bridged by the halogen atom.[58,62]

Very recently Kaplan and Drury[60] have presented additional support for reaction 30 with $n = 1$ and $X = I$ and $Y = I$, Ph_3Sn and Ph_3Ge. Cyclopropane was produced at 285° for reactions with Ph_3SiH, Ph_2SiH_2, or Ph_3GeH initiated with benzoyl peroxide or *t*-butyl peroxide. Cyclopropane was not formed with Y = Br, PhS, or Cl. Kaplan points out that to establish the operation of the S_H2 mechanism (actually S_Hi) it is necessary to demonstrate (i) the intermediacy of $YCH_2CH_2\dot{C}H_2$ as a precursor of cyclopropane and (ii) the non-operation of a mechanism involving radical displacement on the Y of a 1,3-Y-bridged trimethylene species. The absence of bridging for Y = I was shown (see also footnotes 58 and 62) and the abstract implies that for $ICH_2CH_2\dot{C}H_2$ the yield of $ICH_2CH_2CH_2Cl$ relative to cyclopropane increases as the concentration of CCl_4 as solvent is increased. It is concluded

[59] N. J. Friswell and B. G. Gowenlock, *Chem. Comm.*, 277 (1965).

[60] L. Kaplan, (a) *J. Amer. Chem. Soc.*, **89,** 1753 (1967); (b) *ibid.*, **89,** 4566 (1967); (c) *J. Org. Chem.*, **32,** 4059 (1967); (d) *Chem. Comm.*, 754 (1968); (e) *ibid.*, 106 (1969); (f) L. Kaplan and R. Drury, abstract 5.19 of paper presented at the Joint Meeting of the Chem. Soc. and the Royal Inst. Chem. Edinburgh (April 1970).

[61] P. B. Chock and J. Halpern, *J. Amer. Chem. Soc.*, **91,** 582 (1969).

[62] J. F. Garst and J. T. Barbas, *ibid.*, **91,** 3385 (1969).

[63] J. K. Kochi and D. M. Singleton, *J. Org. Chem.*, **33,** 1027 (1968).

[64] See also, D. C. Blomstrom, K. Herbig, and H. E. Simmons, *ibid.*, **30,** 959 (1965).

that the radical leaving ability of Y from $YCH_2CH_2\dot{C}H_2$ decreases in the order $Ph_3Sn > I > Ph_3Ge$.

It is difficult to assess the true significance of this work but, in the first place, it is hard to believe that cyclopropane with ~28 kcal/mole of strain energy[42] could be formed in what appears to be a relatively facile S_Hi reaction (e.g., the reaction of $Ph_3SnCH_2CH_2CH_2Cl$ with 3% Ph_3SiH is reported to give cyclopropane quantitatively). For Y = I the data certainly do not require an S_Hi process, that is,

$$ICH_2CH_2\dot{C}H_2 + Ph_3M^{\cdot} \longrightarrow {}^{\cdot}CH_2CH_2\dot{C}H_2 + Ph_3MI \quad (31)$$
$$\downarrow \Delta$$
$$(M = Si, Ge)$$

For Y = Ph_3Sn and Ph_3Ge the ease of S_H2 reactions at these metals (see Chapter V(C) and V(D)) indicates other possibilities such as reaction 32 followed by 31,

$$Ph_3SnCH_2CH_2CH_2I + R^{\cdot} \rightarrow Ph_3SnR + {}^{\cdot}CH_2CH_2CH_2I \quad (32)$$

where $R^{\cdot}$ is any radical in the system, including the initiating radicals $PhC(O)O^{\cdot}$ and $Bu^tO^{\cdot}$ which would not abstract iodine. Another possibility is an S_Hi reaction at tin,

$$Ph_3SnCH_2CH_2\dot{C}H_2 \longrightarrow$$

[Resonance structures: Ph_3Sn bridged by CH_2–CH_2–${}^{\cdot}CH_2$ $\longleftrightarrow$ $Ph_3Sn\cdot$ with CH_2, CH_2, CH_2 ring $\longleftrightarrow$ Ph_3Sn with ${}^{\cdot}CH_2$–CH_2–CH_2]

$$\downarrow$$
$$Ph_3Sn^{\cdot} + \Delta \quad (33)$$

or even an attack on one of the phenyl rings,

[Structure: Ph_2Sn attached to cyclohexadienyl ring bridged via $CH_2CH_2CH_2$ to ring carbon bearing H] $\longrightarrow$ [Structure: Ph_2Sn= cyclohexadiene bearing $CH_2CH_2\dot{C}H_2$ and H] $\longrightarrow$

$$Ph_3Sn^{\cdot} + {}^{\cdot}CH_2CH_2\dot{C}H_2 \quad (34)$$

In view of the weight of evidence against S_Hi reactions at unstrained sp^3 hybridized carbon by carbon centered radicals, it would seem premature to conclude from Kaplan and Drury's data that such processes can occur with considerable facility to yield the highly strained cyclopropane molecule.

A few other reported examples of S_H2 reactions at sp^3 hybridized carbon are worth noting. Thus, Davidson and Stephenson[65] have explained the formation of tetramethylsilane in the pyrolysis (523–555°) of hexamethydisilane by the reaction,

$$Me_3Si^{\cdot} + Me_3SiSiMe_3 \rightarrow Me_4Si + Me_3\dot{Si}SiMe_2 \quad (35)$$

It was concluded from the absence of methane in the products that methyl radicals cannot have been formed in the system and hence, that the tetramethylsilane cannot have arisen from combination of a methyl with a trimethylsilyl radical. The activation energy for this S_H2 reaction at carbon was estimated to be only 2 kcal/mole greater than for the hydrogen abstraction reaction,

$$Me_3Si^{\cdot} + Me_3SiSiMe_3 \rightarrow Me_3SiH + Me_3SiSiMe_2\dot{C}H_2 \quad (36)$$

The activation energy for the hydrogen abstraction was estimated to be at least 20 kcal/mole. However, in view of the overwhelming evidence from other sources that S_H2 reactions at sp^3 hybridized carbon do not occur at all readily, some other explanation for the formation of tetramethylsilane seems to be required. Possibly, at these high temperatures any methyl radicals formed undergo S_H2 reactions at silicon much more rapidly than they abstract hydrogen to give methane.

$$Me^{\cdot} + Me_3SiSiMe_3 \xrightarrow{\text{fast}} Me_4Si + Me_3Si^{\cdot} \quad (37)$$

$$Me^{\cdot} + Me_3SiSiMe_3 \xrightarrow{\text{slow}} CH_4 + Me_3SiSiMe\dot{C}H_2 \quad (38)$$

The autoxidations of hexaarylethanes and tetraaryldialkylethanes (which for a long time were believed to dissociate spontaneously into free radicals and exist in solution in equilibrium with the radicals) have been shown to be chain processes.[66–72] Hydroperoxides are only obtained in the presence of hydrogen donors such as pyrogallol, under which conditions the chain process is inhibited. The uninhibited chain yields peroxides and the reaction scheme for hexaphenylethane, for example, has been formulated as

$$Ph_3CCPh_3 \rightleftarrows 2Ph_3C^{\cdot} \quad (39)$$

$$Ph_3C^{\cdot} + O_2 \rightarrow Ph_3COO^{\cdot} \quad (40)$$

$$Ph_3COO^{\cdot} + Ph_3CCPh_3 \rightarrow Ph_3COOCPh_3 + Ph_3C^{\cdot} \quad (41)$$

[65] I. M. T. Davidson and I. L. Stephenson, *J. Chem. Soc. (A)*, 282 (1968).
[66] K. Ziegler and L. Ewald, *Ann.*, **504,** 162 (1933).
[67] K. Ziegler, L. Ewald, and A. Seib, *ibid.*, **504,** 182 (1933).
[68] K. Ziegler, *ibid.*, **551,** 127 (1942).
[69] K. Ziegler, A. Seib, F. Knoevenagel, P. Herte, and F. Andreas, *ibid.*, **551,** 150 (1942).
[70] K. Ziegler, R. B. Whitney, and P. Herte, *ibid.*, **551,** 187 (1942).
[71] K. Ziegler and P. Herte, *ibid.*, **551,** 206, 222 (1942).
[72] N. N. Lichtin and G. R. Thomas, *J. Amer. Chem. Soc.*, **76,** 3020 (1954).

Reaction 41 represents an S_H2 reaction at sp^3 hybridized carbon. However, while a homolytic substitution at carbon would certainly appear to be occurring it now seems unlikely that the carbon under attack is sp^3 hybridized. Lankamp, Nauta, and MacLean[73] have shown that diarylmethyl and triarylmethyl radicals dimerize unsymmetrically forming derivatives of 1-methylene-cyclohexa-2,5-diene, while true ethanes (e.g., $Ph_2CHCHPh_2$) do not dissociate at all readily:

$$2Ph_3C^{\cdot} \rightleftharpoons Ph_2C{=}C_6H_5(H)(CPh_3) \qquad (42)$$

The unsymmetrical nature of the dimers has been confirmed by ^{13}C n.m.r spectroscopy.[74] The reaction products indicate that the peroxy radical attacks the methylene carbon.

$$Ph_3COO^{\cdot} + Ph_2C{=}C_6H_5(H)(CPh_3) \longrightarrow Ph_3COOCPh_3 + {}^{\cdot}CPh_3 \qquad (43)$$

The occurrence in autoxidations of S_H2 reactions at carbon have also been postulated by Cullis and co-workers for the gas phase oxidation of trimethylaluminum,[74a]

$$Me_2AlOO^{\cdot} + Al_2Me_6 \rightarrow Me_2AlOOMe + 2Me_2Al^{\cdot} + Me^{\cdot} \qquad (44)$$

$$Me_2AlO^{\cdot} + Al_2Me_6 \rightarrow Me_2AlOMe + 2Me_2Al^{\cdot} + Me^{\cdot} \qquad (45)$$

and triethylindium,[75]

$$Et_2InOO^{\cdot} + Et_3In \rightarrow Et_2InOOEt + Et_2In^{\cdot} \qquad (46)$$

$$EtO^{\cdot} + Et_3In \rightarrow Et_2O + Et_2In^{\cdot} \qquad (47)$$

However, it seems much more probable that these oxidations involve S_H2 reactions by alkoxy and peroxy radicals at the metal rather than at carbon (see Chapter IV).

Although the foregoing account indicates that there are no unequivocal examples of S_H2 reactions in which a free radical attacks an unstrained sp^3 hybridized carbon, there are some examples of attack on strained rings. For instance, Gritter and Wallace[76] have reported that *t*-butoxy radicals attack propylene oxide at 150° to yield, among other products, 5-hydroxy-2-hexanone.[77] This compound is believed to be formed by the following

[73] H. Lankamp, Th. W. Nauta, and C. MacLean, *Tetrahedron Lett.*, 249 (1968).
[74] H. A. Staab, H. Brettschneider, and H. Brunner, *Chem. Ber.*, **103**, 1101 (1970).
[74a] See footnotes 75–77, Chapter IV.
[75] See footnote 107, Chapter IV.
[76] R. J. Gritter and T. J. Wallace, *J. Org. Chem.*, **26**, 282 (1961).
[77] This compound is not formed at 70°; see footnote 37, Chapter V.

sequence of reactions, which includes the S_H2 process, reaction 50.

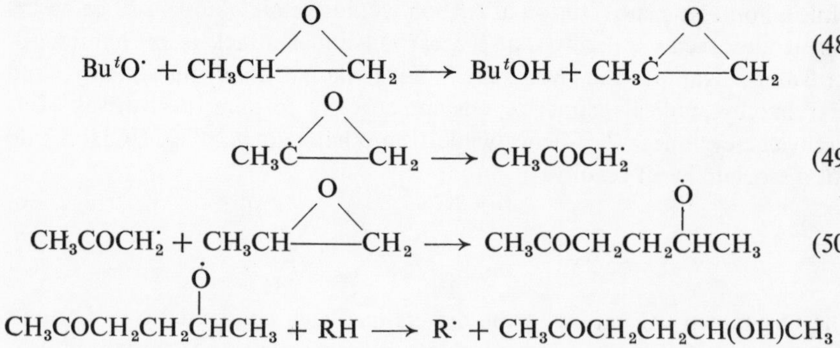

$$Bu^tO^\cdot + CH_3CH\overset{O}{\frown}CH_2 \longrightarrow Bu^tOH + CH_3\dot{C}\overset{O}{\frown}CH_2 \quad (48)$$

$$CH_3\dot{C}\overset{O}{\frown}CH_2 \longrightarrow CH_3COCH_2^\cdot \quad (49)$$

$$CH_3COCH_2^\cdot + CH_3CH\overset{O}{\frown}CH_2 \longrightarrow CH_3COCH_2CH_2\overset{\dot{O}}{C}HCH_3 \quad (50)$$

$$CH_3COCH_2CH_2\overset{\dot{O}}{C}HCH_3 + RH \longrightarrow R^\cdot + CH_3COCH_2CH_2CH(OH)CH_3 \quad (51)$$

A second example of an S_H2 reaction at carbon in a strained ring is implied by Pincock and Torupka's[48] report that 1,3-dehydroadamantane reacts rapidly with oxygen by a free-radical chain process to yield a polymeric 1,3-dioxyadamantane $[–O–C_{10}H_{14}–O–]_n$, presumably by the reaction sequence,

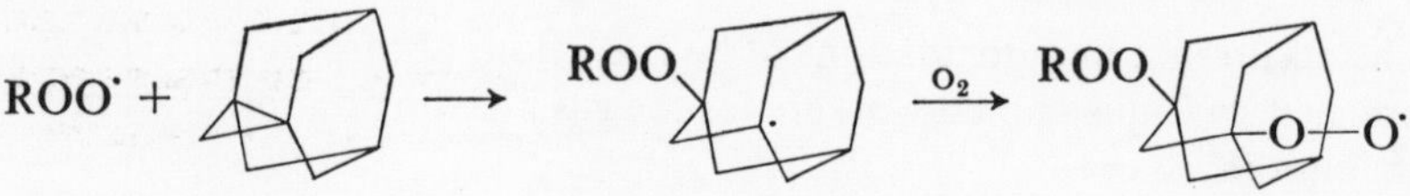

ROO· + [1,3-dehydroadamantane] → ROO–[adamantyl·] $\xrightarrow{O_2}$ ROO–[adamantyl]–O–O· (52)

Yet another example is provided by Jarvis' report[77a] that the cyclopropane ring in dibenzotricyclo $[3.3.0.0^{2,8}]$-3, 6-octadiene is opened with inversion by the trichloromethyl radical in the free radical addition of bromotrichloromethane.

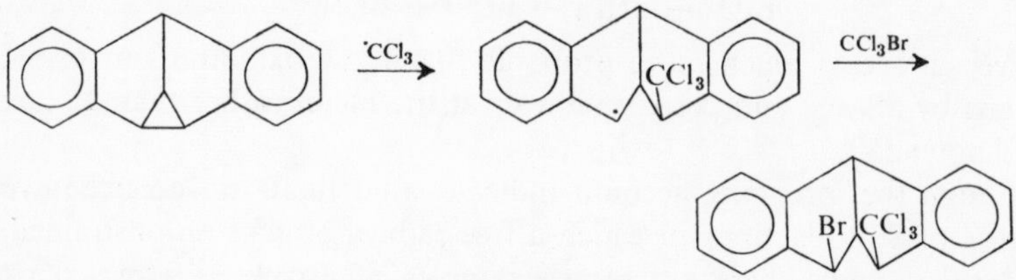

$\xrightarrow{\cdot CCl_3}$ [CCl$_3$] $\xrightarrow{CCl_3Br}$ [Br CCl$_3$] (53)

Homolytic substitutions at unsaturated carbon (olefins, aromatics, ketones, etc.) are quite common. They are more likely to occur in the gas phase than in solution because the reagent concentration is generally lower. The intermediate adduct radical is therefore less likely to be trapped by abstracting a labile atom or by adding to a second unsaturated molecule. In addition, the "hot" adduct radical is less likely to be deactivated in the gas phase and is therefore more likely to undergo an elimination. Among the earliest reports

[77a] B. B. Jarvis. *J. Org. Chem.*, **35**, 924 (1970).

of this type of reaction appear to be those relating to the attack of methyl radicals on the α,β-unsaturated ketones, biacetyl,[78] *trans*-methyl propenyl ketone,[79,80] and crotonaldehyde[80] in the gas phase, e.g.

$$CH_3^{\cdot} + CH_3COCOCH_3 \rightarrow CH_3COCH_3 + CH_3\dot{C}O \qquad (54)$$

$$CH_3^{\cdot} + CH_3CH{=}CHCOCH_3 \rightarrow CH_3CH{=}CHCH_3 + CH_3\dot{C}O \qquad (55)$$

Bentrude and Darnell[81,82] have recently shown that similar reactions take place in solution. Thus,[81] when benzoyl peroxide is thermally decomposed at 80° in deoxygenated cyclohexane or dioxane to which biacetyl has been added, acetyl products can be obtained in up to 70% yields based on the assumed formation of 2 moles of radicals per mole of peroxide decomposed. At high biacetyl concentrations acetophenone is produced, the phenyl radical coming from decarboxylated benzoyloxy radicals.

$$R^{\cdot} + CH_3COCOCH_3 \rightleftarrows CH_3\underset{\underset{R}{|}}{\overset{\overset{\dot{O}}{|}}{C}}\!-\!\overset{\overset{O}{\|}}{C}CH_3 \longrightarrow CH_3COR + CH_3\dot{C}O \qquad (56)$$

The same type of products can also be formed by the photolysis of dilute biacetyl solutions.[82] In this case, the $R^{\cdot}$ radical is generated by hydrogen abstraction from the solvent by the electronically excited diketone.

A comprehensive discussion of homolytic substitution at sp^2 hybridized carbon including an attempt to distinguish between an addition-elimination sequence and a direct substitution for each reaction would be out of place in this review. Instead, it seems more appropriate to describe a specific reaction that has been the subject of recent debate.

The pyrolysis of acetaldehyde has been very thoroughly studied at temperatures from 450 to 550°.[83–92] The overall reaction is a simple chain

[78] F. E. Blacet and W. E. Bell, *Disc. Faraday Soc.*, **14,** 70 (1953).
[79] J. N. Pitts, Jr., R. S. Tolberg, and T. W. Martin, *J. Amer. Chem. Soc.*, **76,** 2843 (1954); **79,** 6370 (1957).
[80] J. N. Pitts, Jr., D. D. Thompson, and R. W. Woolfolk, *ibid.*, **80,** 66 (1958).
[81] W. G. Bentrude and K. R. Darnell, *ibid.*, **90,** 3588 (1968).
[82] W. G. Bentrude and K. R. Darnell, *Chem. Comm.*, 810 (1968).
[83] A. B. Trenwith, *J. Chem. Soc.*, 4426 (1963).
[84] R. W. Dexter and A. B. Trenwith, *ibid.*, 5459 (1964).
[85] M. Eusef and K. J. Laidler, *Can. J. Chem.*, **42,** 1851 (1964).
[86] K. J. Laidler, *Chemical Kinetics*, 2nd ed. McGraw-Hill, New York, 1965, pp. 408–410.
[87] K. J. Laidler and M. T. H. Liu, *Proc. Roy. Soc.* (*London*), **A297,** 365 (1967).
[88] M. T. H. Liu and K. J. Laidler, *Can. J. Chem.*, **46,** 479 (1968).
[89] L. Batt, *J. Chem. Phys.*, **47,** 3674 (1967).
[90] See footnote 42, Chapter V. (p. 139–142).
[91] L. Phillips, *J. Phys. Chem.*, **73,** 2090 (1969).
[92] L. Batt, *ibid.*, **73,** 2091 (1969).

decomposition, the rate of which is proportional to $[CH_3CHO]^{3/2}$. The major products are methane and carbon monoxide,

$$CH_3^{\cdot} + CH_3CHO \rightarrow CH_4 + CH_3\dot{C}O \quad (57)$$

$$CH_3\dot{C}O\ (+\ M) \rightarrow CH_3^{\cdot} + CO\ (+\ M) \quad (58)$$

There are also a number of minor products, hydrogen, acetone, propionaldehyde, ethane, and ethylene. The initial rate of production of acetone is much greater than the initial rate of ethane production.[84] This led to the suggestion that the acetone was formed by the combination of $CH_3\dot{C}O$ and $CH_3^{\cdot}$ radicals and hence should replace $2CH_3^{\cdot} \rightarrow C_2H_6$ as the main chain terminating step. However, this termination is incompatible with the experimental reaction order of 3/2 in acetaldehyde.[85] Laidler and co-workers[86–88] have therefore proposed that the acetone is formed by the attack of $CH_3^{\cdot}$ on the acetaldehyde,

$$CH_3^{\cdot} + CH_3CHO \rightarrow CH_3COCH_3 + H^{\cdot} \quad (59)$$

The rate constant for this reaction at 523° was estimated[87] to be 7×10^3 M^{-1} sec^{-1} and its variation with temperature is given by[88] $\log k_{59} = 7.2 - 12{,}400/2.3RT$ (M^{-1} sec^{-1}). Although reaction 59 is written as a concerted process it is actually considered to occur in two stages.[85–92]

$$CH_3^{\cdot} + CH_3CHO \rightleftarrows (CH_3)_2\overset{H\ *}{C}O^{\cdot} \quad (60)$$

$$(CH_3)_2\overset{H\ *}{C}O^{\cdot} \rightarrow CH_3COCH_3 + H^{\cdot} \quad (61)$$

In radicals, as in normal molecules, C–H bonds are stronger than C–C bonds and hence $k_{-60} \gg k_{61}$. The rate constants and activation enthalpies and entropies of reactions 60, −60, and 61 have been estimated by Liu and Laidler[88] ($E_{60} = 10.6$, $E_{-60} = 17.3$, $E_{61} = 19.1$ kcal/mole) and by Batt[89,92] ($E_{60} = 11.5$, $E_{-60} = 17.2$, $E_{61} = 22\text{–}24$ kcal/mole). Phillips[91] has pointed out that the rate constants for decomposition of the isopropoxy radical (k_{-60} and k_{61}) will be pressure dependent under the conditions of Laidler and Liu's experiments (523°, 195 torr). Batt[92] estimates that $k_{-60} = 10^{10.1}$ sec^{-1} and $k_{61} = 10^{8.7}$ sec^{-1} which means that spontaneous decomposition (favoring acetaldehyde over acetone by a factor of 25) is much faster than quenching of the hot isopropoxy radical unless deactivation occurs on every collision which is unlikely. Batt's calculated rate of acetone production[92] is in excellent agreement with Laidler and Liu's measured rate which implies that most, if not all,[93] of the acetone is produced by a synchronous or

[93] Benson[90] has suggested that acetone may also arise by the sequence,

$$CH_3^{\cdot} + CH_2CO \rightleftarrows CH_3COCH_2^{\cdot}$$

$$CH_3COCH_2^{\cdot} + CH_3CHO \rightarrow CH_3COCH_3 + CH_3\dot{C}O$$

"concerted" substitution. However, the reaction is "concerted" only because of the high temperature and low pressure of the experiments. At lower temperatures or higher pressures the hot isopropoxy radical would be deactivated before scission and could be trapped as isopropyl alcohol by a suitable hydrogen donor. Reaction 60 and like processes should therefore be considered as addition-eliminations rather than as homolytic substitutions.

Neumann, Sommer, and Lind[94] have reported that Et_3SnCN is formed when azo-*bis*(1-cyanocyclohexane) is decomposed in triethyltin hydride. The mechanism was suggested to be the addition of the Et_3SnH across the N=N bond followed by molecular elimination of Et_3SnCN. However, Jackson[95] has suggested a simpler reaction scheme in which the $Et_3Sn^{\cdot}$ radical adds to the nitrile group and the resultant radical undergoes β-scission.

$Et_3Sn^{\cdot}$ + (NC)C₆H₁₀–N=N–C₆H₁₀(CN) → [Et_3Sn–C(=N·)C₆H₁₀–N=N–C₆H₁₀(CN)] → Et_3SnCN + C₆H₁₀·–N=N–C₆H₁₀(CN) (62)

RADICAL FOR ATOM SUBSTITUTIONS

These have been suggested only very occasionally. As examples, we might note Kochi and Buchanan's[96] tentative proposal that aralkyl radicals can substitute the halogen of aralkyl halides,

$$ArCH_2^{\cdot} + ArCH_2Br \rightarrow ArCH_2CH_2Ar + Br\cdot \quad (63)$$

[94] W. P. Neumann, R. Sommer, and H. Lind, *Ann.*, **688**, 14 (1965).
[95] R. A. Jackson, *Adv. Free-Radical Chem.*, **3**, 231 (1969).
[96] J. K. Kochi and D. Buchanan, *J. Amer. Chem. Soc.*, **87**, 853 (1965).

and Thomas' conclusion[97] that the most likely reaction between methyl iodide and a hydroxyl radical is the substitution of the iodine,

$$HO^{\cdot} + CH_3I \rightarrow CH_3OH + I^{\cdot} \quad (64)$$

Reaction 63 appears unlikely, but reaction 64 might indeed occur since it satisfies the requirements mentioned at the beginning of this Chapter. That is, a genuine S_H2 process at an *sp*³ hybridized carbon is only likely to be found among reactions which involve high exothermicity and low steric hindrance. However, in connection with reaction 64 it must be noted that Slaugh[98] has shown that *t*-butoxy radicals do not substitute for the iodine of 1-iodobutane.

$$Bu^tO^{\cdot} + ICH_2CH_2CH_2CH_3 \longrightarrow Bu^tOCH_2CH_2CH_2CH_3 + I^{\cdot} \quad (65)$$

B. SILICON

A silicon atom is, theoretically, more likely to become involved in an S_H2 reaction than a saturated carbon because it has energetically available 3*d* orbitals which would allow a bond to be formed to the incoming radical before an existing bond was broken. However, there are still very few examples of such reactions.

Hanson[99] found that dimethyldichlorosilane undergoes condensation in a high voltage a-c mercury arc to give 25-40% of 2,2,4,4-tetrachloro-2,4-disilapentane together with H_2, CH_4, C_2H_2, C_2H_4, C_2H_6, HCl, and CH_3SiCl_2H. The following free-radical chain mechanism was suggested to occur.

$$CH_3SiCl_2CH_2^{\cdot} + CH_3SiCl_2CH_3 \longrightarrow CH_3SiCl_2CH_2SiCl_2CH_3 + CH_3^{\cdot} \quad (66)$$

$$CH_3^{\cdot} + CH_3SiCl_2CH_3 \longrightarrow CH_4 + CH_3SiCl_2CH_2^{\cdot} \quad (67)$$

Further support for an S_H2 reaction at silicon by an α-silylalkyl radical comes from Urry's report[100] that the photolysis of *bis*(trimethylsilylmethyl)-mercury

[97] J. K. Thomas, *J. Phys. Chem.*, **71,** 1919 (1967).
[98] L. H. Slaugh, *J. Amer. Chem. Soc.*, **83,** 2734 (1969).
[99] J. E. Hanson, *Diss. Abstr.*, **25,** 818 (1964).
[100] G. Urry, *Amer. Chem. Soc. Abstracts 153rd Meeting* (Miami Beach), L63 (1967).

in the presence of tetramethylsilane yields 2,2,4,4-tetramethyl-2,4-disilapentane almost exclusively.

$$\{(CH_3)_3SiCH_2\}_2Hg \xrightarrow{h\nu} Hg + 2(CH_3)_3SiCH_2^{\cdot} \xrightarrow{(CH_3)_4Si} (CH_3)_3SiCH_2Si(CH_3)_3 + CH_3^{\cdot} \quad (68)$$

A homolytic substitution for hydrogen by a siloxy radical has been tentatively suggested to explain the formation of disiloxanes, nitrogen, and an enhanced rate of production of hydrogen when methylsilanes ($MeSiH_3$, Me_2SiH_2, Me_3SiH) are photosensitized in the presence of nitric oxide.[101]

$$Me_3Si^{\cdot} + NO \rightarrow Me_3SiON\colon$$
$$\rightarrow [Me_3SiON{=}NOSiMe_3] \rightarrow 2Me_3SiO^{\cdot} + N_2 \quad (69)$$
$$Me_3SiO^{\cdot} + Me_3SiH \rightarrow Me_3SiOSiMe_3 + H^{\cdot} \quad (70)$$

This substitution for hydrogen seems improbable in view of the fact that with *t*-butoxy radicals, silanes undergo S_H2 reactions at hydrogen not silicon.[102–104]

$$Me_3CO^{\cdot} + Me_3SiH \rightarrow Me_3COH + Me_3Si^{\cdot} \quad (71)$$

The hexamethyldisiloxane and hydrogen found by Nay et al.[101] were probably formed by routes other than that proposed.

Substitution for hydrogen has also been proposed by Ring, Puentes and O'Neal[104a] in the flow pyrolysis of silane at ~500°. From a study of SiH_4-SiD_4 mixtures these workers conclude that a free-radical chain of considerable length is involved. The principal propagating reactions are

$$H^{\cdot} + SiH_4 \rightarrow H_2 + {}^{\cdot}SiH_3 \quad (72)$$
$${}^{\cdot}SiH_3 + SiH_4 \rightarrow Si_2H_6 + H^{\cdot} \quad (73)$$

and termination occurs by

$$2\,{}^{\cdot}SiH_3 \rightarrow Si_2H_6 \quad (74)$$

The activation energy of reaction 73 is estimated to be *c.a.* 13 kcal/mole with a pre-exponential factor of 10^{10} M^{-1} sec^{-1}. The activation energy for the reverse reaction, which is also an S_H2 reaction at silicon, probably has a small activation energy (~3 kcal/mole) and occurs on virtually every collision. However, the above kinetic scheme should be accepted with some reserve since it leads to a termination pre-exponential factor, A_{73}, of only ~10^6

[101] M. A. Nay, G. N. C. Woodall, O. P. Strausz, and H. E. Gunning, *J. Amer. Chem. Soc.*, **87,** 179 (1965).
[102] P. J. Krusic and J. K. Kochi, *ibid.*, **91,** 3938 (1969).
[103] H. Sakurai, A. Hosomi, and M. Kumado, *Bull. Chem. Soc. Japan*, **40,** 1551 (1967).
[104] S. W. Bennett, C. Eaborn, A. Hudson, H. A. Hussein, and R. A. Jackson, *J. Organometal. Chem.*, **16,** P36 (1969).
[104a] M. A. Ring, M. J. Puentes and H. E. O'Neal. *J. Amer. Chem. Soc.*, **92,** 4845 (1970).

M^{-1} sec^{-1}. Such a low value is unlikely to be correct since the analogous coupling of trimethylsilyl radicals occurs on every collision both in solution[104b] and in the gas phase.[104c]

The highly electronegative radical *bis*(trifluoromethyl)nitroxide reacts with silicon tetrachloride at 85° to yield[105] $\{(CF_3)_2NO\}_4Si$. The same product is formed with the halosilanes[105a] SiH_3Br and SiH_2I_2. The mechanism of these reactions has not been studied but an S_H2 process on silicon is a possibility.

$$(CF_3)_2NO^{\cdot} + R_3SiX \rightarrow (CF_3)_2NOSiR_3 + X^{\cdot} \tag{75}$$

A similar reaction occurs with germyl iodide.

The trifluoromethyl radical has been reported to substitute tetramethylsilane at 150° in the gas phase.[106]

$$CF_3^{\cdot} + (CH_3)_4Si \rightarrow CF_3Si(CH_3)_3 + CH_3^{\cdot} \tag{76}$$

Band and Davidson[107] have recently reported that the gas-phase reaction between hexamethyldisilane and iodine at 185–250° is first order in disilane and half order in iodine and gives trimethylsilyl iodide as the only product. A simple free-radical chain process involving an S_H2 reaction at silicon was proposed.

$$I_2 + M \rightleftarrows 2I^{\cdot} + M \tag{77}$$

$$I^{\cdot} + Me_3SiSiMe_3 \rightarrow Me_3SiI + Me_3Si^{\cdot} \tag{78}$$

$$Me_3Si^{\cdot} + I_2 \rightarrow Me_3SiI + I^{\cdot} \tag{79}$$

The rate constant for the S_H2 reaction 78 could be represented by

$$k_{78} = 10^{8.23\pm0.50} \exp(-8100 \pm 1100/RT)\ M^{-1}\ sec^{-1}$$

It was suggested that this reaction is a synchronous process because a stepwise process would be expected to have a lower pre-exponential factor.

Intramolecular homolytic substitutions at silicon have been proposed by several groups of workers. Thus, Razuvaev et al.[108] have shown that the phenoxy radical produced by oxidizing 2,4-di-*t*-butyl-6-trimethylsilylphenol with lead dioxide or alkaline potassium ferricyanide is not stable in oxygen-free inert solvents. The ESR signal and the color disappear rapidly and the carbon-carbon dimer is produced, presumably via the intermediate aryl radical.

[104b] P. T. Frangopol and K. U. Ingold. J. Organometal Chem. **25,** C9 (1970).
[104c] P. Cadman. Private communication.
[105] H. J. Emeléus, *Suomen Kemistilehti* (*B*), **42,** 157 (1969).
[105a] See footnote 17, Chapter IV.
[106] See footnote 50, Chapter IV.
[107] S. J. Band and I. M. T. Davidson, *Trans. Faraday Soc.*, **66,** 406 (1970).
[108] G. A. Razuvaev, N. S. Vasileiskaya, and D. V. Muslin, *J. Organometal. Chem.*, **7,** 531 (1967).

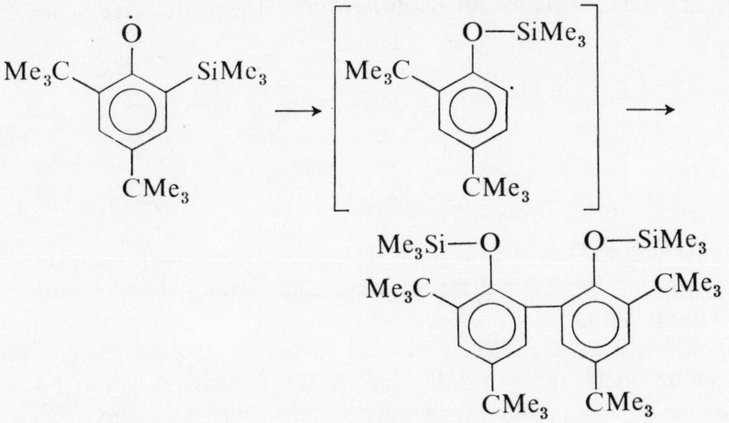

Ȯ
Me_3C $SiMe_3$
CMe_3
→
[O—$SiMe_3$
Me_3C ·
CMe_3]
→
Me_3Si—O O—$SiMe_3$
Me_3C CMe_3
CMe_3 CMe_3

(80)

This behavior is quite different to that of the 2,4,6-tri-*t*-butylphenoxy radical which is completely stable under analogous conditions.

Migration of an *o*-trimethylsilyl group also occurs and the dimer is formed when 2,6-di-(trimethylsilyl)-4-*t*-butyl-, 4-isopropyl-, and 4-trimethylsilyl-phenols are oxidized.[109,110] However, migration does not occur with the 4-methyl compound. In this case the product results from the coupling of two canonical forms of the initial radical.

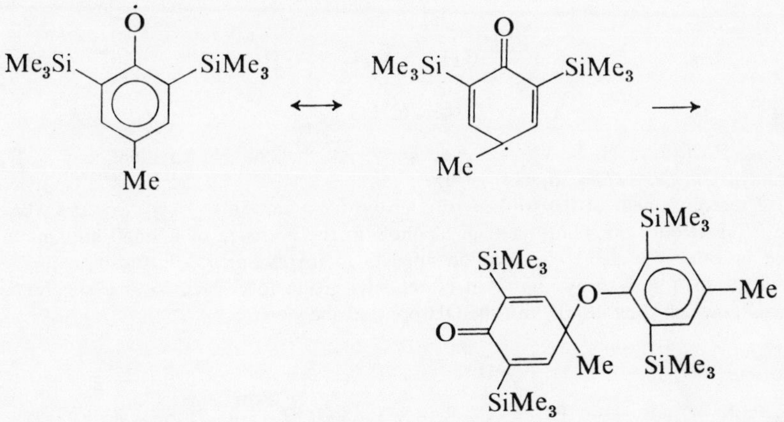

Ȯ
Me_3Si $SiMe_3$
Me
↔
O
Me_3Si $SiMe_3$
Me ·
→
$SiMe_3$
$SiMe_3$
O= O— —Me
Me $SiMe_3$
$SiMe_3$

(81)

The trimethylsilyl group is eliminated and does not migrate to oxygen

[109] G. A. Razuvaev, N. S. Vasileiskaya, and D. V. Muslin, *Dokl. Akad. Nauk SSSR*, **175,** 620 (1967).

[110] G. A. Razuvaev, I. L. Khrzhanovskaya, N. S. Vasileiskaya, and D. V. Muslin, *ibid.*, **177,** 600 (1967).

during the lead dioxide oxidation of 2,6-di-*tert*-butyl-4-trimethylsilylphenol[110a]

$$\text{2,6-}(Me_3C)_2\text{-4-}(SiMe_3)C_6H_2\dot{O} \longrightarrow \text{3,3',5,5'-}(Me_3C)_4\text{-diphenoquinone} \qquad (82)$$

The strong silicon-oxygen bond formed in these 1,3 rearrangements[111] provides the driving force for these intramolecular homolytic substitutions from the ortho position.

The chlorodimethylsilyl radical produced in a mercury arc reacts with 1,3-butadiene to yield (among other products) 1-methyl-1-chloro-1-silacyclopent-3-ene.[113] This compound is apparently formed by a silyl radical addition followed by an S_Hi ring closure with displacement of a methyl radical.

$$(CH_3)_2\dot{S}iCl + CH_2{=}CHCH{=}CH_2$$
$$\downarrow$$
$$[(CH_3)_2SiClCH_2\dot{C}HCH{=}CH_2 \leftrightarrow (CH_3)_2SiClCH_2CH{=}CH\dot{C}H_2]$$
$$\downarrow$$
$$\text{1-chloro-1-methyl-1-silacyclopent-3-ene } (Cl(CH_3)Si(CH_2)_2(HC{=}CH)) + CH_3\cdot \qquad (83)$$

[110a] G. A. Razuvaev, N. S. Vasileiskaya, D. V. Muslin, N. N. Vavilina, and S. N. Uspenskaya. *Zh. Org. Khim.*, **6,** 980 (1970).

[111] A 1,2 rearrangement of the triphenylsilyl group from carbon to oxygen occurs when triphenylsilyl ketones are photolyzed in alcohols in the presence of a small amount of pyridine or other amine.[112] The reaction appears to involve an n-π^* transition in the carbonyl group, followed by migration of the silyl group to oxygen to give a carbene intermediate which then inserts into the OH bond of the alcohol.

$$Ph_3Si{-}\overset{O}{\overset{\|}{C}}{-}R \xrightarrow[\text{(base)}]{h\nu} Ph_3Si{-}\overset{\overset{*}{O}}{\overset{|}{C}}{-}R \longrightarrow Ph_3SiO\ddot{C}R \xrightarrow{R'OH} Ph_3SiO\overset{H}{\overset{|}{\underset{OR'}{\underset{|}{C}}}}R$$

Similar 1,2 rearrangements of silyl groups from silicon to carbon apparently can occur at high temperatures.[95]

$$R_3Si\dot{S}iCH_2R_2 \rightarrow R_3SiCH_2\dot{S}iR_2$$

[112] A. G. Brook and J. M. Duff, *J. Amer. Chem. Soc.*, **89,** 454 (1967).

[113] J. E. Hanson and G. Urry in *A General Study of the Properties of Silyl Radicals* by G. Urry, AD 621131. Avail. CFSTI (1965), *U.S. Govt. Res. Develop. Rept.*, **40,** 31 (1965).

The results of Fish, Kuivila, and Tyminski[114] on the addition of trimethyltin hydride to butadiene are interesting in connection with this silyl radical addition. At 100° the free-radical addition products were the 1,2 adduct, allylcarbinyltrimethyltin (7.1%) and the 1,4 adducts, *cis*-(55.4%) and *trans*-(37.5%) crotyltrimethyltin. On steric grounds, the *trans*-adduct would be expected to predominate. The high yield of the *cis* adduct was ascribed to the formation of a bridged radical in which a *d* orbital of the tin atom serves to delocalize the odd electron. In the absence of the good hydrogen donor

Me_3Sn — or — $Me_3Sn^{\cdot}$ — Me_3Sn — Me_3Sn — Sn Me_3 ⟷ $Sn^{\cdot}$ Me_3 — Me_3SnH — Me_3Sn *trans* — Me_3SnH — Me_3Sn *cis*

(84)

Me_3SnH, the adduct radical might eliminate a methyl radical (as in the silyl case) and give a stannacyclopentene.

C. GERMANIUM

There does not appear to be any authentic examples of S_H2 reactions at germanium. This probably reflects the limited extent of our knowledge of organogermanium chemistry rather than any inherent difficulty in these reactions since S_H2 processes at silicon and tin appear to be firmly established.

The reaction of tetraethylgermane with di-*t*-butyl peroxide at 130° gives

[114] R. H. Fish, H. G. Kuivila, and I. J. Tyminski, *J. Amer. Chem. Soc.*, **89**, 5861 (1967).

7% (mole % based on peroxide decomposed) of *t*-butoxytriethylgermane.[115,116] This is unlikely to be a direct substitution at germanium,

$$Me_3CO^{\cdot} + Et_4Ge \longrightarrow Me_3COGeEt_3 + Et^{\cdot} \quad (85)$$

because ethylene was produced in substantial quantities (25%) whereas ethane was not formed. This argues against the presence of ethyl radicals. The reaction was therefore formulated as,

$$Me_3CO^{\cdot} + Et_4Ge \rightarrow Me_3COH + Et_3GeCH_2CH_2^{\cdot} \quad (86)$$

$$Et_3GeCH_2CH_2^{\cdot} \rightarrow Et_3Ge^{\cdot} + C_2H_4 \quad (87)$$

$$Et_3Ge^{\cdot} + Me_3COOCMe_3 \rightarrow Me_3COGeEt_3 + Me_3CO^{\cdot} \quad (88)$$

In the reaction with benzoyl peroxide at 100° the analogous product, triethylgermyl benzoate, was not identified among the rather complex mixture of reaction products.[115] However, in addition to ethylene (39%) there was some ethane (2%) which argues that the reaction

$$PhC(O)O^{\cdot} + Et_4Ge \rightarrow PhC(O)OGeEt_3 + Et^{\cdot} \quad (89)$$

may have taken place. The presence of 18% 2-benzoyloxyethyltriethylgermane indicates that the 2-triethylgermylethyl radicals induce the decomposition of benzoyl peroxide.

$$Et_3GeCH_2CH_2^{\cdot} + PhC(O)OOC(O)Ph \rightarrow Et_3GeCH_2CH_2OC(O)Ph + PhC(O)O^{\cdot} \quad (90)$$

An S_H2 reaction between $CF_3^{\cdot}$ radicals and tetramethylgermane apparently occurs at 150° in the gas phase.[106]

D. TIN

In contrast to the other elements of Group IVA there are numerous examples of S_H2 reactions at tin. These reactions can be conveniently subdivided into those in which a tin–tin bond is cleaved and those in which a tin–carbon bond is cleaved.

[115] N. S. Vyazankin, E. N. Gladyshev, and G. A. Razuvaev, *Dokl. Akad. Nauk SSSR*, **153** 104 (1963).

[116] With tetraethylsilane[117] and hexaethyldisilane[118] the dimers

$$(Et_3SiC_2H_4)_2 \quad \text{and} \quad (Et_3SiSi(Et)_2C_2H_4)_2$$

are formed.

[117] G. A. Razuvaev, N. S. Vyazankin, and O. S. D'yachkovskaya, *Zhur. Obshch. Khim.*, **32,** 2161 (1962).

[118] N. S. Vyazankin, G. A. Razuvaev, and O. A. Kruglaya, *Izv. Akad. Nauk, SSSR, Otd. Khim. Nauk*, 2008 (1962).

CLEAVAGE OF THE Sn–Sn BOND

In 1959 Kaesz, Phillips, and Stone[119] reported that the UV irradiation of a gaseous mixture of hexamethylditin and trifluoromethyl iodide at room temperature proceeds according to the overall equation,

$$Me_3SnSnMe_3 + CF_3I \xrightarrow{h\nu} Me_3SnCF_3 + Me_3SnI \tag{91}$$

An analogous product is formed with perfluoroethyl iodide. Subsequently, Clark and Willis[120] suggested a bimolecular mechanism for this reaction. This was later shown to be incorrect and the following free-radical chain was proposed.[121]

$$Me_3SnSnMe_3 \xrightarrow{h\nu} 2Me_3Sn^{\cdot} \tag{92}$$

$$Me_3Sn^{\cdot} + CF_3I \longrightarrow Me_3SnCF_3 + I^{\cdot} \tag{93}$$

$$I^{\cdot} + Me_3SnSnMe_3 \longrightarrow Me_3SnI + Me_3Sn^{\cdot} \tag{94}$$

However, as Jackson[95] has pointed out, the liberation of an iodine atom in an S_H2 reaction at carbon, is much less probable than the liberation of $CF_3^{\cdot}$ in an S_H2 reaction at iodine. The most probable mechanism involves an S_H2 reaction at tin by the $CF_3^{\cdot}$ radical, for example,

$$Me_3Sn^{\cdot} + CF_3I \rightarrow Me_3SnI + CF_3^{\cdot} \tag{95}$$

$$CF_3^{\cdot} + Me_3SnSnMe_3 \rightarrow Me_3SnCF_3 + Me_3Sn^{\cdot} \tag{96}$$

The facility with which fluoroalkyl radicals cleave tin–tin bonds has been amply demonstrated by Clark and co-workers.[122,123] The UV irradiation of hexamethylditin with tetrafluoroethylene in the gas phase at 25–75° gave telomers of the type $Me_3Sn(CF_2CF_2)_nSnMe_3$.

$$Me_3Sn^{\cdot} + CF_2{=}CF_2 \rightarrow Me_3SnCF_2CF_2^{\cdot} \tag{97}$$

$$Me_3SnCF_2CF_2^{\cdot} + (n-1)CF_2{=}CF_2 \rightarrow Me_3Sn(CF_2CF_2)_n^{\cdot} \tag{98}$$

$$Me_3Sn(CF_2CF_2)_n^{\cdot} + Me_3SnSnMe_3 \rightarrow Me_3Sn(CF_2CF_2)_nSnMe_3 + Me_3Sn^{\cdot} \tag{99}$$

The products also include polyfluorotrimethyltin compounds,

$$Me_3Sn(CF_2CF_2)_nH$$

and trimethyltin fluoride. The former is presumably formed by an S_H2 reaction at hydrogen rather than at tin, that is,

$$Me_3Sn(CF_2CF_2)_n^{\cdot} + Me_3SnSnMe_3 \rightarrow Me_3Sn(CF_2CF_2)_nH + Me_3SnSn(Me_2)CH_2^{\cdot} \tag{100}$$

[119] H. D. Kaesz, J. R. Phillips, and F. G. A. Stone, *Chem. Ind.* (*London*), 1409 (1959).
[120] H. C. Clark and C. J. Willis, *J. Amer. Chem. Soc.*, **82,** 1888 (1960).
[121] R. D. Chambers, H. C. Clark, and C. J. Willis, *Chem. Ind.* (*London*), 76 (1960).
[122] M. A. A. Beg and H. C. Clark, *ibid.*, 140 (1962).
[123] H. C. Clark, J. D. Cotton, and J. H. Tsai, *Can. J. Chem.*, **44,** 903 (1966).

The trimethyltin fluoride is believed to arise from an α-fluorine atom elimination from the initial adduct radical which is followed by an S_H2 reaction of the fluorine on the ditin compound.

$$Me_3SnCF_2CF_2^{\cdot} \rightarrow Me_3SnCF{=}CF_2 + F^{\cdot} \quad (101)$$

$$F^{\cdot} + Me_3SnSnMe_3 \rightarrow Me_3SnF + Me_3Sn^{\cdot} \quad (102)$$

The photoreaction of hexamethylditin with perfluoropropene[123] gives $Me_3SnCF_2CF(CF_3)SnMe_3$ and $Me_3SnCF_2CF(CF_3)H$. The latter product was the only isomer of hexafluoropropyltrimethyltin which indicates that the attack of the tin radical occurs exclusively at the $F_2C{=}$ group.

$$Me_3Sn^{\cdot} + F_2C{=}CFCF_3 \rightarrow Me_3SnCF_2\dot{C}FCF_3 \quad (103)$$

This direction of attack is consistent with the nucleophilic character of trialkyltin radicals. Hexamethylditin also forms telomers with CF_2CFH (predominant addition to $=CFH$) and CF_2CH_2 (addition to $=CH_2$). However, with CF_2CFCl the products were Me_3SnCF_2CFHCl, Me_4Sn, Me_3SnF, and Me_3SnCl, no *bis*(trimethyltin) adduct being formed. With CF_2CFBr the products formed by irradiation at room temperature were perfluorvinyltrimethyltin, Me_4Sn, Me_3SnF, and Me_3SnBr. The following radical chain sequence is indicated.

$$Me_3Sn^{\cdot} + CF_2{=}CFBr \rightarrow [Me_3SnCFBrCF_2^{\cdot}] \rightarrow Me_3SnCF{=}CF_2 + Br^{\cdot} \quad (104)$$

$$Br^{\cdot} + Me_3SnSnMe_3 \rightarrow Me_3SnBr + Me_3Sn^{\cdot} \quad (105)$$

There have been a number of publications by Razuvaev and co-workers on the oxidation of compounds with tin–tin bonds by peroxides.[118,124–129] Thus, hexaethylditin reacts with benzoyl peroxide at room temperature to yield, in benzene, triethyltin benzoate.[124] Since neither reactant decomposes appreciably at room temperature their ready reaction was explained by the formation of an unstable intermediate cyclic complex.

[124] G. A. Razuvaev, N. S. Vyazankin, and O. A. Shchepetkova, *Zhur. Obshch. Khim.*, **30**, 2498 (1960).
[125] G. A. Razuvaev, O. A. Shchepetkova, and N. S. Vyazankin, *ibid.*, **31**, 3401 (1961).
[126] G. A. Razuvaev, N. S. Vyazankin, and O. A. Shchepetkova, *ibid.*, **31**, 3762 (1961).
[127] I. B. Rabinovich, V. I. Tel'noi, P. M. Nickolaev, and G. A. Razuvaev, *Dokl. Akad. Nauk SSSR*, **138**, 852 (1961).
[128] G. A. Razuvaev, Yu. I. Dergunov, and N. S. Vyazankin, *ibid.*, **145**, 347 (1962).
[129] G. A. Razuvaev, N. S. Vyazankin, and O. A. Shchepetkova, *Tetrahedron*, **18**, 667 (1962).

$$Et_3SnSnEt_3 + Ph\overset{O}{\overset{\|}{C}}O O\overset{O}{\overset{\|}{C}}Ph \longrightarrow \left[\begin{array}{c} Et_3Sn \text{---} SnEt_3 \\ \vdots \qquad \vdots \\ O \qquad \vdots \\ \| \qquad \vdots \\ Ph\text{---}C\text{==}O\text{---}O\text{---}C\text{---}Ph \\ \qquad\qquad\qquad \| \\ \qquad\qquad\qquad O \end{array}\right] \longrightarrow \begin{array}{c} Et_3Sn \quad SnEt_3 \\ | \qquad | \\ O + O\text{---}C\text{---}Ph \\ | \qquad\qquad \| \\ Ph\text{---}C\text{=}O \qquad O \end{array} \quad (106)$$

The possibility of a free-radical chain process involving S_H2 reactions at tin and oxygen seems to have been overlooked.

$$Et_3Sn^\cdot + Ph\overset{O}{\overset{\|}{C}}O O\overset{O}{\overset{\|}{C}}Ph \longrightarrow Ph\overset{O}{\overset{\|}{C}}OSnEt_3 + Ph\overset{O}{\overset{\|}{C}}O^\cdot \quad (107)$$

$$Ph\overset{O}{\overset{\|}{C}}O^\cdot + Et_3SnSnEt_3 \longrightarrow Ph\overset{O}{\overset{\|}{C}}OSnEt_3 + Et_3Sn^\cdot \quad (108)$$

Such a chain might be initiated by a direct reaction between the peroxide and ditin compound or by a reaction of the latter with traces of oxygen. Evidence favoring the formation of some free radicals in the reaction comes from the observation[124] that there was a 17.5% yield of triethyltin chloride when the reaction was carried out in carbon tetrachloride. Furthermore,[124] acrylonitrile was rapidly polymerized by benzoyl peroxide and the ditin in CCl_4. At 90° the reaction of 0.025 mole of benzoyl peroxide with 0.05 mole hexaethylditin in 0.15 mole CCl_4 yields[128] (in moles per mole of peroxide) 0.40 CO_2, 0.46 C_2H_6, 13.9 C_2H_4, 1.09 n-C_4H_{10}, 29.4 $CHCl_3$, 15.6 Et_3SnCl, 8.4 Et_2SnCl_2, and tar. This is clearly a chain reaction yielding triethyltin radicals but the high yield of ethylene and chloroform point to an attack on the ethyl groups of the hexaethylditin by a trichloromethyl radical. Presumably the $CCl_3^\cdot$ radical is less able to cleave the tin–tin bond by an attack on tin than the benzoyloxy radical both for steric reasons and because the reaction will be much less exothermic. The difference between $CCl_3^\cdot$ and $CF_3^\cdot$ radicals must presumably be ascribed chiefly to steric effects. Ditin compounds are even less reactive towards simple alkyl radicals than towards the perhaloalkyl radicals.[124,130]

Hexaethylditin is also cleaved at room temperature by acetyl benzoyl

[130] G. A. Razuvaev, N. S. Vyazankin, E. N. Gladyshev, and I. A. Borodavko, *Zhur. Obshch. Khim.*, **32**, 2154 (1962).

peroxide, dicyclohexyl percarbonate, nitrosoacetanilide and lead tetra-acetate.[124] The products are quite consistent with free radical chains involving S_H2 reactions at tin by oxy radicals. This ditin compound is not, however, cleaved by 2-cyanopropyl radicals formed from azoisobutyronitrile at room temperature.[124] Hexaalkylditins do not react with di-*t*-alkyl peroxides such as di-*t*-butyl peroxide below the temperature at which the peroxide cleaves to alkoxy radicals (~130°).[118,131] Alkoxytrialkyltins are formed indicating a cleavage of the tin–tin bond,

$$R'O^{\cdot} + R_3SnSnR_3 \rightarrow R'OSnR_3 + R_3Sn^{\cdot} \qquad (109)$$

The saturated hydrocarbon RH and the corresponding olefin R(−H) are also formed together with some R_2. These products imply that tin–carbon bonds are also cleaved under these conditions.

$$R'O^{\cdot} + R_3SnSnR_3 \rightarrow R_3SnSn(R_2)OR' + R^{\cdot} \qquad (110)$$

Acetone is not formed from *t*-butoxy radicals which indicates that ditins are quite efficient alkoxy radical traps. The presence of *t*-butanol implies that some of the olefin is formed by hydrogen abstraction from the β-position of the alkyl groups.

$$R'O^{\cdot} + R_3SnSnR_3 \rightarrow R'OH + R_3SnSn(R_2)CH_2\dot{C}H\text{—} \rightarrow R_3Sn\dot{S}nR_2 + CH_2{=}CH\text{—} \qquad (111)$$

The tin radicals formed in these reactions appear to condense together to give dark-red nondistillable polymeric tin compounds.

The photoinitiated autoxidation of neat hexaethylditin and of solutions in nonane is a free-radical chain process at temperatures above about 20° and the principal initial product at these temperatures is *bis*(triethyltin) peroxide.[132] A triethyltinperoxy radical is presumably involved in carrying the chain.

$$Et_3SnOO^{\cdot} + Et_3SnSnEt_3 \rightarrow Et_3SnOOSnEt_3 + Et_3Sn^{\cdot} \qquad (112)$$

$$Et_3Sn^{\cdot} + O_2 \rightarrow Et_3SnOO^{\cdot} \qquad (113)$$

The peroxide is reduced to oxide by unreacted hexaethylditin, this reaction yielding some free radicals.

$$Et_3SnOOSnEt_3 + Et_3SnSnEt_3 \rightarrow Et_3SnOSnEt_3 + Et_3Sn^{\cdot} + Et_3SnO^{\cdot} \qquad (114)$$

Minor products of the oxidation include diethyltin oxide, ethoxytriethyltin, ethanol, acetaldehyde, and water. It is believed that some pentaethyldistannyl ethyl peroxide is formed in the photoinitiation.

[131] W. P. Neumann, K. Rubsamen, and R. Sommer, *Chem. Ber.*, **100,** 1063 (1967).
[132] Yu. A. Aleksandrov and B. A. Radbil', *Zhur. Obshch. Khim.*, **36,** 543 (1966).

CLEAVAGE OF THE Sn–C BOND

Numerous examples of S_H2 reactions at tin which result in the cleavage of a tin–carbon bond have been reported by Razuvaev in his studies of the tetra-alkyltin-peroxide reaction and by Neumann in his studies of the trialkyltin hydride-peroxide reaction. Thus, the reaction of tetraethyltin with benzoyl peroxide at 95° for 16 hours under free radical conditions yields triethyltin benzoate (0.66 moles per mole of peroxide), diethyltin dibenzoate (0.37), ethylene (0.55), ethane (0.26), butane (0.02), CO_2 (0.20), and ethyl benzoate (high yield).[133] Similar products are formed with triethyltin chloride, bromide, and benzoate. The products indicate that ethyl radicals are formed in the reaction and a reasonable reaction scheme is

$$PhC(O)O^{\cdot} + Et_4Sn \rightarrow PhC(O)OSnEt_3 + Et^{\cdot} \tag{115}$$

$$Et^{\cdot} + PhC(O)OOC(O)Ph \rightarrow PhC(O)OEt + PhC(O)O^{\cdot} \tag{116}$$

However, the reaction with acetyl benzoyl peroxide[117,133] gave triethyltin acetate (0.42), triethyltin benzoate (0.43), C_2H_4 (0.54), C_2H_6 (0.13), C_4H_{10} (0.02), CO_2 (0.61), and CH_4 (0.48). The rapid decarboxylation of acetoxy radicals rules out the formation of triethyltin acetate by an S_H2 reaction at tin by an acetoxy radical. This led to the suggestion that reaction occurs at least in part by way of an intermediate complex.

$$\begin{array}{ccc} Et_3Sn & & Et \\ \vdots & & \vdots \\ O & & \overset{O}{\|} \\ \| & & \\ MeCO & & O\text{—}C\text{—}Ph \end{array}$$

To the present reviewers it seems far more likely that the formation of triethyltin acetate is due to an S_H2 reaction of a triethyltin radical at the peroxidic oxygen.[134]

$$Et_3Sn^{\cdot} + \begin{array}{l} O\text{—}\overset{O}{\overset{\|}{C}}\text{—}Me \\ | \\ O\text{—}\underset{O}{\underset{\|}{C}}\text{—}Ph \end{array} \begin{array}{l} \nearrow Et_3SnO\overset{O}{\overset{\|}{C}}Me + Ph\overset{O}{\overset{\|}{C}}O^{\cdot} \\ \\ \searrow Et_3SnO\overset{O}{\overset{\|}{C}}Ph + (Me\overset{O}{\overset{\|}{C}}O^{\cdot}) \longrightarrow Me^{\cdot} + CO_2 \end{array} \tag{117}$$

[133] G. A. Razuvaev, O. S. D'yachkova, N. S. Vyazankin, and O. A. Shchepetkova, *Dokl. Akad. Nauk SSSR*, **137**, 618 (1961).

[134] K. Rubsamen, W. P. Neumann, R. Sommer, and U. Frommer, *Chem. Ber.*, **102**, 1290 (1969).

These reactions are known to proceed rapidly (see Chapter VII). The tin radical could be formed by abstraction of the β-hydrogen, that is,

$$R^{\cdot} + Et_4Sn \rightarrow RH + Et_3SnCH_2\dot{C}H_2 \rightarrow Et_3Sn^{\cdot} + C_2H_4 \qquad (118)$$

or in the S_H2 reaction between the benzoyloxy radical and the tetraethyltin, reaction 115. The probable intermediacy of some trialkyltin radicals in these reactions is indicated by the fact that trialkyltin halides are produced when the reactions are carried out in halogenated solvents.[128–130] For example,[128] heating 0.40 mole of tetraethyltin with 0.47 mole CCl_4 and 0.01 mole of benzoyl peroxide at 75–80° for 12 hr gave 29 moles (per mole of peroxide) triethyltin chloride. Other products of particular interest include 13.7 moles C_2H_4, 0.95 moles C_2H_6, and 0.20 moles n-C_4H_{10}. The last two products strongly suggest the intermediacy of ethyl radicals.

The reaction of tetraethyltin with di-*t*-butylperoxide at 145° for 16 hr[117] gives *t*-butoxytriethyltin (1.12 moles per mole of peroxide decomposed) ethylene (1.25), ethane (0.50), and *t*-butanol (0.82). This reaction may, in part, involve an S_H2 reaction at tin with the substitution of *t*-butoxy for ethyl and/or an S_H2 reaction at oxygen with the substitution of triethyltin for *t*-butoxy. However the high yield of ethylene and *t*-butanol indicates that many *t*-butoxy radicals abstract hydrogen from the tin compound.

$$Me_3CO^{\cdot} + (C_2H_5)_4Sn \rightarrow Me_3COH + (C_2H_5)_3SnCH_2\dot{C}H_2 \rightarrow (C_2H_5)_3Sn^{\cdot} + C_2H_4 \qquad (119)$$

A reasonable route to the *t*-butoxytriethyltin involving the initial dimerization of triethyltin radicals is suggested by the work of Neumann, Rubsamen, and Sommer[131] who showed that the decomposition of di-*t*-butyl peroxide at 130° is not induced by four times the molar quantity of tributyltin hydride. *Tert* butanol and hexabutylditin were produced quantitatively. However, equimolar quantities of the tin hydride and the peroxide yield *t*-butoxytributyltin which suggests that the overall process should be formulated as,

$$Me_3CO^{\cdot} + Bu^n_3SnH \rightarrow Me_3COH + Bu^n_3Sn^{\cdot} \qquad (120)$$

$$Bu^n_3Sn^{\cdot} + Bu^n_3Sn^{\cdot} \rightarrow Bu^n_3SnSnBu^n_3 \qquad (121)$$

$$Me_3CO^{\cdot} + Bu^n_3SnSnBu^n_3 \rightarrow Me_3COSnBu^n_3 + Bu^n_3Sn^{\cdot} \qquad (122)$$

A similar scheme is clearly possible with tetraethyltin with the tin radicals formed by reaction 119.

As was mentioned above, the reaction of hexaalkylditins R_6Sn_2 with *t*-butyl peroxide at 130° yields not only *t*-butoxytrialkyltin (Sn–Sn cleavage, reaction 112), but also the hydrocarbons RH, R(–H), and R_2 (Sn–C cleavage).[118,131]

The reaction of trialkyltin hydrides, R_3SnH, with dibenzoyl peroxide and diacyl peroxides yields principally the trialkyltin ester, $R_3SnOC(O)R'$.[131] However, there is always some RH hydrocarbon (e.g., ethane from triethyltin hydride) together with some $R_2Sn(OC(O)R')_2$. Both of these products are, presumably, the result of S_H2 cleavage of a tin–carbon bond, that is,

$$R'C(O)O^{\cdot} + R_3SnOC(O)R' \rightarrow R_2Sn(OC(O)R')_2 + R^{\cdot} \quad (123)$$

$$R^{\cdot} + R_3SnH \rightarrow RH + R_3Sn^{\cdot} \quad (124)$$

Although the reactions of organotin compounds with peroxides are rather complex there seems to be no doubt that S_H2 processes at tin of the general type

$$R'O^{\cdot} + RSn\!\!\overset{/}{\underset{\backslash}{-}} \longrightarrow R'OSn\!\!\overset{/}{\underset{\backslash}{-}} + R^{\cdot} \quad (125)$$

occur with reasonable facility.

The photo-initiated autoxidation of tetraalkyltin compounds is a free-radical chain process yielding alkylperoxytrialkyltin compounds as one of the principal products.[135,136] It seems likely that the following reactions are important in propagating the chain.

$$R^{\cdot} + O_2 \rightarrow ROO^{\cdot} \quad (126)$$

$$ROO^{\cdot} + R_4Sn \rightarrow ROOSnR_3 + R^{\cdot} \quad (127)$$

Additional products are formed by hydrogen abstractions from the alkyl groups and by photolysis of the peroxytin compound.

The reaction of halogens with tetraalkyltins

$$X_2 + R_4Sn \rightarrow RX + R_3SnX \quad (128)$$

may occur by either an electrophilic substitution at a saturated carbon atom or by a homolytic substitution at tin.[137] The relative importance of the two processes depends on the experimental conditions and on the dielectric constant, polarizability, and nucleophilicity of the solvent. In polar media, the solvent itself acts as the nucleophilic catalyst and the reaction can be represented as,

$$R_4Sn + solv \rightleftarrows R_4Sn^{-}solv^{+} \xrightarrow{X_2} [\overset{\delta-}{X_2} \cdots R \cdots \overset{\delta-}{R_3Sn}solv^{+}]^{\ddagger} \rightarrow R_3SnX + RX \quad (129)$$

[135] Yu. A. Aleksandrov, B. A. Radbil', and V. A. Shushunov, *Zh. Obshch. Khim.*, **37**, 208 (1967).
[136] Yu. A. Aleksandrov and B. A. Radbil', *ibid.*, **37**, 2345 (1967).
[137] S. Boué, M. Gielen, and J. Nasielski, *J. Organometal. Chem.*, **9**, 443, 461, 481 (1967).

In less active solvents, the halogen acts as the nucleophile,

$$R_4Sn + X_2 \longrightarrow R_4Sn^{-}X_2^{+} \longrightarrow \left[R_3Sn \overset{X}{\underset{R}{\diamond}} X \right]^{\ddagger} \longrightarrow R_3SnX + RX$$

$$R_4Sn^{-}X_2^{+} \xrightarrow{X_2} [\overset{\delta-}{X_2} \cdots R \cdots \overset{\delta-}{R_3}SnX_2^{+}]^{\ddagger} \longrightarrow R_3SnX + RX \qquad (130)$$

In relatively nonpolar solvents such as chlorobenzene, and under the influence of light, the reaction is a free radical chain which is inhibited by oxygen and hydroquinone.

$$Br_2 \xrightarrow[(410\ nm)]{h\nu} 2Br^{\cdot} \qquad (131)$$

$$Br^{\cdot} + R_4Sn \rightarrow R_3SnBr + R^{\cdot} \qquad (132)$$

$$R^{\cdot} + Br_2 \rightarrow RBr + Br^{\cdot} \qquad (133)$$

The rate of the S_H2 reaction 132 increases with increasing stability of the displaced radical. For example, with Me_3SnR' the relative rates for different R' are: *iso*-propyl (5.5) > ethyl (3.7) = *n*-butyl (3.7) > *n*-propyl (3.35) > methyl (1). For tetraalkyltins the relative rates are *iso*-propyl (10.0) > ethyl (5.5) > *n*-propyl (4.0) ≈ *n*-butyl (3.8) > methyl (1), which indicates that the reaction is subject to steric hindrance. Some evidence that the homolytic substitution may involve an intermediate with a lifetime sufficient for it to react with a second molecule of tetraalkyltin was obtained in competitive experiments. The suggested mechanism involves a stepwise S_H2 reaction.

$$Br^{\cdot} + R_4Sn \rightleftarrows (R_4Sn^{-}Br^{+})^{\cdot} \rightarrow R_3SnBr + R^{\cdot} \qquad (134)$$

$$(R_4Sn^{-}Br^{+})^{\cdot} + R'_4Sn \rightarrow R_4Sn + R'_3SnBr + R'^{\cdot} \qquad (135)$$

Additional support for a free-radical mechanism for the bromination and iodination of tetraalkyltins comes from studies on an optically active (1-methyl-2,2-diphenylcyclopropyl) trimethyltin.[137a] In CCl_4 the 1-bromo- (or iodo)-1-methyl-2,2-diphenylcyclopropane product has only a small degree of retention of configuration which argues in favor of an intermediate alkyl radical and hence of an S_H2 process at tin. The overall reaction with iodine can be represented as

$$\text{(1-SnMe}_3\text{-1-Me-2,2-Ph}_2\text{-cyclopropane)} \xrightarrow[CCl_4]{I_2} \text{(1-I-1-Me-2,2-Ph}_2\text{-cyclopropane)} + \text{(1-SnMe}_2\text{I-1-Me-2,2-Ph}_2\text{-cyclopropane)} + Ph_2CHC(CH_3){=}CH_2 \qquad (136)$$

	Starting tin compound	Iodide	$SnMe_2I$ product	$Ph_2CHC(CH_3)=CH_2$
Yield		42%	41%	4%
Optical purity	69%	2.3%	69%	

[137a] K. Sisido, T. Miyanisi, T. Isida and S. Kozima, *ibid*, **23**, 117 (1970).

The reaction of tetraalkyltin compounds with perfluoroalkyl iodides yields small quantities of $R_3SnC_nF_{2n+1}$ presumably by an S_H2 reaction in which a carbon radical displaces carbon from tin.[119] Trifluoromethyl radicals produced by the photolysis of hexafluoroacetone undergo an S_H2 reaction with tetramethyltin.[106]

$$CF_3^{\cdot} + (CH_3)_4Sn \rightarrow CF_3Sn(CH_3)_3 + CH_3^{\cdot} \quad (137)$$

This reaction has a rate constant of $1 \times 10^4\ M^{-1}\ sec^{-1}$ at 150°.

A more interesting example of this type of reaction has recently been reported by Sato and Moritani.[138] The irradiation of an ethereal solution of 1,1-diphenyl-1-stannacyclohepta-2,6-diene (*trans-trans* isomer) **1** resulted in its smooth conversion to dimer **2** (34%) and polymer (56%). The same products are formed in 2-propanol and in benzene. The polymer was obtained by reaction in refluxing benzene in the dark in the presence of azo-*bis*-isobutyronitrile. The following reaction scheme was proposed.

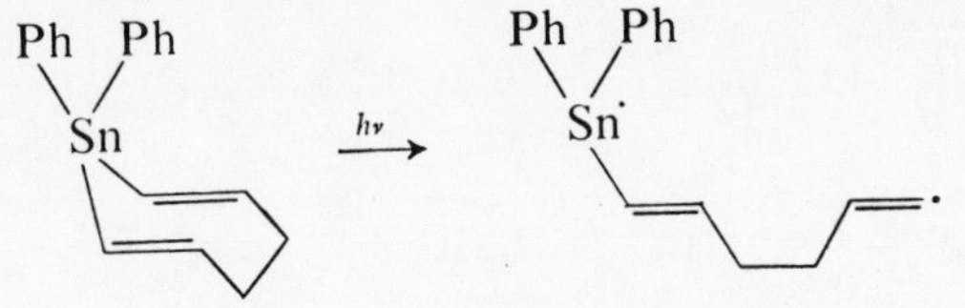

(138)

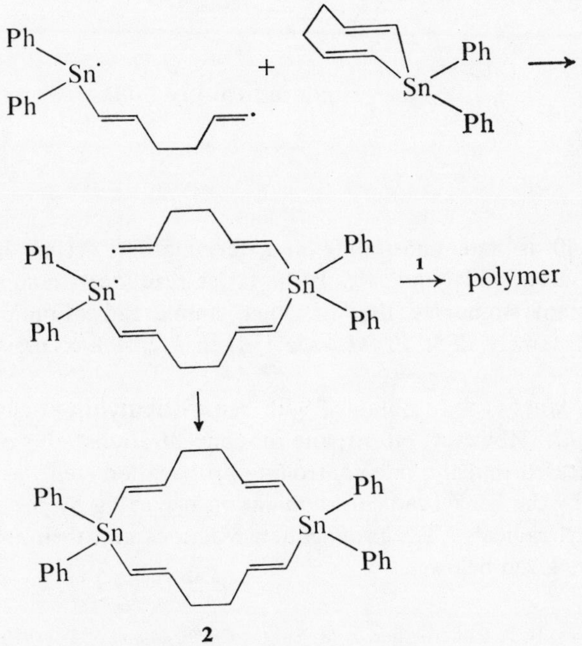

(139)

[138] T. Sato and I. Moritani, *Tetrahedron Lett.*, 3181 (1969).

The reaction of *N*-bromosuccinimide with tetrabutyltin and with benzyltributyltin in acetone at 35° is a free-radical chain process which can be initiated with *t*-butyl hyponitrite and completely inhibited with galvinoxyl.[139] The succinimidyl radical enters into an S_H2 reaction at tin displacing a butyl radical from tetrabutyltin and a benzyl radical from benzyltributyltin. Butyl bromide (or benzyl bromide) are formed quantitatively together with *N*-tributylstannyl succinimide. Only one group is cleaved even in the presence of a three-fold excess of *N*-bromosuccinimide. The reaction kinetics with tetrabutyltin are consistent with the following propagation and termination steps.

$$(CH_2CO)_2N^{\cdot} + Bu_4Sn \longrightarrow (CH_2CO)_2N{-}SnBu_3 + Bu^{\cdot} \qquad (140)$$

$$Bu^{\cdot} + (CH_2CO)_2NBr \longrightarrow BuBr + (CH_2CO)_2N^{\cdot} \qquad (141)$$

$$2\,(CH_2CO)_2N^{\cdot} \longrightarrow \text{nonradical products} \qquad (142)$$

Reaction 140 is rate controlling and termination occurs by the self-reaction of succinimidyl radicals. If the latter reaction is assumed to have a rate constant similar to that of other simple radical-radical reactions in solution, namely, $2 \times 10^9\ M^{-1}\ \text{sec}^{-1}$, then $k_{140} \approx 8 \times 10^3\ M^{-1}\ \text{sec}^{-1}$ at 35°.

The same kinetics were obtained with benzyltributyltin at concentrations below 0.25 *M*. However, at organotin concentrations above 0.5 *M* the kinetics indicated that the rate controlling propagation step was abstraction of bromine by the benzyl radical, termination occurring by the self-reaction of two benzyl radicals. The propagation reactions and their estimated rate constants are given below.

[139] A. G. Davies, B. P. Roberts, and J. M. Smith, *Chem. Comm.*, 557 (1970).

$$\text{(succinimidyl)N}^{\cdot} + PhCH_2SnBu_3 \xrightarrow{1.2 \times 10^5\ M^{-1}\ sec^{-1}} \text{(succinimidyl)N—}SnBu_3 + PhCH_2^{\cdot} \quad (143)$$

$$PhCH_2^{\cdot} + \text{(succinimidyl)NBr} \xrightarrow{6 \times 10^5\ M^{-1}\ sec^{-1}} PhCH_2Br + \text{(succinimidyl)N}^{\cdot} \quad (144)$$

Kupchik and Lanigan[140] had previously reported that *N*-bromosuccinimide reacts with tetraphenyltin in refluxing carbon tetrachloride to give bromobenzene. Some evidence for a rapidly hydrolyzed *N*-triphenylstannylsuccinimide was found. More recently, a similar reaction between *N*-bromosuccinimide and trimethyl-*p*-tolyltin has been reported in which the *p*-tolyl group was selectively cleaved.[141] For neither of these reactions was the mechanism discussed. A free-radical chain reaction similar to that for the tetraalkyltin compounds is certainly possible. However, a heterolytic (electrophilic) cleavage of the arylcarbon–tin bond is perhaps more likely in view of a report[142] that *N*-bromosuccinimide and dibenzyldiphenyltin refluxed for 30 min in $CHCl_3$ give bromobenzene, succinimide, and from the residue, dibenzyltin dichloride. A free-radical mechanism would be expected to have given benzyl bromide since the benzyl radical should be displaced from the tin more readily than a phenyl radical.

Tetra-*n*-butyltin reacts slowly with thiophenol at 100° in the presence of an initiator to give phenylthiotri-*n*-butyltin.[143] An S_H2 attack on tin by the phenylthiyl radical seems a likely mechanism.

E. LEAD

Organolead compounds are less stable than organotin compounds and consequently their known chemistry is less extensive. However there are a number of examples of S_H2 reactions at lead.

[140] E. J. Kupchik and T. Lanigan, *J. Org. Chem.*, **27**, 3661 (1962).
[141] J. C. Maire, R. Prosperini, and J. Van Rietschoten, *J. Organometal. Chem.*, **21**, P41 (1970).
[142] G. A. Razuvaev and V. Fetyukova, *Zh. Obshch. Khim.*, **21**, 1010 (1951).
[143] A. G. Davies and J. M. Smith, unpublished results.

Hexaethyldilead reacts at room temperature with oxygen and acyl peroxides,[144] and with lead tetraacetate and nitrosoacetanilide[124] in almost exactly the same way as hexaethylditin. Since hexaethyldilead (and the other reagents) are thermally stable at this temperature it seems reasonable to postulate radical chain mechanisms for these reactions, one step of which is an S_H2 reaction by an oxy radical at lead with displacement of triethyllead.

$$RO^{\cdot} + Et_3PbPbEt_3 \rightarrow ROPbEt_3 + Et_3Pb^{\cdot} \qquad (145)$$

Tetraalkyllead compounds react with perfluoroalkyl iodides under the influence of heat or ultraviolet light to yield compounds of the type $R_3PbC_nF_{2n+1}$.[119] By analogy with the alkyl tin and ditin reactions with these iodides, the reaction with lead can probably be formulated as,

$$CF_3^{\cdot} + Me_4Pb \rightarrow Me_3PbCF_3 + Me^{\cdot} \qquad (146)$$

$$Me^{\cdot} + CF_3I \rightarrow MeI + CF_3^{\cdot} \qquad (147)$$

The reaction of 0.005 mole benzoyl peroxide with 0.10 mole tetraethyllead at 80° yields (in moles per mole of peroxide): CO_2 (0.04); C_2H_4 (0.38); C_2H_6 (0.92); n-C_4H_{10} (0.26); and $(C_2H_5)_3PbOC(O)C_6H_5$ (0.60).[117,133] The high yields of ethane, butane, and triethyllead benzoate indicate an S_H2 reaction at lead with displacement of an ethyl radical.

$$PhC(O)O^{\cdot} + Et_4Pb \rightarrow Et_3PbOC(O)Ph + Et^{\cdot} \qquad (148)$$

The reaction of *t*-butoxy radicals (produced by the thermal decomposition of *t*-butyl peroxyoxalate) with tetraethyl lead yields ethyl radicals.[144a] These radicals were trapped with 2-methyl-2-nitrosopropane and the resulting *t*-butyl ethyl nitroxide radical was identified by e.s.r. spectroscopy.

$$Me_3CO^{\cdot} + PbEt_4 \rightarrow Me_3COPbEt_3 + Et^{\cdot} \qquad (149)$$

$$Et^{\cdot} + Me_3CNO \rightarrow Me_3CN(O^{\cdot})Et \qquad (150)$$

The photo-initiated autoxidation of tetraethyllead yields some ethylperoxytriethyllead.[135] The chain length is less than unity at temperatures below 50° but increases with an increase in temperature. The reaction

$$EtOO^{\cdot} + Et_4Pb \rightarrow EtOOPbEt_3 + Et^{\cdot} \qquad (151)$$

must be fairly slow and have quite a high activation energy.

Tetramethyllead and tetraphenyllead react with *N*-bromosuccinimide in toluene at room temperature to yield trimethyllead and triphenyllead succinimide almost quantitatively.[145] The exothermic reaction is preceded by an

[144] Yu. A. Aleksandrov, T. G. Brilkina, A. A. Kvasov, G. A. Razuvaev, and V. A. Shushunov, *Dokl. Akad. Nauk SSSR*, **129**, 321 (1959).

[144a] K. Torssell. *Tetrahedron*, **26**, 2759 (1970).

[145] B. C. Pant and W. E. Davidson, *J. Organometal. Chem.*, **19**, P3 (1969).

induction period which strongly suggests the occurrence of a free-radical chain with an S_H2 reaction at lead as one propagating step, for example,

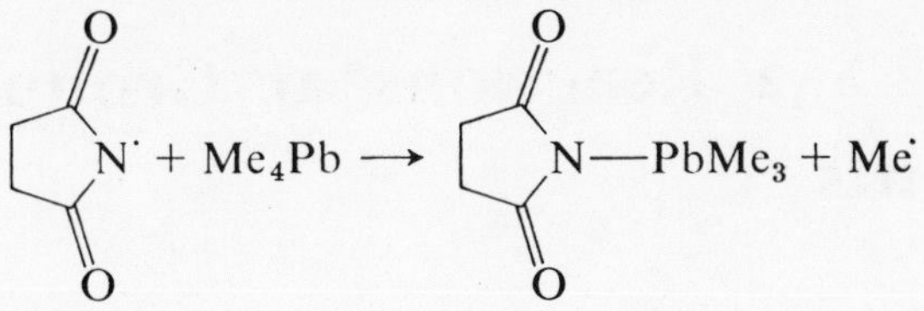

$$\text{(succinimidyl)N}^{\cdot} + Me_4Pb \longrightarrow \text{(succinimidyl)N—PbMe}_3 + Me^{\cdot} \quad (152)$$

$$Me^{\cdot} + \text{(succinimidyl)N—Br} \longrightarrow MeBr + \text{(succinimidyl)N}^{\cdot} \quad (153)$$

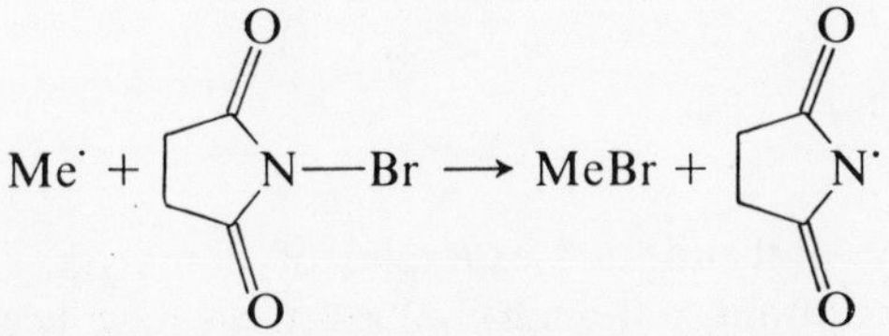

Reaction did not occur with *N*-chlorosuccinimide but it seems possible that it would occur in the presence of a source of free radicals.

The reactions of a wider variety of tetraorganolead compounds with *N*-bromosuccinimide have since been examined.[145a] The compounds studied included Me_4Pb, Et_4Pb, Bu_4Pb, Ph_4Pb, $(p\text{-}CH_3C_6H_4)_4Pb$, and $(2C_4H_3S)_4Pb$ together with a number of unsymmetrically substituted compounds. There is a high degree of selectivity between the different groups of the unsymmetrically substituted leads. The following orders of ease of cleavage were observed: (i), *p*-tolyl > phenyl > alkyl > benzyl and (ii), 2-thienyl > phenyl > benzyl. The occurrence of an induction period with tetramethyl lead implies a free-radical chain process in this case. However, the easier cleavage of aryl than alkyl groups is inconsistent with an S_H2 displacement of the aryl groups. It seems likely that the aryl groups are cleaved heterolytically and the alkyl groups homolytically.

[145a] B. C. Pant. *ibid.*, **24,** 697 (1970).

VI S_H2 Reactions at Group VA Elements

A. NITROGEN

There are few, if any, authenticated examples of homolytic substitutions occurring at a nitrogen atom, although S_H2 reactions have been proposed many times in the past.

The interaction of hydrogen atoms with ammonia has been studied on numerous occasions,[1–5] and the results of early work on the kinetics of the photochemical decomposition of ammonia were even interpreted in terms of an *intermediate* species $\dot{N}H_4$.[2]

In 1936 Farkas and Melville[5] concluded that reaction 1 was responsible for the mercury-photosensitized exchange occurring between ammonia (or ND_3) and deuterium (or H_2) above 300°, an activation energy of 11 ± 1 kcal/mole and an A-factor of $2 \times 10^{10}\ M^{-1}\ sec^{-1}$ were assigned to the reaction.

$$D^{\cdot} + NH_3 \rightarrow DNH_2 + H^{\cdot} \tag{1}$$

Later, Melville and Bolland confirmed these values.[6] The quantum yield for the exchange was found to be greater than unity, which was consistent with the occurrence of a chain reaction formed by the incursion of 1 followed by 2.

$$H^{\cdot} + D_2 \rightarrow HD + D^{\cdot} \tag{2}$$

These results were quoted by Steacie[7,8] without comment on the validity of

[1] E. Boehm and K. F. Bonhoeffer, *Z. physik. Chem.*, **119**, 385, 474 (1926).
[2] L. Farkas and P. Harteck, *ibid.*, **B25**, 257 (1934).
[3] H. S. Taylor and J. C. Jungers, *J. Chem. Phys.*, **2**, 452 (1934).
[4] K. H. Geib and E. W. R. Steacie, *Z. Physik. Chem.*, **B29**, 215 (1935).
[5] A. Farkas and H. W. Melville, *Proc. Roy. Soc.* (*London*), **A157**, 625 (1936).
[6] H. W. Melville and J. L. Bolland, *ibid.*, **A160**, 384 (1937).
[7] E. W. R. Steacie and M. Szwarc, *J. Chem. Phys.*, **19**, 1309 (1951).
[8] E. W. R. Steacie, "*Atomic and Free-radical Reactions*," 2nd ed., Reinhold, New York (1954).

the proposed mechanism, which involved bimolecular homolytic substitution at nitrogen, reaction 1. However, Steacie[8] pointed out that a similar conclusion regarding the mechanism[4] of the exchange between methane (isoelectronic with ammonia) and deuterium atoms was in error and, in particular, that the chain mechanism shown in reactions 3 and 4 could account for the results.

$$D^{\cdot} + CH_4 \rightarrow {}^{\cdot}CH_3 + HD \quad (3)$$

$$^{\cdot}CH_3 + D_2 \rightarrow CH_3D + D^{\cdot} \quad (4)$$

In fact, an activation energy of 37 kcal/mole has been calculated for the alternative S_H2 displacement 5 (see Chapter V).

$$D^{\cdot} + CH_4 \rightarrow CH_3D + H^{\cdot} \quad (5)$$

Thus, it seems that the exchange mechanism shown in equation 1 will be slow compared to the alternative chain process, in which both 6 and 7 are approximately thermoneutral.[9]

$$D^{\cdot} + NH_3 \rightarrow HD + {}^{\cdot}NH_2 \quad (6)$$

$$^{\cdot}NH_2 + D_2 \rightarrow NH_2D + D^{\cdot} \quad (7)$$

More recently[10] the interaction of hydrogen atoms, generated by microwave discharge in molecular hydrogen, with ammonia was studied in a flow system at 150°. However, the results throw no light on the mechanism of the exchange reaction, since deuterium was not used and only the decrease in ammonia concentration was monitored.

A similar problem arises for the interaction of hydrogen atoms with hydrazine.[11–13]

$$H^{\cdot} + H_2N{-}NH_2 \rightarrow \tfrac{1}{2}N_2 + H_2 + NH_3 \quad (8)$$

Two alternative mechanisms were considered and the one involving homolytic displacement at nitrogen was favored by analogy with the then-accepted mechanism for deuterium atom exchange with ammonia.[12]

$$H^{\cdot} + H_2N{-}NH_2 \rightarrow NH_3 + \dot{N}H_2 \quad (9)$$

$$\dot{N}H_2 \rightarrow \tfrac{1}{2}N_2 + H_2 \quad (10)$$

$$H^{\cdot} + H_2N{-}NH_2 \rightarrow H\dot{N}{-}NH_2 + H_2 \quad (11)$$

$$H\dot{N}{-}NH_2 \rightarrow \tfrac{1}{2}N_2 + NH_3 \quad (12)$$

[9] In contrast, atom abstraction from molecular hydrogen by the phosphino radical would be about 14 kcal/mole endothermic. The activation energy for $^{\cdot}PH_2 + D_2 \rightarrow PH_2D + D^{\cdot}$ cannot be less than this figure.

[10] M. Schiavello and G. G. Volpi, *J. Chem. Phys.*, **37**, 1510 (1962).

[11] E. O. Wiig, *J. Amer. Chem. Soc.*, **59**, 827 (1937) and references cited.

[12] E. A. B. Birse and H. W. Melville, *Proc. Roy. Soc.* (*London*), **A175**, 164 (1940).

[13] S. N. Foner and R. L. Hudson, *J. Chem. Phys.*, **29**, 442 (1958).

The activation energy for the equivalent displacement on carbon, reaction 13, has been calculated to be about 30 kcal/mole.[14]

$$H^{\cdot} + CH_3—CH_3 \rightarrow CH_4 + {}^{\cdot}CH_3 \tag{13}$$

However, in the case of hydrazine, the displacement would be much more exothermic[15] because of the weak N–N bond broken, (similar displacements at oxygen in the isoelectronic peroxides are quite common).

The question of the occurrence of the homolytic substitution reaction 9 was settled by Schiavello and Volpi,[9] who showed that NH_2D was not a product of the interaction of deuterium atoms with hydrazine in the temperature range 25–150°. These authors discussed the overall process in terms of reactions 11, 14, 15, and 16.

$$2H\dot{N}—NH_2 \rightarrow 2NH_3 + N_2 \tag{14}$$

$$H^{\cdot} + H\dot{N}—NH_2 \rightarrow H_2 + N_2H_2 \tag{15}$$

$$N_2H_2 \rightarrow N_2 + H_2 \tag{16}$$

They derived the following values for k_{11} and k_{14}:

$$k_{11} = 3.5 \times 10^8 e^{-200/RT}\ M^{-1}\ \text{sec}^{-1}$$

$$k_{14} \geqslant 3 \times 10^9\ M^{-1}\ \text{sec}^{-1} \text{ at } 150°$$

Rice and co-workers studied the pyrolysis of propylamine at 650° in the hope of identifying an induced decomposition of the amine by attack of methyl radicals produced as secondary products of the pyrolysis.[16]

$$Me^{\cdot} + n\text{-}C_3H_7NH_2 \rightarrow MeNH_2 + C_3H_7^{\cdot} \tag{17}$$

No methylamine was formed and it was concluded that homolytic displacement at nitrogen did not occur. Methylamine was one of the products of the reaction of hydrogen atoms, produced by the discharge tube method, with azomethane at elevated temperatures, whereas at low temperatures (27°) dimethylhydrazine was the sole product.[17] It was concluded that the methylamine formed at 110° was produced by a homolytic displacement at nitrogen by hydrogen atoms. An activation energy of 8 kcal/mole was attributed to reaction 18, which would be about 35 kcal/mole exothermic.

$$H^{\cdot} + \underset{H}{MeN}—\underset{H}{NMe} \longrightarrow MeNH_2 + Me\dot{N}H \tag{18}$$

[14] See footnote 3, Chapter V.

[15] $D(H_2N\text{–}H) = 103$ kcal/mole; $D(H_2N\text{–}NH_2) = 56$ kcal/mole (see footnote 46, Chapter V).

[16] F. O. Rice, W. D. Walters, and P. M. Ruoff, *J. Chem. Phys.*, **8**, 259 (1940).

[17] H. Henkin and H. A. Taylor, *ibid.*, **8**, 1 (1940).

However, as in the case of hydrazine itself, abstraction of the hydrogen bound to nitrogen in the substituted hydrazine would be expected to be a very facile process.

Snipes and Schmidt[18] have reported that the radical $CH_3\dot{C}HCOOH$ is formed by exposure of crystalline alanine to γ rays and to hydrogen atoms from a microwave discharge tube. They suggest that the hydrogen atoms abstract the amino group,

$$H^{\cdot} + H_3C\text{—}\underset{\substack{|\\H}}{\overset{\substack{NH_2\\|}}{C}}\text{—}COOH \longrightarrow H_3C\text{—}\underset{\substack{|\\H}}{\overset{\substack{H\\ \vdots \\ H\cdots N\cdots H\\|}}{C}}\text{—}COOH \longrightarrow$$

$$H_3C\text{—}\underset{\substack{|\\H}}{\dot{C}}\text{—}COOH + NH_3 \tag{19}$$

However, hydrogen atom abstraction from the CH_3 group would seem to provide a more likely route to the reported radical, that is,

$$H^{\cdot} + CH_3CH(NH_2)COOH \rightarrow H_2 + {}^{\cdot}CH_2CH(NH_2)COOH \tag{20}$$

$${}^{\cdot}CH_2CH(NH_2)COOH \rightarrow CH_2{=}CHCOOH + {}^{\cdot}NH_2 \tag{21}$$

$$H^{\cdot} + CH_2{=}CHCOOH \rightarrow CH_3\dot{C}HCOOH \tag{22}$$

A rather special case is presented by tetrafluorohydrazine which participates in a variety of free radical reactions, as might be expected from its low N–N bond strength (19.8 kcal/mole).[19] Even at room temperature some difluoroamino radicals are present in equilibrium with the parent hydrazine, and if a radical is generated in the presence of tetrafluorohydrazine a difluoroamino compound is generally isolated. For example, if an azoalkane is photolyzed with tetrafluorohydrazine the alkyldifluoroamine is produced.[20]

$$RN{=}NR \xrightarrow{h\nu} 2R^{\cdot} + N_2 \tag{23}$$

$$R^{\cdot} + {}^{\cdot}NF_2 \rightarrow RNF_2 \tag{24}$$

However, it is conceivable that at low temperatures a homolytic displacement at nitrogen might compete favorably with reaction 24, but such a process does not appear to have been identified at present.

$$R^{\cdot} + F_2N\text{—}NF_2 \rightarrow RNF_2 + {}^{\cdot}NF_2 \tag{25}$$

[18] W. Snipes and J. Schmidt, *Radiation Res.*, **29,** 194 (1966).
[19] C. B. Colburn, *Chem. Brit.*, **2,** 336 (1966).
[20] R. C. Petry and J. P. Freeman, *J. Amer. Chem. Soc.*, **83,** 3912 (1961).

Alkanes react with nitric acid in the gas phase to yield nitroalkanes along with aldehydes, ketones, and their various further oxidation products, a reaction first studied extensively by Hass and co-workers.[21] One of the theories of mechanism considered during this work was that free radicals, thermally produced from the nitrating agent, or from unstable oxidation products, attack the alkane to produce an alkyl radical, which engages in a chain reaction.[22,23]

$$RH + \text{initiator} \rightarrow R^{\cdot} \tag{26}$$

$$R^{\cdot} + HONO_2 \rightarrow RNO_2 + \dot{O}H \tag{27}$$

$$RH + \dot{O}H \rightarrow R^{\cdot} + H_2O \tag{28}$$

The first propagating step of the chain, reaction 27, implies S_H2 attack of the alkyl radical on nitrogen, expelling an hydroxy radical.

The kinetics of the vapor-phase nitration of methane by nitric acid at 349.5° were studied by Godfrey, Hughes, and Ingold.[24] It was shown that the major reaction occurring was a free-radical chain process of short chain length. The proposed mechanism involved bimolecular homolytic attack of a methyl radical on nitric acid as part of the chain-propagating sequence.

$$H\dot{O} + CH_4 \rightarrow H_2O + \dot{C}H_3 \tag{29}$$

$$\dot{C}H_3 + HONO_2 \rightarrow CH_3NO_2^* + \dot{O}H \tag{30}$$

(* Includes CH_3NO_2 and CH_3ONO, the latter being unstable under the reaction conditions.)

A further reaction between methyl radicals and nitric acid was suggested, formally homolytic displacement at oxygen.

$$\dot{C}H_3 + HONO_2 \rightarrow CH_3OH + \dot{N}O_2 \tag{31}$$

In order to explain the formation of dimeric nitrosomethane in the reaction between methyl radicals, from acetyl peroxide, and *s*-butyl nitrite in the liquid phase at 73°, Kharasch, Meltzer, and Nudenburg[25] suggested the occurrence of reaction 32.

$$Me^{\cdot} + Bu^sONO \rightarrow MeNO + Bu^sO^{\cdot} \tag{32}$$

A similar reaction was proposed by Gray for the decomposition of *t*-butyl nitrite induced by methyl radicals in the vapor or liquid phase.[26,27] Phillips

[21] H. B. Hass, E. B. Hodge, and B. M. Vanderbilt, *Ind. Eng. Chem.*, **28,** 339 (1936).
[22] H. B. Hass and J. A. Paterson, *ibid.*, **30,** 67 (1938).
[23] R. F. McCleary and E. F. Degering, *ibid.*, **30,** 64 (1938).
[24] T. S. Godfrey, E. D. Hughes, and C. K. Ingold, *J. Chem. Soc.*, 1063 (1965).
[25] M. S. Kharasch, T. H. Meltzer, and W. Nudenburg, *J. Org. Chem.*, **22,** 37 (1957).
[26] P. Gray, *Chem. and Ind.* (*London*), 120 (1960).
[27] P. Gray and P. Rathbone, *Proc. Chem. Soc.*, 316 (1960).

and co-workers[28,29] have studied the reaction of methyl radicals, from the pyrolysis of di-*t*-butyl peroxide, with a series of alkyl nitrites in the vapor-phase at 160–180°. On the basis of the low yields of methane and ethane together with the high yields of alcohol derived from the nitrite they concluded that homolytic displacement on nitrogen was occurring and they pointed out that reaction 33 would be 19 kcal/mole exothermic.

$$\mathrm{Me^{\cdot} + MeONO \rightarrow MeNO + MeO^{\cdot}} \tag{33}$$

Alkoxy radicals add to C-nitroso compounds to give alkoxyalkylnitroxides which are unstable, losing an alkyl radical and forming an alkyl nitrite.[30,31]

$$\mathrm{Bu^{t}O^{\cdot} + RNO \longrightarrow Bu^{t}O\underset{\underset{\displaystyle O^{\cdot}}{|}}{N}R} \tag{34}$$

$$\mathrm{Bu^{t}O\underset{\underset{\displaystyle O^{\cdot}}{|}}{N}R \longrightarrow Bu^{t}ONO + R^{\cdot}} \tag{35}$$

This may be considered as the stepwise homolytic displacement of an alkyl radical from nitrogen by an alkoxy radical. Reactions 32 and 33 might also proceed by way of an alkoxyalkylnitroxide, the reverse of reaction 35 constituting the initial step, followed by the reverse of reaction 34.

When mixtures of hydrogen peroxide and methylamine, dimethylamine or diethylamine were photolyzed at 77°K the only species observed by e.s.r. spectroscopy were alkylamino radicals or dialkylamino radicals.[32,32a] These radicals were presumably formed by abstraction of hydrogen from the amines by hydroxyl radicals.

$$\mathrm{HO^{\cdot} + HNRR' \longrightarrow H_2O + {}^{\cdot}NRR'} \tag{36}$$

$$R = \mathrm{Me}, R' = \mathrm{H};\ R = R' = \mathrm{Me};\ R = R' = \mathrm{Et}.$$

However, when triethylamine and hydrogen peroxide were irradiated the spectrum of the ethyl radical was observed. When illumination ceased the

[28] B. Jest and L. Phillips, *ibid.*, 73 (1960).
[29] B. Bromberger and L. Phillips, *J. Chem. Soc.*, 5302 (1961).
[30] A. Mackor, Th. A. J. W. Wajer, Th. J. deBoer, and J. D. W. vanVoorst, *Tetrahedron Lett.*, 385 (1967).
[31] A. Mackor, Th. A. J. W. Wajer, and Th. J. deBoer, *Tetrahedron*, **24**, 1623 (1968).
[32] V. I. Mal'tsev and A. A. Petrov, *Zhur. Org. Chem.*, **3**, 216 (1967).
[32a] cf. However, W. C. Danen and T. T. Kensler, *J. Amer. Chem. Soc.*, **92**, 5235 (1970).

ethyl radical spectrum decayed and the spectrum due to the $CH_3\dot{C}HNEt_2$ radical appeared, evidently the result of attack of the ethyl radical on triethylamine.

$$Et^{\cdot} + CH_3CH_2NEt_2 \rightarrow EtH + CH_3\dot{C}HNEt_2 \quad (37)$$

The mechanism by which the ethyl radical is generated is not known, but a direct homolytic displacement at nitrogen by the hydroxyl radical seems unlikely. One possibility is that transfer of an electron from the triethylamine to the hydroxyl radical occurs to give the amine radical cation and a hydroxide anion which then react together to generate the ethyl radical.

$$HO^{\cdot} + NEt_3 \rightarrow HO^- + \dot{\overset{+}{N}}Et_3 \rightarrow HONEt_2 + Et^{\cdot} \quad (38)$$

The photo reduction of aromatic carbonyl compounds by amines of low ionization potential occurs by electron transfer to the electron deficient excited carbonyl group.[33]

$$Ar_2CO^* + Et_3N \rightarrow Ar_2\dot{C}{-}\overset{-}{O} + Et_3\overset{\cdot+}{N} \quad (39)$$

$$Ar_2\overset{-}{C}{-}\dot{O} + CH_3CH_2\overset{\cdot+}{N}Et_2 \rightarrow Ar_2\dot{C}{-}OH + CH_3\dot{C}HNEt_2 \quad (40)$$

The reaction of benzoyl azide with tributyltin hydride is a free-radical chain process which includes attack of the tin radical at nitrogen.[34]

$$Bu_3Sn^{\cdot} + PhC(O){-}N{-}\overset{-}{N}{\equiv}\overset{+}{N} \rightarrow PhC(O){-}N{-}SnBu_3 + N_2 \quad (41)$$

$$Bu_3SnH + PhC(O){-}\dot{N}{-}SnBu_3 \rightarrow PhC(O)NHSnBu_3 + Bu_3Sn^{\cdot} \quad (42)$$

However, reaction 41 may well involve a rate controlling electron transfer. The reaction of phenyl azide with triorganotin hydrides[34a] or with organosilicon hydrides[34b] probably proceeds by a similar chain sequence.

B. PHOSPHORUS

In general, phosphorus compounds prefer to react by ionic routes, utilizing the nucleophilic reactivity of the lone pair of electrons in tervalent compounds or the electrophilicity of the phosphorus atom in quinquevalent derivatives.

[33] R. S. Davidson and R. Wilson, *J. Chem. Soc.* (*B*), 71 (1970).

[34] M. Frankel, D. Wagner, D. Gertner, and A. Zilkha, *J. Organometal. Chem.*, **2,** 518 (1967).

[34a] H. Schumann and S. Ronecker, *J. Organometal. Chem.*, **23,** 451 (1970).

[34b] F. A. Carey and C. W. Hsu, *Tetrahedron Letters*, 3885 (1970).

However, radical mechanisms are also quite common, and seem likely to become of increasing importance.[35–39]

The presence of unoccupied low energy *d*-orbitals makes valence shell expansion of second (and higher) row elements possible. Thus tervalent phosphorus compounds unlike those of nitrogen, can react with a free radical to increase the coordination number to four and give a species with nine valence electrons.

$$Y^{\cdot} + PX_3 \rightleftharpoons Y\dot{P}X_3 \tag{43}$$

The formation of such a *phosphoranyl radical* was first suggested in 1957.[40–42] Loss of $\dot{X}$ from the intermediate phosphoranyl radical would result in the overall homolytic substitution of X for Y by the S_H2 (stepwise) mechanism.

$$Y\dot{P}X_3 \rightarrow YPX_2 + X^{\cdot} \tag{44}$$

If $Y\dot{P}X_3$ represents only a transition state the mechanism would be designated S_H2 (synchronous). In fact it is in the free-radical chemistry of phosphorus compounds that there is most evidence for the stepwise process. In addition to the α-scission pathway for decomposition of the intermediate phosphoranyl radical, shown in equation 44, $Y\dot{P}X_3$ may also undergo β-scission, mainly on account of the strength of the P=O and P=S bonds in quinquevalent phosphorus compounds. For example, when di-*t*-butyl peroxide or dicumyl peroxide are allowed to react with triethyl phosphite (either thermally or photochemically) the products are triethyl phosphate and hydrocarbon mixtures arising from alkyl radical dimerization and disproportionation.[43]

$$RO^{\cdot} + P(OEt)_3 \rightarrow RO\dot{P}(OEt)_3 \tag{45}$$

$$RO\dot{P}(OEt)_3 \rightarrow R^{\cdot} + O{=}P(OEt)_3 \tag{46}$$

$$R = Bu^t \text{ or } PhCMe_2$$

Processes of this type, which do not involve homolytic substitution at the phosphorus atom, are not the subject of this Review, but discussion of them will be included when it aids the understanding of the S_H2 reaction.

[35] J. I. G. Cadogan, *Quart. Rev.*, **16,** 226 (1962).
[36] R. F. Hudson, *Structure and Mechanism in Organophosphorus Chemistry*, Academic Press, London, (1965) Ch. 9.
[37] C. Walling and M. S. Pearson, *Topics Phosphorus Chem.*, **3,** 1 (1966).
[38] A. J. Kirby and S. G. Warren, *The Organic Chemistry of Phosphorus*, Elsevier, Amsterdam, 1967.
[39] J. I. G. Cadogan, *Adv. Free-Radical Chem.*, **2,** 203 (1968).
[40] F. Ramirez and N. McKelvie, *J. Amer. Chem. Soc.*, **79,** 5829 (1957).
[41] C. Walling and R. Rabinowitz, *ibid.*, **79,** 5326 (1957).
[42] G. Kamai and F. M. Kharrasova, *Zhur. Obshch. Khim.*, **27,** 953 (1957).
[43] C. Walling and R. Rabinowitz, *J. Amer. Chem. Soc.*, **81,** 1243 (1959).

Substitution by Oxygen-Centered Radicals

The major products of the autoxidation of trialkylphosphines in hydrocarbon solvents at room temperature are trialkylphosphine oxide and alkyl dialkylphosphinate in approximately equimolar amounts.[44] Dialkyl alkylphosphonate and trialkyl phosphate are also formed to a small extent. Buckler[44] found that the autoxidation could be inhibited by small quantities of diphenylamine or hydroquinone and he proposed that the reaction occurred by the free-radical chain mechanism shown in equations 47–55.

$$RO^{\cdot} + R_3P \rightarrow R^{\cdot} + R_3PO \tag{47}$$
$$RO^{\cdot} + R_3P \rightarrow R^{\cdot} + R_2POR \tag{48}$$
$$RO_2^{\cdot} + R_2POR \rightarrow RO^{\cdot} + R_2P(O)OR \tag{49}$$
$$RO^{\cdot} + R_2POR \rightarrow R^{\cdot} + R_2P(O)OR \tag{50}$$
$$RO^{\cdot} + R_2POR \rightarrow R^{\cdot} + RP(OR)_2 \tag{51}$$
$$RO_2^{\cdot} + RP(OR)_2 \rightarrow RO^{\cdot} + RP(O)(OR)_2 \tag{52}$$
$$RO^{\cdot} + RP(OR)_2 \rightarrow R^{\cdot} + RP(O)(OR)_2 \tag{53}$$
$$RO^{\cdot} + RP(OR)_2 \rightarrow R^{\cdot} + P(OR)_3 \tag{54}$$
$$RO_2^{\cdot} + P(OR)_3 \rightarrow RO^{\cdot} + O{=}P(OR)_3 \tag{55}$$
$$R = Bu^n, \text{ cyclohexyl.}$$

The relative amounts of phosphine oxide and phosphinite ester (which is further oxidized to the isolated phosphinate ester) would be determined by the competition of reactions 47 and 48. In support of this proposition it was shown that when di-*t*-butyl peroxide was thermally (130°) decomposed in the presence of an excess of tributylphosphine, *t*-butyl dibutylphosphinite and tributylphosphine oxide were formed in the ratio 4:1.

$$Bu^tO^{\cdot} + PBu_3 \rightarrow Bu^tOPBu_2 + Bu^{\cdot} \tag{56}$$
$$Bu^tO^{\cdot} + PBu_3 \rightarrow O{=}PBu_3 + Bu^{t\cdot} \tag{57}$$

A subsequent kinetic study of the autoxidation[45] gave results in agreement with Buckler's mechanism. The oxidation reaction 57 is thermodynamically more favorable than the S_H2 displacement 56, as can be seen from the following bond energy data.[46–48]

[44] See footnote 3, Chapter I.

[45] M. B. Floyd and C. E. Boozer, *J. Amer. Chem. Soc.*, **85**, 984 (1963).

[46] See footnote 102, Chapter III.

[47] S. W. Benson and R. Shaw, Chapter 2 in "*Organic Peroxides*" Ed. D. Swern, **1**, Wiley, New York, 1970.

[48] S. B. Hartley, W. S. Holmes, J. K. Jacques, M. F. Mole, and J. C. McCoubrey, *Quart. Rev.*, **17**, 204 (1963).

$D(Bu_3P{=}O)$	137	$\bar{D}$[P—O in $(EtO)_3P$]	92
$D(Bu^t{-}O^\cdot)$	89	$\bar{D}$(P—C in Et_3P)	62
difference	48 kcal/mole	difference	30 kcal/mole

This led Buckler[44] to suggest a direct displacement mechanism [S_H2 (synchronous)] for step 56 to account for the kinetic product control. Since $D(Bu^tO{-}O\cdot)$ is only 57 kcal/mole,[47] phosphine oxide production will be some 50 kcal/mole more exothermic than alkyl radical displacement in the reaction between an alkylperoxy radical and a trialkylphosphine.

More recently the e.s.r. spectrum of the displaced alkyl radical has been observed when *t*-butoxy radicals, generated photolytically from di-*t*-butyl peroxide, react with trialkylphosphines while in the cavity of the spectrometer.[49]

$$Bu^tO^\cdot + PR_3 \rightarrow Bu^tOPR_2 + R^\cdot \tag{58}$$

$$R = Me, Et, Bu^i, Pr^i, \text{cyclohexyl}.$$

In the case of trimethylphosphine the intensity of the methyl radical spectrum was lower than that of the radical from the higher alkyls, and a second spectrum attributed to an intermediate phosphoranyl radical was observed.

$$(CH_3)_3CO^\cdot + (CH_3)_3P \rightarrow (CH_3)_3CO\dot{P}(CH_3)_3 \tag{59}$$

$$(CH_3)_3CO\dot{P}(CH_3)_3 \rightarrow (CH_3)_3COP(CH_3)_2 + CH_3^\cdot \tag{60}$$

No spectrum due to the *t*-butyl radical was observed in these reactions which were carried out at a lower temperature than that used by Buckler[44] in his preparative experiment. If the thermally and photolytically generated *t*-butoxy radicals behave similarly, reaction 56 must have a lower activation energy than reaction 57, although the activation energies must be low for both reactions.

The e.s.r. spectrum of $(CH_3)_3CO\dot{P}(CH_3)_3$ indicated that one methyl group was distinct from the other two and a trigonal-bipyramidal configuration was proposed, similar to that suggested for other phosphoranyl radicals, $\dot{P}F_4$,[50,51] $\dot{P}Cl_4$,[52] and $Me_2\dot{P}Cl_2$.[52] Thus, it would appear that an intermediate phosphoranyl radical is indeed formed in the interaction of an alkoxy radical with a tervalent phosphorus compound and that this intermediate may undergo competing α-scission or β-scission.

[49] See footnote 2, Chapter I.
[50] See footnote 35, Chapter I
[51] See footnote 36, Chapter I.
[52] See footnote 37, Chapter I.

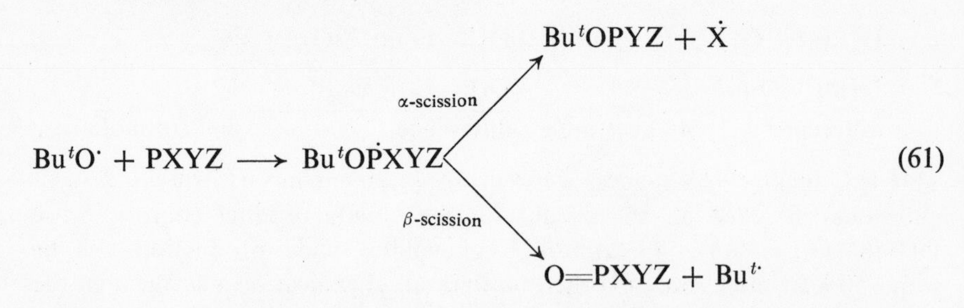

(61)

Which of the two pathways is followed is critically dependent on the groups, X, Y, and Z, the solvent and the temperature. Perhaps the increase in rate of β-scission compared to α-scission in more polar solvents noted by Buckler[44] is due to the relatively more polar transition state for β-scission (compare the effect of solvent on the rate of β-scission of *t*-alkoxy radicals).[53] With trialkyl phosphites ($X = Y = Z = OR$) the only products are those of β-scission, usually the trialkyl phosphate[43] and the *t*-butyl radical,[54–56] although when triallyl phosphite was used, the more stable allyl radical was lost.[54]

$$Bu^tO^{\cdot} + P(OCH_2CH{=}CH_2)_3 \longrightarrow Bu^tO\dot{P}(OCH_2CH{=}CH_2)_3$$
$$\downarrow$$
$$CH_2{=}CH\dot{C}H_2 + Bu^tOP(OCH_2CH{=}CH_2)_2 \quad (P{=}O) \qquad (62)$$

When a dilute solution of triphenyl phosphite and di-*t*-butyl peroxide in cyclopropane was irradiated at $-85°$ the e.s.r. spectrum of the phenoxy radical was observed.[54]

$$Bu^tO^{\cdot} + P(OPh)_3 \rightarrow Bu^tO\dot{P}(OPh)_3 \rightarrow Bu^tOP(OPh)_2 + \dot{O}Ph \qquad (63)$$

Here the stability of the phenoxy radical allows P–O cleavage of the phosphoranyl radical, a reaction which had been proposed earlier on the basis of chemical studies.[57–59] A similar homolytic displacement generating a 2,4,6 tri-*t*-butylphenoxy radical may be responsible for the efficiency, compared with trialkyl phosphites, of **1** as an inhibitor of polypropylene autoxidation,[60]

[53] C. Walling and P. J. Wagner, *J. Amer. Chem. Soc.* **86,** 3368 (1964).
[54] See footnote 2, Chapter I.
[55] A. Hudson and H. A. Hussain, *J. Chem. Soc.* (*B*), 793 (1969).
[56] J. Q. Adams and H. G. Ingersoll, *Abstracts*, *157th National Meeting of the American Chemical Society*, Minneapolis, Minn., April 13, 1969, PETR-002.
[57] See footnote 16, Chapter III.
[58] W. G. Bentrude, *Tetrahedron Lett.*, 3543 (1965).
[59] W. H. Starnes, Jr., and N. P. Neureiter, *J. Org. Chem.*, **32,** 333 (1967).
[60] P. I. Levin, *Zhur. Fiz. Khim.*, **38,** 672 (1964).

while trialkyl phosphites cause much more rapid heterolytic decomposition of chain-initiating hydroperoxides than does **1**.[61]

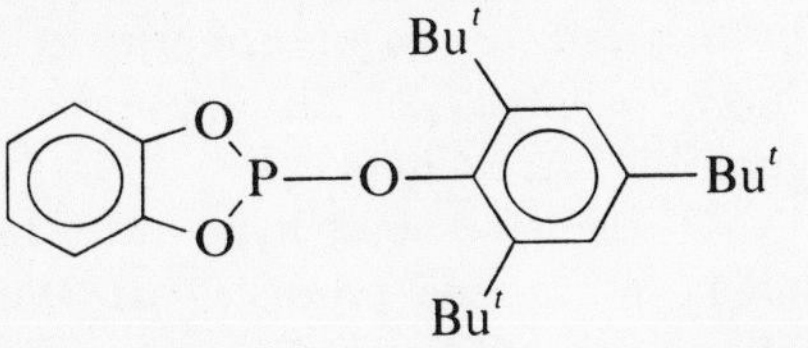

Bu^t

O P—O— —Bu^t **1**

O

Bu^t

A possible mechanism for the formation of polymeric products from the reaction of certain cyclic phosphites with *t*-butoxy radicals involves α-scission (i.e., displacement) of an alkoxy radical from phosphorus.[58]

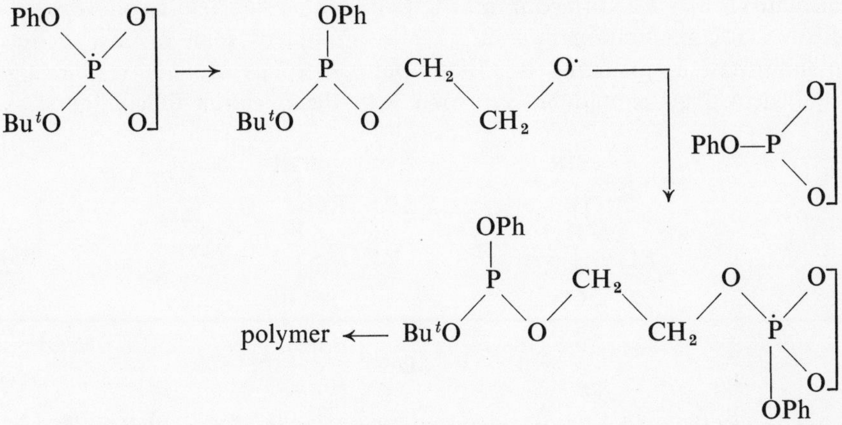

PhO O

$\dot{P}$ ⟶ OPh P CH$_2$ O$^\cdot$

Bu^{t}O O Bu^{t}O O CH$_2$

PhO—P (O, O)

OPh

P CH$_2$ O O

polymer ⟵ Bu^{t}O O CH$_2$ $\dot{P}$

O

OPh

(64)

Relief of ring strain was thought to account for the preferential loss of an alkoxy radical rather than the more stable phenoxy radical.

When triphenylphosphine (X = Y = Z = Ph) reacts with *t*-butoxy radicals, β-scission of the phosphoranyl radical occurs to give triphenylphosphine oxide in high yield.[44] The fact that phenyl radicals are not displaced from phosphorus could be rationalized in terms of the increased P–C(aryl) bond strength as compared with the P–C(alkyl) bond. However, in some recent studies on the stereochemistry of these reactions, Bentrude allowed optically active methylpropylphenylphosphine to react with *t*-butoxy radicals from the thermolysis of di-*t*-butyl hyponitrite at 69° in acetonitrile, and found that methylpropylphenylphosphine oxide was formed with retention of configuration.[62] The polar solvent would confer a preference for β-scission of the

[61] P. A. Kirpichnikov, N. A. Mukmeneva, A. N. Pudovik, and N. S. Kolyubakina, *Dokl. Akad. Nauk. SSSR*, **164**, 1050 (1965).
[62] See footnote 38, Chapter I.

phosphoranyl radical but perhaps the presence of a phenyl group attached to phosphorus may increase the rate of β-scission compared with α-scission [the latter would give *t*-butyl methylphenylphosphinite, $Bu^tOP(Me)Ph$]. The transfer of oxygen from *t*-butoxy radicals to phosphorus in the geometrically isomeric pair of 2-methoxy-5-*t*-butyl-1,3,2-dioxaphosphorinans also occurs with retention of configuration[62] (see Chapter I).

The thermolysis of ^{14}C-labeled di-*t*-butyl hyponitrite in the presence of tri-*t*-butyl phosphite at 65° in benzene produced tri-*t*-butyl phosphate in almost quantitative yield and approximately 75% of the product phosphate was ^{14}C-labeled. This result was interpreted in terms of the essentially *irreversible* formation of a phosphoranyl radical intermediate in a manner which allowed statistical scrambling of the label.[63] If an unsymmetrical intermediate such as **2** is formed it must have a lifetime sufficient to allow equilibration of configurationally nonequivalent groups by some process such as pseudorotation, assuming that the *t*-butoxy groups are not reactionally equivalent. This should be compared with the retention of configuration

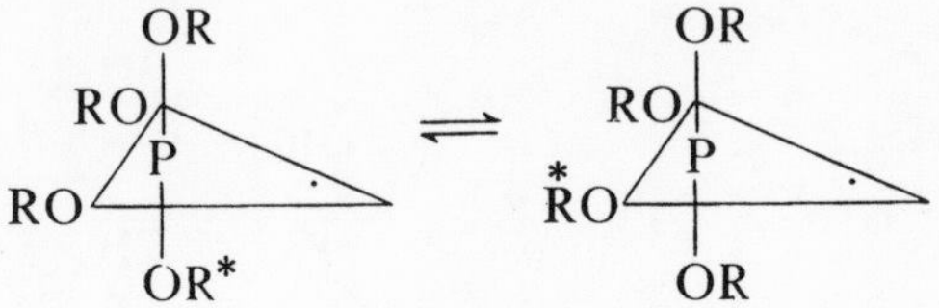

OR, RO, RO, OR* around P· $\rightleftharpoons$ OR, RO, R*O, OR around P· (65)

2

$R = Bu^t$

observed in the oxidation of stereoisomeric phosphines[62] and phosphites[62] by the *t*-butoxy radical discussed above, where pseudorotation in a phosphoranyl intermediate would be expected to lead to some inversion. Bentrude[63] concludes that the configuration, configurational stability and lifetime of phosphoranyl radicals depends on the nature of the substituents around phosphorus.

The reaction of an alkoxy radical at a tervalent phosphorus center is extremely rapid and by studying the competition with hydrogen abstraction from cyclohexane, Walling and Pearson[57] determined a series of reactivities of phosphorus compounds towards the *t*-butoxy radical at 130°. Thus, triethyl phosphite and triphenylphosphine were found to be 600–800 and 7 times, respectively, as reactive as cyclohexane towards the *t*-butoxy radical. However, by direct competition for the *t*-butoxy radical between pairs of phosphorus compounds, the following relative reactivities were obtained:

[63] See footnote 39, Chapter I.

$(EtO)_3P:Ph_3P = 1.6 \pm 0.2$; $Bu_3P:Ph_3P = 2.46 \pm 0.25$ and $(EtO)_3P:Bu_3P = 0.83 \pm 0.07$. The result from the direct competition between triethyl phosphite and triphenylphosphine is very different from that obtained from the indirect study utilizing the competition with hydrogen abstraction from cyclohexane [$(EtO)_3P:Ph_3P = ca.\ 100$]. Walling and Pearson[57] considered that the *indirect* result was more correct. However, Starnes and Neureiter[59] found that the rate of deoxgenation of the *t*-butoxy radical by triphenylphosphine was at least 400 ± 200 times faster than hydrogen abstraction from *n*-heptane at 100°. As the differences in competing hydrocarbon and temperature would not be expected to have a large effect, this result is in fairly good agreement with the *direct* comparison between triethyl phosphite and triphenylphosphine obtained by Walling and Pearson, with triphenylphosphine only slightly less reactive than triethyl phosphite towards *t*-butoxy radicals.

In contrast to their behavior towards tervalent phosphorus compounds, when *t*-butoxy radicals react with a trialkylphosphine oxide[54] or a trialkyl phosphate[55] they do so by abstraction of hydrogen from C–H bonds and do not attack the phosphorus center.

$$(Me_2CH)_3PO + Bu^tO^{\cdot} \longrightarrow Me_2\dot{C}\overset{\overset{\displaystyle O}{\|}}{P}(CHMe_2)_2 + Bu^tOH \qquad (66)$$

$$(MeO)_3PO + Bu^tO^{\cdot} \longrightarrow H_2\dot{C}O\overset{\overset{\displaystyle O}{\|}}{P}(OMe)_2 + Bu^tOH \qquad (67)$$

Similarly, hydrogen abstraction was observed when hydroxyl radicals were allowed to react with dialkylphosphonates and dialkyl alkylphosphonates in a flow system using e.s.r. spectroscopy to identify the radical formed.[64] With the hydroxyl radical, even trishydroxymethylphosphine apparently underwent only side chain abstraction.[64] A suggested route[65] for the production of alcohol in the reaction of alkoxy radicals with trialkylphosphine oxides, in which the addition of the former to the latter is likened to the rapid addition of radicals to nitroso compounds to give nitroxides, seems unlikely.

$$\begin{array}{l} RO^{\cdot} + R'_3PO \longrightarrow R'_3P(OR)O^{\cdot} \xrightarrow{\text{solvent}} R'_3P(OR)OH \\ \qquad\qquad\qquad\qquad\qquad\qquad\qquad\qquad \rightleftharpoons \\ \qquad\qquad\qquad\qquad\qquad\qquad\qquad R'_3PO + ROH \end{array} \qquad (68)$$

Although α-scission of a phosphoranyl radical, leading to overall homolytic displacement, can occur in favorable cases, the species formed by addition of an alkylperoxy radical to a tervalent phosphorus compound would

[64] E. A. C. Lucken, *J. Chem. Soc.* (*A*), 1354, 1357 (1966).
[65] J. H. Boyer and J. D. Woodyard, *J. Org. Chem.*, **33**, 3329 (1958).

be expected to undergo β-scission exclusively because of the low O–O bond strength.[47]

$$ROO^{\cdot} + PX_3 \rightarrow [ROO\dot{P}X_3] \rightarrow RO^{\cdot} + O{=}PX_3 \quad (69)$$

It is not clear if a discrete intermediate is involved or whether $[ROO\dot{P}X_3]$ is only a transition state, but the overall reaction has been suggested many times in the thermal and photolytically initiated free-radical chain autoxidation of phosphines[44,45] and phosphites.[43,57,66–68]

The oxidative phosphonation of hydrocarbons with phosphorus trichloride was first described by Clayton and Jensen.[69] To a first approximation, saturated hydrocarbons give alkylphosphonyl dichlorides by the overall reaction,

$$RH + 2PCl_3 + O_2 \rightarrow RP(O)Cl_2 + P(O)Cl_3 + HCl \quad (70)$$

The mechanism of the chlorophosphonation reaction, which competes with the oxidation of phosphorus trichloride, was extensively investigated by Mayo, Durham, and Griggs[70] and later by Flurry and Boozer[71] whose free-radical chain mechanism was in essential agreement with that of the former workers. Mayo's formulation[70] involved no homolytic displacements at phosphorus. It was proposed that alkoxy radicals reacted with phosphorus trichloride to give phosphorus oxychloride.

$$RO^{\cdot} + PCl_3 \rightarrow RO\dot{P}Cl_3 \rightarrow R^{\cdot} + O{=}PCl_3 \quad (71)$$

Flurry and Boozer[71] found that the chlorophosphonation reaction was not spontaneous, as Mayo had thought, if the reactants were sufficiently pure. They suggested that initiation might be brought about by the reaction of a trace of hydroperoxide with the phosphorus trichloride.

$$PCl_3 + ROOH \rightarrow Cl_2POOR + HCl \quad (72)$$

$$Cl_2POOR \rightarrow Cl_2PO^{\cdot} + {}^{\cdot}OR \quad (73)$$

$$RO^{\cdot} + PCl_3 \rightarrow ROPCl_2 + Cl^{\cdot} \quad (74)$$

In the presence of chlorides of quinquevalent phosphorus, 72 and 73 were replaced by the more rapid reactions 75 and 76, both radicals produced in the latter being capable of displacing a chlorine atom from phosphorus trichloride.

$$P(O)Cl_3 + ROOH \rightarrow P(O)(OOR)Cl_2 + HCl \quad (75)$$

$$P(O)(OOR)Cl_2 \rightarrow P(O)(O^{\cdot})Cl_2 + {}^{\cdot}OR \quad (76)$$

[66] K. Smeykal, H. Baltz, and H. Fisher, *J. Prakt. Chem.*, **22**, 186 (1963).
[67] J. I. G. Cadogan, M. Cameron-Wood, and W. R. Foster, *J. Chem. Soc.*, 2549 (1963).
[68] J. B. Plumb and C. E. Griffin, *J. Org. Chem.*, **28**, 2908 (1963).
[69] J. O. Clayton and W. L. Jensen, *J. Amer. Chem. Soc.*, **70**, 3880 (1948).
[70] F. R. Mayo, L. Y. Durham, and K. S. Griggs, *ibid.*, **85**, 3156 (1963).
[71] R. L. Flurry and C. E. Boozer, *J. Org. Chem.*, **31**, 2076 (1966).

The importance of reaction 74 relative to 71 was shown by the fact that when di-*t*-butyl peroxide was thermally decomposed at 125° in the presence of phosphorus trichloride only a small amount of phosphorus oxychloride was formed but phosphite esters were detected.

Homolytic displacement at the P–P bonds in white phosphorus (tetrahedral P_4) may possibly occur during the autoxidation of the element[72,73] or its reaction with di-*t*-butyl peroxide.[74] Benzene solutions of mixtures of phosphorus and olefins absorb oxygen to give polymeric products containing P–C bonds in a free-radical chain reaction.[75] However the mechanisms of these reactions are only poorly defined at the present time.

Substitution by Sulfur-Centered Radicals

By analogy with the reaction of alkoxy radicals, alkylthiyl radicals would be expected to form phosphoranyl radicals by addition to tervalent phosphorus compounds.

$$RS^{\cdot} + PX_3 \rightarrow RS\dot{P}X_3 \tag{77}$$

In theory the phosphoranyl radical could next undergo α-scission or β-scission to give overall homolytic substitution at or oxidation of the phosphorus.

$$RS\dot{P}X_3 \xrightarrow{\alpha\text{-scission}} RSPX_2 + \dot{X} \tag{78a}$$

$$RS\dot{P}X_3 \xrightarrow{\beta\text{-scission}} S{=}PX_3 + \dot{R} \tag{78b}$$

No examples of the S_H2 reaction 78a have been reported. Reaction 78b always occurs in accord with the weaker single bonds which sulfur (compared with oxygen) forms to carbon and phosphorus, coupled with the relative strength of the P=S bond. The overall result of reactions 77 and 78b forms a step in the free-radical chain mechanism for the

[72] F. S. Dainton and H. M. Kimberly, *Trans. Faraday Soc.*, **46,** 629 (1950).
[73] P. W. Schenk and H. Vietzke, *Z. Anorg. Allgem. Chem.*, **326,** 152 (1963).
[74] W. E. Garwood, L. A. Hamilton, and F. M. Seger, *Ind. Eng. Chem.*, **52,** 401 (1960).
[75] C. Walling, F. R. Stacey, S. E. Jamison, and E. S. Huyser, *J. Amer. Chem. Soc.*, **80,** 4543, 4546 (1958).

desulfurization of thiols or disulfides by trialkylphosphines[76] or phosphites[41,43,76] first proposed by Walling and Rabinowitz.[41,43]

$$RS^{\cdot} + PX_3 \xrightarrow{(77)} RS\dot{P}X_3 \xrightarrow{(78b)} R^{\cdot} + S{=}PX_3 \qquad (79)$$

$$R^{\cdot} + RSH \rightarrow RH + RS^{\cdot} \qquad (80)$$

or,

$$R^{\cdot} + RSSR \rightarrow RSR + RS^{\cdot} \qquad (80a)$$

$$X = OR', R, Ph.$$

Competitive studies indicated that the order of reactivity of trivalent phosphorus compounds towards thiyl radicals was

$$Bu_3P > (EtO)_3P > Ph_3P > (PhO)_3P$$

and that the rate coefficient for reaction 79 ($R = Bu^n$, $X = OEt$) is $\sim 10^8\ M^{-1}\ sec^{-1}$ at 70°.[57] Burkhart[77] has made use of the reaction of thiols with triethyl phosphite to determine the rate of self-reaction of the alkyl radicals, $R^{\cdot}$, under conditions where this is the major chain-terminating process.

The occurrence of reaction 79 has been demonstrated directly by the observation of the e.s.r. spectrum of the *t*-butyl radical when a solution containing di-*t*-butyl disulfide and triisopropylphosphine was irradiated with u.v. light in the cavity of the spectrometer.[54] No isopropyl radical, the product of homolytic displacement at phosphorus, was detected. Bentrude[62] has shown that the oxidation of optically active methylpropylphenylphosphine or of mixtures of *cis*- and *trans*-2-methoxy-5-*t*-butyl-1,3,2-dioxaphosphorinans with *t*-butylthiyl radicals involves retention of configuration at the phosphorus center, the same result as was obtained in the oxidation with *t*-butoxy radicals. Further evidence for a phosphoranyl radical intermediate in reaction 79 was provided by Walling, Basedow, and Savas[76] who obtained a small amount of toluene from the reaction of butanethiol with benzyl diethyl phosphite.

$$BuS^{\cdot} + PhCH_2OP(OEt)_2 \longrightarrow BuS{-}\overset{\displaystyle OEt}{\underset{\displaystyle OCH_2Ph}{\dot{P}}}{-}OEt \longrightarrow BuS\underset{\displaystyle \| \atop \displaystyle O}{P}(OEt)_2 + PhCH_2^{\cdot} \qquad (81)$$

In this case, scission of the $PhCH_2$–O bond is more favorable than loss of the butyl radical because of the stability of the benzyl radical which is formed. It seems likely that an α-scission of the intermediate phosphoranyl radical

[76] C. Walling, O. H. Basedow, and E. S. Savas, *ibid.*, **82,** 2181 (1960).
[77] R. D. Burkhart, *J. Phys. Chem.*, **70,** 605 (1966); *J. Amer. Chem. Soc.*, **90,** 273 (1968).

might occur in the reaction of a benzylphosphine with, for example, the methylthiyl radical leading to overall homolytic displacement.

$$\mathrm{MeS^{\cdot} + R_2PCH_2Ph \longrightarrow MeS\overset{R}{\underset{R}{P^{\cdot}}}CH_2Ph \longrightarrow MeSPR_2 + \dot{C}H_2Ph} \quad (82)$$

Substitution by Carbon-Centered Radicals

Phosphoranyl radicals may also be formed by addition of a carbon-centered radical to a tervalent phosphorus compound and, as before, α-scission or β-scission of the intermediate can subsequently occur. This reaction appears to be very rapid. For example, Bridger and Russell[78] obtained only a trace of chlorobenzene after decomposition of phenylazotriphenylmethane in a mixture of carbon tetrachloride and triphenylphosphine in the molar ratio 4.02:1. The oxidation of a series of alkyl diphenylphosphinites by phenyl radicals (from phenylazotriphenylmethane in benzene) gave very little concurrent phenylation of the solvent.[79]

$$\mathrm{Ph_2POR + Ph^{\cdot} \rightarrow Ph_3\dot{P}OR \rightarrow Ph_3PO + R^{\cdot}} \quad (83)$$

$$\mathrm{R = Me, Pr^{\mathit{i}}, C_6H_{11}CH_2, PhCH_2}$$

In addition to the phenyl radicals[79,80] methyl and isopropyl radicals would also bring about reaction 83 (for R = Me). However, benzyl radicals gave no oxidized phosphorus compounds and *t*-butyl radicals only served to initiate the chain reaction of methyl diphenylphosphinite with methyl radicals.[79]

slow:

$$\mathrm{Bu^{\mathit{t}\cdot} + MeOPPh_2 \rightarrow Bu^{\mathit{t}}\dot{P}Ph_2(OMe) \rightarrow Bu^{\mathit{t}}P(O)Ph_2 + Me^{\cdot}} \quad (84)$$

fast:

$$\mathrm{Me^{\cdot} + MeOPPh_2 \rightarrow Me\dot{P}Ph_2(OMe) \rightarrow MeP(O)Ph_2 + Me^{\cdot}} \quad (85)$$

A similar reaction can be brought about using dimethylamino radicals produced by photolysis of tetramethyltetrazene.[81] The failure of *t*-butyl radicals to oxidize tervalent phosphorus compounds had been noted previously.[43,80]

The addition of carbon-centered radicals to trivalent phosphorus appears to be reversible as well as rapid. Mayo, Durham, and Griggs[70] noted that

[78] R. F. Bridger and G. A. Russell, *ibid.*, **85**, 3754 (1963).
[79] R. S. Davidson, *Tetrahedron*, **25**, 3383 (1969).
[80] W. Bentrude, J.-J. L. Fu, and C. E. Griffin, *Tetrahedron Lett.*, 6033 (1968).
[81] R. S. Davidson, *ibid.*, 3029 (1968).

although phosphorus trichloride is half as reactive as oxygen towards cyclohexyl radicals, an efficient trap, in this case oxygen, for the phosphoranyl radical must be present.

$$R^{\cdot} + PCl_3 \rightleftharpoons R\dot{P}Cl_3 \tag{86}$$

$$R\dot{P}Cl_3 + O_2 \rightarrow RP(Cl_3)OO^{\cdot} \tag{87}$$

The reversal of reaction 86 should occur more readily than the loss of a chlorine atom from the phosphoranyl radical on account of the greater strength of the P–Cl bond compared with the P–C bond.[82] If loss of a chlorine atom did occur, the overall result would be homolytic substitution at phosphorus. Such a process has been suggested[83] to account for the direct alkylation of phosphorus trichloride by methane and ethane during short residence times at 575–600°. The reaction was catalyzed by oxygen and inhibited by propene indicating a chain process.

$$PCl_3 \rightarrow {}^{\cdot}PCl_2 + Cl^{\cdot} \tag{88}$$

$$Cl^{\cdot} + CH_4 \rightarrow CH_3^{\cdot} + HCl \tag{89}$$

$$CH_3^{\cdot} + PCl_3 \rightarrow CH_3PCl_2 + Cl^{\cdot} \tag{90}$$

A similar reaction can be achieved by the electron or γ-irradiation of mixtures of hydrocarbons and phosphorus trichloride.[84,85]

Homolytic substitution at phosphorus should be more likely in the reaction of alkyl radicals with phosphorus tribromide or phosphorus triiodide,[82] but there is also a possibility that halogen abstraction would occur, as has been suggested as a step in the reaction of olefins with phosphorus trichloride.[86]

$$\gt C{=}C\lt + {}^{\cdot}PCl_2 \longrightarrow Cl_2P\overset{|}{\underset{|}{C}}{-}C\lt^{\cdot} \tag{91}$$

$$Cl_2P\overset{|}{\underset{|}{C}}{-}C\lt^{\cdot} + PCl_3 \longrightarrow Cl_2P\overset{|}{\underset{|}{C}}{-}\overset{|}{\underset{|}{C}}{-}Cl + {}^{\cdot}PCl_2 \tag{92}$$

However, recent work on the olefin addition reaction favors the intermediacy of phosphoranyl radicals, with both $Cl_2P^{\cdot}$ and $Cl^{\cdot}$ as chain carriers.[87–89]

[82] $\bar{D}$(P–X) in PX_3 in kcal/mole[48]; $\bar{D}$(P–Cl) = 76; $\bar{D}$(P–Br) = 62 and $\bar{D}$(P–I) = 44.
[83] J. A. Pianfetti and L. D. Quin, *J. Amer. Chem. Soc.*, **84**, 851 (1962).
[84] A. Henglein, *Intern. J. Appl. Radiation and Isotopes*, **8**, 156 (1960).
[85] E. I. Babkina and I. V. Vereshinskii, *Zh. Obshch. Khim.*, **37**, 513 (1967); **38**, 1772 (1968).
[86] M. S. Kharasch, E. V. Jensen, and W. H. Urry, *J. Amer. Chem. Soc.*, **67**, 1864 (1945).
[87] L. A. Errede and W. A. Pearson, *ibid.*, **83**, 954 (1961).
[88] B. Fontal and H. Goldwhite, *Chem. Comm.*, 111 (1965).
[89] J. R. Little and P. F. Hartman, *J. Amer. Chem. Soc.*, **88**, 96 (1966).

$$R^{\cdot} + PCl_3 \longrightarrow R\dot{P}Cl_3 \begin{cases} \nearrow RCl + {}^{\cdot}PCl_2 \\ \searrow RPCl_2 + Cl^{\cdot} \end{cases} \tag{93}$$

Photolysis of iodo- and bromo-benzenes in the presence of trialkyl phosphites results in the formation of dialkyl phenylphosphonates in good yield.[90] It was suggested that this and similar reactions with trialkyl phosphites and triaryl phosphites involved the addition of phenyl radicals to give intermediate phosphoranyl radicals.[91]

$$PhI \xrightarrow{h\nu} Ph^{\cdot} + I^{\cdot} \tag{94}$$

$$Ph^{\cdot} + P(OR)_3 \longrightarrow (RO)_3\dot{P}Ph \xrightarrow[\text{or } I_2]{I^{\cdot}} [PhP(OR)_3]^+I^- \downarrow RI + PhP(O)(OR)_2 \tag{95}$$

Such mechanisms should be viewed with caution, however, since it appears that a nucleophilic photoreaction can occur between an electronically excited aryl halide and a variety of nucleophiles.[80,92] This nucleophilic photoreaction, rather than the alternative homolytic displacement of halogen by the phenyl radical, may account for the reaction of iodobenzene with phosphorus trihalides induced by UV irradiation.[93]

$$PhI + PX_3 \rightarrow PhPX_2 \quad X = Br, Cl \tag{96}$$

When triphenylphosphine in benzene-ethanol solvent is irradiated with UV light, ethyldiphenylphosphine is isolated as a major product.[94] It was thought to arise by the (thermodynamically unfavorable) homolytic substitution of a phenyl group by an ethyl radical.

$$Et^{\cdot} + PPh_3 \rightleftharpoons Et\dot{P}Ph_3 \rightleftharpoons EtPPh_2 + Ph^{\cdot} \tag{97}$$

The ethyl radical could arise from the interesting reaction of a diphenylphosphino radical (from the photolysis of triphenylphosphine) with ethanol.[95]

$$EtOH + {}^{\cdot}PPh_2 \longrightarrow Et\text{—}\overset{+}{O}(H)\text{—}\dot{\bar{P}}Ph_2 \longrightarrow Et\dot{O} + HPPh_2 \tag{98}$$

$$EtO^{\cdot} + PPh_3 \longrightarrow Et^{\cdot} + O{=}PPh_3 \tag{99}$$

[90] C. E. Griffin and J. B. Plumb, *J. Org. Chem.*, **27**, 4711 (1962).
[91] J. B. Plumb, R. Obrycki, and C. E. Griffin, *ibid.*, **31**, 2455 (1966).
[92] J. A. Barltrop, N. J. Bunce, and A. Thompson, *J. Chem. Soc.* (*C*), 1142 (1967).
[93] See footnote 14, Chapter IV.
[94] M. L. Kaufman and C. E. Griffin, *Tetrahedron Lett.*, 769 (1965).
[95] R. S. Davidson, R. A. Sheldon, and S. Tripett, *J. Chem. Soc.*, 722 (1966).

Polyhalogenomethanes, especially carbon tetrachloride and bromotrichloromethane, may react with tervalent phosphorus compounds by homolytic and heterolytic routes depending on the nature of the reactants and on the conditions.[39] Even when colligation of a polyhalomethyl radical and a tervalent phosphorus compound leads to a phosphoranyl radical, there appear to be few, if any, unambiguous examples of the α-scission of the latter to give an overall homolytic substitution at phosphorus.

$$Cl_3C^{\cdot} + PX_3 \rightarrow Cl_3C\dot{P}X_3 \rightarrow Cl_3CPX_2 + \dot{X} \qquad (100)$$

However, a free-radical mechanism may be at work in the reactions between triphenylphosphine or chlorodiphenylphosphine and trifluoroiodomethane which give diphenyltrifluoromethylphosphine,[96] Ph_2PCF_3.

The thermal alkylation of white phosphorus by alkyl and aryl halides has also been attributed to free-radical reactions.[97] Initial radical attack at the apices of the tetrahedron of the P_4 molecule was considered to cause P–P bond rupture and this may or may not involve a phosphoranyl intermediate. A correlation was found between the temperature at which the reaction would occur and the stability of the carbon radical involved.[97] Thus benzyl chloride reacted at a higher temperature (330°) than triphenylmethyl chloride (225°).

Organophosphorus polymers have been obtained after UV irradiation of solutions of white phosphorus in alkyl halides and halobenzenes.[98] White phosphorus and carbon tetrachloride react during 104 hr at 157° to give trichloromethylphosphorus dichloride in 28% yield. The reaction can be initiated at lower temperatures with visible light or ^{60}Co-radiation.[99] These reactions may involve carbon radical attack at phosphorus displacing a phosphorus centered radical. A similar S_H2 process may be involved in the addition of tetramethyldiphosphine to olefins,[100] and, more probably, in the reactions between tetraphenyldiphosphine[96] or tetraphenylcyclotetraphosphine,[101] $(PhP)_4$, and trifluoroiodomethane to give diphenyltrifluoromethylphosphine or phenyl-bis(trifluoromethyl)phosphine, respectively. These reactions, which were brought about thermally or by the action of UV light, probably involve attack of the trifluoromethyl radical at phosphorus with resulting P–P bond cleavage.

Trifluoromethyl radicals, from the decomposition of benzoyl peroxide in

[96] M. A. A. Beg and H. C. Clark, *Can. J. Chem.*, **40,** 283 (1962).
[97] K. A. Petrov, V. V. Smirnov, and V. I. Emel'yanov, *Zh. Obshch. Khim.*, **31,** 3027 (1961).
[98] M. Ya. Kraft and V. P. Parini, *Dokl. Akad. Nauk, SSSR.*, **77,** 57 (1951); *Sb. Statei Obshch. Khim., Akad. Nauk, SSSR*, **1,** 716 (1953).
[99] V. D. Perner and A. Henglein, *Z. Naturforsch*, **17B,** 703 (1962).
[100] A. B. Burg, *J. Amer. Chem. Soc.*, **83,** 2226 (1961).
[101] M. A. A. Beg and H. C. Clark, *Can. J. Chem.*, **39,** 564 (1961).

the presence of fluoroform, react with white phosphorus in the gas phase or in solution to form the cyclic tetraphosphine, **3**.[102]

$$\begin{array}{ccc} CF_3P & — & PCF_3 \\ | & & | \\ CF_3P & — & PCF_3 \end{array} \qquad \mathbf{3}$$

Similarly, the reaction of trifluoroiodomethane or of heptafluoropropyl iodide with white phosphorus at 200–220° is thought to involve radical intermediates.[103–105]

$$R^FI + P_4 \rightarrow R^F_nPI_{3-n} \qquad (101)$$

$$R^F = CF_3, \quad n = 0 - 3; \qquad R^F = C_3F_7, \quad n = 1 \text{ or } 2.$$

Homolytic displacement at P–P bonds may also occur in the reaction at 220° of tetrafluoroethylene and other fluorinated unsaturated compounds with red phosphorus in the presence of iodine.[106,107]

Substitution by Nitrogen-Centered Radicals

There appear to be no examples of homolytic substitution by nitrogen-centered radicals at a phosphorus atom. Davidson[81] has reported the oxidation of alkyl diphenylphosphinites by dimethylamino radicals generated by the photolysis of tetramethyltetrazene.

$$Me_2N^\cdot + Ph_2POR \rightarrow Ph_2\dot{P}(OR)NMe_2 \rightarrow Ph_2P(O)NMe_2 + R^\cdot \qquad (102)$$

$$R^\cdot + Ph_2POR \rightarrow Ph_2\dot{P}(OR)R \rightarrow Ph_2P(O)R + R^\cdot \qquad (103)$$

$$R = \text{Me, Et, cyclohexyl, } n\text{-octyl, crotyl, cinnamyl, bornyl.}$$

These reactions involve β-scission of the phosphoranyl intermediate. Photolysis of a mixture of tetramethyltetrazene and a trialkylphosphine in the cavity of an e.s.r. spectrometer did not yield the spectrum of the alkyl radical, indicating that reaction 104 must be slow in contrast to the corresponding displacement by *t*-butoxy radicals.[108]

$$Me_2N^\cdot + PR_3 \rightarrow Me_2N\dot{P}R_3 \rightarrow Me_2NPR_2 + R^\cdot \qquad (104)$$

$$R = Bu^n, Bu^s.$$

This result is in accord with the P–O and P–N bond strengths.[109]

[102] W. H. Watson, *Texas J. Sci.*, **11**, 471 (1950).

[103] F. W. Bennett, H. J. Emeléus, and R. N. Haszeldine, *J. Chem. Soc.*, 1565 (1953).

[104] H. J. Emeléus and J. D. Smith, *ibid.*, 375 (1959).

[105] H. J. Emeléus, *Angew. Chem. Intern. Edn. Eng.*, **1**, 129 (1962).

[106] C. G. Krespan, *J. Amer. Chem. Soc.*, **83**, 3432 (1961).

[107] C. G. Krespan and C. M. Langkammerer, *J. Org. Chem.*, **27**, 3584 (1962).

[108] See footnote 78, Chapter III.

[109] $\overline{D}$(P–X) in PX_3 in kcal/mole[48]; $\overline{D}(P\text{–}OMe) = 91$; $\overline{D}(P\text{–}NMe_2) = 70$.

The reaction of *N*-halo-amines and *N*-halo-amides with tervalent phosphorus compounds occurs by a polar mechanism.[110,111] This contrasts with the corresponding reaction with trialkylboranes which involves, at least in part, homolytic substitution at boron by the nitrogen-centered radical formed by halogen abstraction from the *N*-halo compound.

The reaction of nitric oxide with tervalent compounds of phosphorus leads to oxidation to the quinquevalent state and formation of nitrous oxide.[112–114] The mechanism has been pictured as an initial coordination of nitric oxide to phosphorus followed by reaction of the phosphoranyl adduct with a further molecule of nitric oxide. Whether a P–N bond[113] or a P–O bond[112] is formed initially, it is agreed that no cleavage of groups originally attached to phosphorus occurs from the intermediate, that is no homolytic substitution at phosphorus.

Substitution by Other Radicals

In 1937 Melville and Bolland[6] investigated the mercury-photosensitized exchange reaction of deuterium with phosphine and of hydrogen with trideuteriophosphine at pressures of 100–500 mmHg and temperatures from 20 to 600°. Exchange occurred by two mechanisms: 105 predominated at room temperature while 106 was the major process occurring at high temperatures.

$$\left.\begin{aligned} PH_3 + Hg^* &\rightarrow PH_2^{\cdot} + H^{\cdot} + Hg \\ D_2 + Hg^* &\rightarrow 2D^{\cdot} + Hg \\ D^{\cdot} + PH_2^{\cdot} &\rightarrow DPH_2 \end{aligned}\right\} \tag{105}$$

$$\left.\begin{aligned} D_2 + Hg^* &\rightarrow 2D^{\cdot} + Hg \\ D^{\cdot} + PH_3 &\rightarrow DPH_2 + H^{\cdot} \\ H^{\cdot} + D_2 &\rightarrow HD + D^{\cdot} \end{aligned}\right\} \tag{106}$$

The sequence shown in 106 contains a step involving homolytic displacement at phosphorus which might well proceed by way of an intermediate phosphoranyl radical, $D\dot{P}H_3$. A similar symmetrical atom exchange could occur whenever halogen atoms are generated in the presence of phosphorus trihalide.[115]

$$Cl^{\cdot} + PCl_3 \rightleftharpoons \dot{P}Cl_4 \tag{107}$$

[110] B. Miller, *Topics Phosphorus Chem.*, **2,** 133 (1965).
[111] R. E. Highsmith and H. H. Sisler, *Inorg. Chem.*, **8,** 1029 (1969) and earlier papers.
[112] See footnote 74, Chapter III.
[113] L. P. Kuhn, J. O. Doali, and C. Wellman, *J. Amer. Chem. Soc.*, **82,** 4792 (1960).
[114] M. Halman and L. Kugel, *J. Chem. Soc.*, 3272 (1962).
[115] D. P. Wyman, J. Y. C. Wang, and W. R. Freeman, *J. Org. Chem.*, **28,** 3173 (1963).

In the preceding sections the facility with which tervalent phosphorus compounds add on free radicals to give phosphoranyl intermediates has been noted. An analogous reaction does not seem to occur readily with quinquevalent phosphorus, and there appears to be no unambiguous example of homolytic displacement from phosphorus in its higher oxidation state.

$$Y^{\cdot} + PX_5 \rightarrow Y\dot{P}X_5 \rightarrow YPX_4 + X^{\cdot} \quad (108)$$

Chlorine abstraction occurs readily from phosphorus pentachloride which has been used to chlorinate hydrocarbons by a free-radical chain mechanism in which chlorine atoms appear to be the main hydrogen-abstracting species, reactions 114 and 115 playing only a minor role.[115,116]

$$PCl_5 \rightleftharpoons PCl_3 + Cl_2 \quad (109)$$

$$Cl_2 \rightleftharpoons 2Cl^{\cdot} \quad (110)$$

$$R^{\cdot} + PCl_5 \rightarrow RCl + {}^{\cdot}PCl_4 \quad (111)$$

$${}^{\cdot}PCl_4 \rightleftharpoons PCl_3 + Cl^{\cdot} \quad (112)$$

$$Cl^{\cdot} + RH \rightarrow HCl + R^{\cdot} \quad (113)$$

$${}^{\cdot}PCl_4 + RH \rightarrow HPCl_4 + R^{\cdot} \quad (114)$$

$$HPCl_4 \rightarrow HCl + PCl_3 \quad (115)$$

The rate of the isotopic exchange between PCl_5 and $^{36}Cl_2$ in the dark was not increased by conducting the reaction in ordinary laboratory light.[117] The exchange was thought to proceed by attack of molecular chlorine and the radical-chain mechanism shown in equations 116 and 117 does not appear to occur.

$$\dot{C}l^* + PCl_5 \rightarrow \overset{*}{C}l\dot{P}Cl_5 \rightarrow \overset{*}{C}lPCl_4 + Cl^{\cdot} \quad (116)$$

$$Cl^{\cdot} + Cl_2^* \rightarrow Cl{-}Cl^* + \dot{C}l^* \quad (117)$$

C. ARSENIC, ANTIMONY, AND BISMUTH

Few reactions of the compounds of arsenic, antimony, or bismuth have been shown to occur by way of a bimolecular homolytic substitution mechanism. Based on available evidence, several reactions of the compounds of these elements seem likely to proceed by homolytic routes, but so far these have not been the subject of detailed mechanistic studies.

[116] For example, E. C. Kooyman, and G. C. Vegter, *Tetrahedron*, **4**, 382 (1958).
[117] J. J. Downs and R. E. Johnson, *J. Amer. Chem. Soc.*, **77**, 2098 (1955).

Substitution by Oxygen-Centered Radicals

The ease of autoxidation, sometimes the spontaneous inflammability, of the trialkylarsines, trialkylstibines and trialkylbismuthines has been known for many years. Claims have been made that the oxidation of alkylarsines by air leads to alkylarsenic oxides of the type R_3AsO,[118] but in a number of cases it has since been found that this oxidation is accompanied by cleavage of the C–As bonds.[119,120] No detailed study of the autoxidation of the trialkylarsines has been made, but it seems very likely that the products and mechanism will be similar to those proposed for the autoxidation of trialkylphosphines.[44] Such a mechanism would involve attack of an alkylperoxy radical at the arsenic center, perhaps to give an intermediate analogous to a phosphoranyl radical, resulting in oxidation of the arsine to arsine oxide and the formation of an alkoxy radical.

$$ROO^{\cdot} + AsR_3 \rightarrow [ROO\dot{A}sR_3] \rightarrow RO^{\cdot} + O{=}AsR_3 \qquad (118)$$

The alkoxy radical would then attack the arsine to expel an alkyl radical or form the arsine oxide, reactions exactly analogous to those undergone by trialkylphospines.

$$RO^{\cdot} + AsR_3 \longrightarrow [RO\dot{A}sR_3] \xrightarrow{\alpha\text{-scission}} ROAsR_2 + R^{\cdot} \qquad (119a)$$

$$[RO\dot{A}sR_3] \xrightarrow{\beta\text{-scission}} R^{\cdot} + O{=}AsR_3 \qquad (119b)$$

The autoxidation of As–As bonded compounds, such as arsenobenzene,[121] $[(PhAs)_5]$, and tetramethyldiarsine (cacodyl), probably involve similar homolytic displacement reactions at arsenic.

The interaction of *t*-butoxy radicals, from the photolysis of di-*t*-butyl peroxide or the thermolysis of di-*t*-butyl hyponitrite, with triethylarsine, triethylstibine and triethylbismuthine has been investigated by carrying out the reactions in the cavity of an e.s.r spectrometer.[54,122] In all cases the only spectrum detected was that of the ethyl radical. Under the conditions of these

[118] A. E. Goddard, "Derivatives of Arsenic" Vol. XI, Pt. II in *Textbook of Inorganic Chemistry*, by J. N. Friend. C. Griffin, London, 1930.
[119] G. J. Burrows, *J. Proc. Roy. Soc. N.S. Wales*, **68,** 72 (1934).
[120] J. Dudonov and H. Medox, *Chem. Ber.*, **68,** 1254 (1935).
[121] F. F. Blicke and F. D. Smith, *J. Amer. Chem. Soc.*, **52,** 2946 (1930).
[122] See footnote 92, Chapter III.

experiments, reaction 120b was not occurring to an appreciable extent in competition with reaction 120a

$$Bu^tO^{\cdot} + MEt_3 \longrightarrow \underset{\mathbf{4}}{[Bu^tO\dot{M}Et_3]} \begin{cases} \nearrow Bu^tOMEt_2 + Et^{\cdot} & (120a) \\ \searrow Bu^{t\cdot} + OMEt_3 & (120b) \end{cases}$$

$$M = As, Sb, Bi.$$

The decrease in stability of the M=O bond in the quinquevalent oxides, OMR_3, on going down Group VA from phosphorus to bismuth should cause a steady increase in the rate of reaction 120a relative to that of 120b, indeed the latter process would not be expected at all for bismuth. The stability of the four-coordinate species **4** would also be expected to decrease on going from arsenic to bismuth and a switch from the S_H2 (stepwise) to the S_H2 (synchronous) mechanism should occur on proceeding down the Group.

The autoxidation of the trialkylstibines and trialkylbismuthines has been studied in more detail than that of the corresponding arsines. Tripropylstibine oxide and the compound Pr_3SbO, Sb_2O_3 were isolated from the autoxidation of neat tripropylstibine.[123] Even tribenzylstibine, which might be expected to undergo ready homolytic cleavage of the C–Sb bond, was reported[124] to yield the oxide, $(PhCH_2)_3SbO$. The gas-phase autoxidation of trimethylstibine and triethylstibine was studied briefly by Bamford and Newitt[125] who found that the latter compound, which oxidized much more rapidly than trimethylstibine, gave small quantities of acetaldehyde but no gaseous products condensible in liquid air. The autoxidation of tributylbismuthine has also been shown to yield small quantities of aldehyde.[126] However, Callingaert, Soroos, and Hnizda[127] have reported that acetaldehyde is not a product of the oxidation of triethylbismuth by oxygen. These authors found diethyl peroxide, ethanol, diethyl ether and ethylene as the organic products of the autoxidation at -25 to $+25°$. When oxygen was allowed to react with an excess of triethylbismuthine at -60 to $-50°$, a clear viscous liquid was formed without evolution of gas. When this liquid was allowed to warm, a vigorous reaction ensued with the liberation of heat and gas and the separation of a solid. It was concluded that a compound Et_3BiO_2 was formed initially. This then reacted with triethylbismuthine to

[123] W. J. Dyke and W. J. Jones, *J. Chem. Soc.*, 1921 (1930).
[124] I. P. Tsukervanik and D. Smirnov, *Zhur. Obshch. Khim.*, **7,** 1527 (1937).
[125] C. H. Bamford and D. M. Newitt, *J. Chem. Soc.*, 695 (1946).
[126] H. Gilman, H. L. Yablunky, and A. C. Svigoon, *J. Amer. Chem. Soc.*, **61,** 1170 (1939).
[127] G. Callingaert, H. Soroos, and H. Hnizda, *ibid.*, **64,** 392 (1942).

give diethylethoxybismuthine. The further oxidation of Et_2BiOEt gave rise to all the liquid products by the decomposition of a compound $Et_2Bi(O_2)OEt$ into free radicals.[127]

The homolytic chain nature of the autoxidation of the trialkyls of antimony and bismuth has been unequivocally demonstrated by the inhibition of the reaction by trace amounts of free-radical scavengers such as galvinoxyl,[128] phenothiazine,[128] 2,2,6,6-tetramethylpiperid-4-one-1-oxyl[128] and diphenylamine.[129] In this way, the autoxidation of the following compounds has been shown to proceed by a free-radical chain mechanism[128]: Me_3Sb, Et_3Sb, Me_2SbPh, Et_2SbPh, Me_3Bi, and Et_3Bi. Thus it appears that the autoxidation of the trialkyls of Group VA elements from phosphorus down to bismuth proceeds by a similar homolytic mechanism which involves the interaction of alkylperoxy and alkoxy radicals at the metal or metalloid atom. Alkoxy radicals will bring about displacement of alkyl radicals from the organometallic compounds, whilst alkylperoxy radicals generally cause oxidation of the Group VA element to its quinquevalent state. An exception may be found in the case of the reaction of an alkylperoxy radical with a trialkylbismuthine where homolytic substitution probably occurs to give a dialkylalkylperoxy bismuthine which undergoes subsequent decomposition by O–O homolysis.[128]

$$ROO^{\cdot} + BiR_3 \rightarrow ROOBiR_2 + R^{\cdot} \qquad (121)$$

The triaryl derivatives of arsenic, antimony and bismuth, like those of phosphorus, are much more stable towards autoxidation than the corresponding trialkyl compounds. The inhibiting effect which triphenylstibine and triphenylbismuthine have on the autoxidation of styrene and of other organic compounds[130] seems worthy of further investigation.

Bis(trifluoromethyl)nitroxide has been shown to undergo a variety of hydrogen abstracting and free-radical scavenging reactions. For example, it will react with toluene to yield *O*-benzyl bis(trifluoromethyl)hydroxylamine.[131]

$$(CF_3)_2NO^{\cdot} + PhCH_3 \rightarrow (CF_3)_2NOH + Ph\dot{C}H_2 \qquad (122)$$

$$(CF_3)_2NO^{\cdot} + Ph\dot{C}H_2 \rightarrow (CF_3)_2NOCH_2Ph \qquad (123)$$

Tris(trifluoromethyl)arsine undergoes a stepwise replacement of trifluoromethyl groups when caused to react with bis(trifluoromethyl)nitroxide, the first two groups being replaced at room temperature whilst the third is cleaved at 70°.[132]

[128] A. G. Davies and S. C. W. Hook, unpublished results.
[129] R. V. Winchester, M.Sc. thesis, University of Auckland, New Zealand (1966).
[130] C. Moureau, C. Dufraisse, and M. Dadoche, *Compt. Rend.*, **187,** 1092 (1928).
[131] H. G. Ang, *Chem. Comm.*, 1320 (1968).
[132] H. G. Ang and K. F. Ho, *J. Organometal. Chem.*, **19,** P19 (1969).

$$2(CF_3)_2NO^{\cdot} + As(CF_3)_3 \xrightarrow[\text{temp.}]{\text{room}} (CF_3)_2NOAs(CF_3)_2 + (CF_3)_2NOCF_3$$

$$(CF_3)_2NOAs(CF_3)_2 \xrightarrow[\text{room temp.}]{2(CF_3)_2NO\cdot} [(CF_3)_2NO]_2AsCF_3 + (CF_3)_2NOCF_3 \quad (124)$$

$$[(CF_3)_2NO]_2AsCF_3 \xrightarrow[70^\circ]{2(CF_3)_2NO\cdot} (CF_3)_2NOCF_3 + [(CF_3)_2NO]_3As$$

The problem of mechanism was not discussed, but the initial step is presumably addition of one nitroxide molecule to the arsine.

$$(CF_3)_2NO^{\cdot} + As(CF_3)_3 \rightarrow \underset{\mathbf{5}}{[(CF_3)_2NO\dot{A}s(CF_3)_3]} \quad (125)$$

The important question is whether **5** undergoes α-scission to give a trifluoromethyl radical, reaction 126, or whether it reacts with a second molecule of nitroxide to give an arsonium salt **6** which then loses *O*-trifluoromethyl bis(trifluoromethyl)hydroxylamine, reaction 129.

$$[(CF_3)_2NO\dot{A}s(CF_3)_3] \rightarrow (CF_3)_2NOAs(CF_3)_2 + \dot{C}F_3 \quad (126)$$

$$(CF_3)_2N\dot{O} + \dot{C}F_3 \rightarrow (CF_3)_2NOCF_3 \quad (127)$$

$$[(CF_3)_2NO\dot{A}s(CF_3)_3] + (CF_3)_2N\dot{O} \rightarrow \underset{\mathbf{6}}{[(CF_3)_2NOAs(CF_3)_3]^+\bar{O}N(CF_3)_2} \quad (128)$$

$$[(CF_3)_2NOAs(CF_3)_3]^+\bar{O}N(CF_3)_2 \rightarrow (CF_3)_2NOAs(CF_3)_2 + (CF_3)_2NOCF_3 \quad (129)$$

A decision between two such alternative mechanisms, one involving an arsonium intermediate the other involving overall homolytic substitution at arsenic, is often required in the reactions of Group VA compounds in general. Seldom is the experimental evidence compelling although the "onium" intermediate is normally preferred. The reaction of bis(trifluoromethyl)nitroxide presents a rather special case, however, as it is itself a free radical and thus the formation of the radical species **5** is almost certain. The overall mechanism will be determined by the competition of reactions 126 and 128 and which process predominates in this case is not known. In general, the mechanism of the reaction of bis(trifluoromethyl)nitroxide with tervalent compounds of Group VA elements will depend critically upon the particular element involved and the groups attached to it. Related to the above reaction is the formation of a small quantity of trifluoronitrosomethane when tris(trifluoromethyl)arsine and nitric oxide are heated together at 70°.[133]

[133] W. R. Cullen, *Can. J. Chem.*, **41**, 317 (1963).

Substitution by Carbon-Centered Radicals

As part of his classic work with free radicals, Paneth studied the interaction of methyl and ethyl radicals with mirrors of arsenic, antimony, and bismuth.[134] With a cold arsenic mirror, tetraalkyldiarsine was the major product together with some trialkylarsine, whereas with a heated mirror pentaalkylpentaarsine $[(RAs)_5]$ predominated. This indicates that the diarsine was formed from $R_2As^{\cdot}$ radicals while the pentaarsine originated from an RAs: species. The formation of these products can be looked upon as occurring by homolytic substitution at the As–As bonds in elemental arsenic. This interpretation becomes of little value when applied to the formation of the alkyls of antimony and bismuth by attack of methyl and ethyl radicals on these more metallic elements. It seems that methylene radicals also react with an arsenic mirror, though in this case the products have not been identified.[135]

In 1862, Cahours showed that heating methyl iodide with elemental arsenic gave tetramethylarsonium iodide as the major product.[136] Heating *amorphous* arsenic with methyl iodide at 100° gave mainly diiodomethylarsine and the same form of the element gave arsenic trichloride and hexachloroethane with carbon tetrachloride at 160°.[137]

$$\frac{2}{n}As_n + 6CCl_4 \xrightarrow{160^\circ} 3C_2Cl_6 + 2AsCl_3 \quad (130)$$

Much later it was shown that many other alkyl halides would react in a similar way to methyl iodide and that the reaction could be catalyzed by copper.[138] Clearly the reaction could involve alkyl radical attack on elemental arsenic.

The interaction of alkyl halides with trialkylarsines or trialkylstibines leads to the formation of tetraalkylarsonium or tetraalkylstibonium halides, respectively. However when triisobutylbismuthine was heated with methyl iodide or ethyl iodide at 150°, no quaternary iodide was obtained but instead, trimethylbismuthine and triethylbismuthine.[139] This reaction probably occurs by way of an unstable tetralkylbismuthonium iodide, but a free-radical process of short chain length is also possible.

$$Et^{\cdot} + BiBu^i_3 \rightarrow EtBiBu^i_2 + Bu^{i\cdot} \quad (131)$$

$$Bu^{i\cdot} + EtI \rightarrow Bu^iI + Et^{\cdot} \quad (132)$$

[134] F. A. Paneth and H. Loliet, *J. Chem. Soc.*, 366 (1935)
[135] F. O. Rice and A. L. Glasebrook, *J. Amer. Chem. Soc.*, **56**, 2381 (1934).
[136] A. Cahours, *Ann.*, **122**, 192 (1862).
[137] V. Auger, *Comp. Rend.*, **145**, 808 (1907).
[138] L. Maier, E. G. Rochow, and W. C. Fernelius, *Inorg. Chem.*, **16**, 213 (1961).
[139] A. Marquardt, *Chem. Ber.*, **21**, 2035 (1888).

The cleavage by alkyl halides of the As–As bond in tetralkyldiarsines[140–142] might also in certain circumstances (e.g., at high temperatures) occur in part by a radical chain process.

The reaction of perfluoroalkyl iodides, in particular trifluoroiodomethane, with elemental arsenic, antimony, and bismuth or with the alkyls and perfluoroalkyls of these elements has been studied extensively, although elucidation of the reaction mechanism has seldom been an objective.[143,144] Attack on arsenic by thermally produced trifluoromethyl radicals and iodine atoms was proposed to account for compounds of the type $(CF_3)_nAsI_{3-n}$ ($n = 1$–3) obtained from trifluoroiodomethane and elemental arsenic at 235°.[145] Trifluoromethyl radicals (from the pyrolysis of hexafluoroacetone) gave a compound thought to be tris(trifluoromethyl)bismuthine when allowed to react with a bismuth mirror.[146]

Trifluoroiodomethane reacts with trimethylphosphine, trimethylarsine, and trimethylstibine at room temperature to give the quaternary tetramethylonium iodides and dimethyl(trifluoromethyl)phosphine, dimethyl(trifluoromethyl)arsine, or dimethyl(trifluoromethyl)stibine, respectively.[147]

$$CF_3I + 2Me_3M \rightarrow CF_3MMe_2 + [Me_4M]^+I^- \qquad (133)$$

$$M = P, As, Sb.$$

A polar mechanism was suggested involving an intermediate quaternary trimethyl(trifluoromethyl) iodide which could either breakdown or react further.

$$CF_3I + MMe_3 \longrightarrow [CF_3MMe_3]^+I^- \xrightarrow{MMe_3} Me_4M^+I^- + Me_2MCF_3 \qquad (134a)$$

$$[CF_3MMe_3]^+I^- \longrightarrow MeI + Me_2MCF_3 \qquad (134b)$$

$$MeI \xrightarrow{Me_3M} [Me_4M]^+I^-$$

$$M = P, As, Sb.$$

140 A. Cahours and A. Rieche, *Ann.*, **92,** 361 (1854).
141 W. Steinkopf and G. Schwen, *Chem. Ber.*, **54,** 1437 (1921).
142 W. Steinkopf, S. Schmidt, and P. Smie, *ibid.*, **59,** 1463 (1926).
143 R. E. Banks and R. N. Haszeldine, *Adv. Inorg. Chem. Radiochem.*, **3,** 337 (1961).
144 W. R. Cullen, *Adv. Organometal. Chem.*, **4,** 145 (1966).
145 E. G. Walaschewski, *Chem. Ber.*, **86,** 272 (1953).
146 T. N. Bell, B. J. Pullman, and B. O. West, *Australian J. Chem.*, **16,** 722 (1963).
147 R. N. Haszeldine and B. O. West, *J. Chem. Soc.*, 3631 (1956).

Later, it was found that free ethyl iodide remained at the end of the reaction of trifluoroiodomethane with triethylstibine thus supporting the mechanism shown in equation 134b.[148] Trifluoroiodomethane reacts in a similar manner with trimethylbismuthine, although now no quaternary iodide is formed.[149]

$$Me_3Bi + CF_3I \xrightarrow{100^\circ} Me_2BiCF_3 + MeI \tag{135}$$

Cullen[144,150] has suggested that the cleavage of the M–C bond (M = P, As, Sb, Bi) by perfluoroalkyl iodides may occur by way of a homolytic mechanism and not via a quaternary intermediate particularly when high temperatures and less basic organometallic compounds (e.g., Me_2AsI) are involved.

$$Me_2AsI + CF_3I \rightarrow MeAs(I)CF_3 + MeI \tag{136}$$

When a mixture of tris(trifluoromethyl)arsine and methyl iodide was exposed to UV light at room temperature, products of the type $(CF_3)_nMe_{3-n}As$ were produced (mainly $n = 2$) as well as some trifluoroiodomethane and hexafluoroethane.[151] In the absence of irradiation no reaction occurred at 140°, but heating at 235° for 24 hours brings about a reaction similar to that initiated by UV light. A homolytic mechanism for this reaction was discounted because of the absence of methane and 1,1,1-trifluoroethane, expected to be formed from free methyl radicals in the system. Instead an unstable arsonium intermediate was proposed.

$$(CF_3)_3As + MeI \xrightarrow[\text{or } \Delta]{h\nu} \left[(CF_3)_3As \begin{smallmatrix} \cdot\cdot\, Me \\ \cdot\cdot\, I \end{smallmatrix} \right] \longrightarrow (CF_3)_2AsMe + CF_3I \tag{137}$$

However, if tris(trifluoromethyl)arsine is a good trap for methyl radicals the absence of methane and trifluoroethane is not surprising and the alternative homolytic mechanism is a definite possibility.

$$MeI \xrightarrow{h\nu} Me^\cdot + I^\cdot \tag{138}$$

$$Me^\cdot + As(CF_3)_3 \rightarrow [Me\dot{A}s(CF_3)_3] \tag{139}$$

$$[Me\dot{A}s(CF_3)_3] \rightarrow MeAs(CF_3)_2 + \dot{C}F_3 \tag{140}$$

$$\dot{C}F_3 + I^\cdot \text{ (or } I_2) \rightarrow CF_3I \tag{141}$$

[148] B. J. Pullman and B. O. West, *Australian J. Chem.*, **17,** 30 (1964).
[149] T. N. Bell, B. J. Pullman, and B. O. West, *Proc. Chem. Soc.*, 224 (1962).
[150] W. R. Cullen, *Can. J. Chem.*, **40,** 426 (1962).
[151] H. J. Emeléus, R. N. Haszeldine, and E. G. Walaschewski, *J. Chem. Soc.*, 1552 (1953).

The existence of a four-coordinate intermediate radical $[Me\dot{A}s(CF_3)_3]$ would be quite likely in the case of the arsine reactions.[152]

The As–As bond in tetramethyldiarsine is readily cleaved by trifluoroiodomethane at room temperature,[153] a process which calls to mind the similar cleavage of the Sn–Sn bond in hexamethylditin (see Chapter V). The As–As bonds in arsenobenzene undergo a similar cleavage when heated with an excess of trifluoroiodomethane at 115° or at a lower temperature when the reactants are exposed to UV light.[154] The products are phenylbis(trifluoromethyl) arsine, trifluoromethyl(iodo)phenylarsine, and phenyldiiodoarsine.

$$(PhAs)_6 + 6CF_3I \rightarrow 2PhAs(CF_3)_2 + 2PhAs(CF_3)I + 2PhAsI_2 \quad (142)$$

Tetramethyldiarsine does not react at once with tetrafluoroethylene at 20°, but when the two are irradiated with UV light reaction does occur to give mainly the 1:4 adduct.[154]

$$Me_2AsAsMe_2 + 4CF_2{=}CF_2 \rightarrow Me_2As(C_2F_4)_4AsMe_2 \quad (143)$$

Finally, tetramethyldiarsine reacts with hexafluorobutyne to give approximately equal amounts of the *cis*-butene and *trans*-butene derivative.[154]

$$Me_2AsAsMe_2 + CF_3C{\equiv}CCF_3 \rightarrow Me_2As(CF_3)C{=}C(CF_3)AsMe_2 \quad (144)$$

Clearly there is a possibility of radical mechanisms being involved in these reactions and S_H2 displacement by perfluoroalkyl radicals at the As–As bonds in the compounds discussed above would be required.

The interaction of aryldiazonium salts with alkali metal arsenites or arsenic trichloride, often in the presence of cuprous salts or copper catalysts, comprise the Bart and Scheller reactions respectively, for the formation of arylcarbon-arsenic bonds.[144] Compounds containing phenyl groups bonded to arsenic or antimony can be formed by the decomposition of benzenediazonium chloride by calcium carbonate in the presence of the respective element.[155] These reactions may involve free aryl radicals and in certain cases homolytic substitution processes might take place.

It has been suggested that useful amounts of dichlorocyclohexylarsine can be obtained when a solution of arsenic trichloride in cyclohexane is irradiated with 3.3 MeV electrons from a Van der Graff generator.[84] In this system cyclohexyl radicals will be generated and part of the product may result from a homolytic substitution reaction at arsenic.

$$cyclo\text{-}C_6H_{11}^{\cdot} + AsCl_3 \rightarrow cyclo\text{-}C_6H_{11}AsCl_2 + Cl^{\cdot} \quad (145)$$

[152] N. J. Friswell and B. G. Gowenlock, *Adv. in Free Radical Chem.*, **2,** 1 (1967).
[153] W. R. Cullen, *Can. J. Chem.*, **38,** 439 (1960).
[154] W. R. Cullen and N. K. Hota, *ibid.*, **42,** 1123 (1964).
[155] W. A. Waters, *J. Chem. Soc.*, 864 (1939).

Huff and Perry[156] have determined the chain transfer constants for a number of organometallic compounds in the radical polymerization of styrene (100°), methyl methacrylate (60°), and acrylonitrile (60°). The chain transfer constants evaluated for tributylstibine were: styrene (5.8×10^{-3}), methylmethacrylate ($<1 \times 10^{-4}$), and acrylonitrile (11.1). The significance of these results is not clear.

Substitution by Sulfur-Centered Radicals

The cleavage of all three Bi–C bonds in triethylbismuthine by hydrogen sulphide was reported as early as 1854.[157]

$$2Et_3Bi + 3H_2S \rightarrow Bi_2S_3 + 6EtH \quad (146)$$

Many years later, Gilman and Nelson[157a,157b] investigated the reaction of a number of organic compounds containing weakly acidic hydrogen atoms with a series of organometallic compounds with a view to using analytically the volume of gaseous alkane (RH) liberated to estimate the organometallic compound.

$$R_nM + mHX \rightarrow R_{n-m}MX_m + mRH \quad (147)$$

$$M = K, Na, Li, Ca, Mg, Zn, Al, Cd, B, Pb, Hg, Bi, Sn, Si.$$

In general, Gilman and Nelson found that there was a broad correlation between the acidity of the active hydrogen compound studied and the ease with which cleavage of the metal–carbon bond took place. However, the reactivity of organolead, organobismuth, and organomercury compounds toward –SH and –SeH[158] groups was anomalous since these organometallic compounds were cleaved more readily by thiols and selenols than by the more acidic carboxylic acids. When a triarylbismuthine was heated with *one* molar equivalent of benzenethiol at 100° or at reflux in chloroform, benzene or toluene solvent, *two* C–Bi bonds were cleaved; the unreacted triarylbismuthine was recovered and no Ar_2BiSPh or $(PhS)_3Bi$ was detected.[159,160]

$$2PhSH + Ar_3Bi \rightarrow ArBi(SPh)_2 + 2ArH \quad (148)$$

$$Ar = Ph, p\text{-}MeC_6H_4, p\text{-}ClC_6H_4$$

At a higher temperature,[160,161] for example in refluxing xylene,[160] triarylthiobismuthines could be obtained, although tri-1-naphthylbismuthine was not

156 See footnote 94, Chapter III.
157 F. J. Dünhaupt, *Ann.*, **92,** 371 (1854).
157a See footnote 96, Chapter III.
157b See footnote 97, Chapter III.
158 M. S. Kharasch, *U.S. Patent*, 1,858,958 (1932); *Chem. Abstr.*, **26,** 3806 (1932).
159 H. Gilman and H. L. Yale, *Chem. Rev.*, **30,** 281 (1942).
160 H. Gilman and H. L. Yale, *J. Amer. Chem. Soc.*, **73,** 2880 (1951).
161 See footnote 168, Chapter III.

cleaved at all by benzenethiol even in boiling xylene. The difficulty of cleavage of a 1-naphthyl group from bismuth was also shown by the selective reaction of 1-naphthyldiphenylbismuthine with benzenethiol in refluxing chloroform to give 1-naphthylbis(phenylthiyl)bismuthine.[160]

$$1\text{-NpBiPh}_2 + 2\text{PhSH} \rightarrow 1\text{-NpBi(SPh)}_2 + 2\text{PhH} \qquad (149)$$

In contrast, it was reported[160] that triphenylstibine was not cleaved by benzenethiol after 80 min at 100°.

An explanation of the "anomalous" behavior observed in the reactions of –SH and –SeH containing compounds with organo-antimony and -bismuth compounds has been recently provided by Davies and Hook.[162] These authors showed that the compounds R_3M (M = Sb, Bi) react with benzenethiol to give phenylthio compounds $R_nM(SPh)_{3-n}$ and that the reactions could be initiated by di-*t*-butyl hyponitrite, and sometimes inhibited by galvinoxyl, phenothiazine or 2,2,6,6-tetramethylpiperid-4-one-1-oxyl, known free-radical scavengers. It was concluded that the cleavage of the M–C bond took place by a homolytic chain mechanism, where the S–M bond is formed by bimolecular homolytic substitution by the phenylthiyl radical at the metal center.

$$\text{PhS}^{\cdot} + \text{MR}_3 \rightarrow \text{PhSMR}_2 + \text{R}^{\cdot} \qquad (150)$$

$$\text{R}^{\cdot} + \text{PhSH} \rightarrow \text{RH} + \text{PhS}^{\cdot} \qquad (151)$$

$$\text{M} = \text{Sb, Bi}; \qquad \text{R} = \text{Me, Et, Ph.}$$

A second and sometimes a third M–C bond was cleaved by a homolytic displacement reaction analogous to 150. The results are summarized in Table I.

Table I Products from the Reaction between Benzenethiol and Organoantimony and Organobismuth Compounds in Benzene Solution[162]

Reactant	Reaction Conditions		Products
	Temperature (°C)	Time (hr)	
Me_3Sb	50	6	$MeSb(SPh)_2$
Et_3Sb	35	168	$Sb(SPh)_3$
Ph_3Sb	50	15	Ph_2SbSPh
Me_3Bi	35	24	$MeBi(SPh)_2$
Et_3Bi	35	16	$EtBi(SPh)_2$
Ph_3Bi	35	5	$PhBi(SPh)_2$

[162] A. G. Davies and S. C. W. Hook, *J. Chem. Soc.* (*B*), 735 (1970).

Reactions 150 and 151 form the propagating steps of a free-radical chain process, for which, in the absence of an added initiator, the initiating radicals may be formed by oxidation of the benzenethiol or of the organometallic compound by adventitious traces of oxygen. Lack of chain initiation or the presence of an unrecognized inhibitor probably account for the unreactivity of triphenylstibine towards benzenethiol reported by Gilman and Yale.[160]

Alkylthiyl radicals interact with tervalent phosphorus compounds to give an intermediate phosphoranyl radical which usually undergoes subsequent β-scission of the C–S bond.

$$RS^{\cdot} + PX_3 \rightarrow RS\dot{P}X_3 \quad (77)$$

$$RS\dot{P}X_3 \xrightarrow[\beta\text{-scission}]{} R^{\cdot} + S{=}PX_3 \quad (78b)$$

Whether or not a similar intermediate is involved in the reactions discussed above is a moot point. However, if one is, then the α-scission products which are observed are in accordance with the increased strength of the M–S bond (M = Sb, Bi) and the decreased M–C bond energies relative to the corresponding quantities for phosphorus compounds.

$$RS^{\cdot} + MR'_3 \longrightarrow RS\dot{M}R'_3 \xrightarrow{\beta\text{-scission}} R^{\cdot} + SMR'_3 \quad (152a)$$

$$RS\dot{M}R'_3 \xrightarrow{\alpha\text{-scission}} RSMR'_2 + R'^{\cdot} \quad (152b)$$

The reaction between a trialkylarsine and a thiol does not appear to have been reported and this reaction would be of interest as it might reveal competition between 152a and 152b (M = As). There is no evidence that the reaction between a disulphide and a tetraalkyldiarsine[163,164] does not proceed by a heterolytic mechanism, particularly at low temperatures (20°), and it should be noted that trivalent phosphorus compounds have been shown to react with disulphides by both polar and free-radical processes.

In 1923 it was shown[165] that thiocyanogen oxidizes triphenylarsine and triphenylstibine to pentavalent derivatives whilst with triphenylbismuthine thiocyanogen brings about Bi–C cleavage to give diphenylthiocyanatobismuthine (Ph_2BiSCN). More recent work has shown that the addition of

[163] W. R. Cullen, P. S. Dhaliwal, and W. B. Fox, *Inorg. Chem.*, **3**, 1332 (1964).
[164] W. R. Cullen and P. S. Dhaliwal, *Can. J. Chem.*, **45**, 379 (1967).
[165] See footnote 173, Chapter III.

thiocyanogen to tervalent organic arsines or organic stibines leads to M-N (M = As, Sb) bonded products.[166]

$$R_3M + (SCN)_2 \rightarrow R_3M(NCS)_2 \quad (153)$$

$$M = As, Sb; \quad R = Me, Ph.$$

However, it was confirmed that triphenylbismuthine and thiocyanogen gave Ph_2BiSCN with Bi-C cleavage.[166]

$$Ph_3Bi + (SCN)_2 \rightarrow Ph_2BiSCN + PhSCN \quad (154)$$

The difference in bonding of the SCN group to the metal in the products of reactions 153 and 154 could be accounted for by assuming different mechanisms for the addition and substitution processes. The addition probably takes place by a heterolytic mechanism similar to that usually assumed for the addition of halogens to tervalent phosphorus compounds. On the other hand, a substitution 154 might occur by a free-radical chain mechanism analogous to that established for the interaction of thiols and triphenylbismuthine (see above).

$$Ph^{\cdot} + (SCN)_2 \rightarrow PhSCN + S\dot{C}N \quad (155)$$

$$S\dot{C}N + Ph_3Bi \rightarrow Ph_2BiSCN + Ph^{\cdot} \quad (156)$$

Substitution by Nitrogen-Centered Radicals

The only example of homolytic substitution by a nitrogen-centered free radical at the metal center in a compound of arsenic, antimony or bismuth appears to be the displacement of an ethyl radical from triethylstibine by the dimethylamino radical. When a toluene solution of triethylstibine and tetramethyltetrazene was irradiated with UV light at $-81°$ in the cavity of an e.s.r. spectrometer, the spectrum of the ethyl radical was observed.[167]

$$Me_2NN{=}NNMe_2 \xrightarrow{h\nu} 2Me_2N^{\cdot} + N_2 \quad (156)$$

$$Me_2N^{\cdot} + SbEt_3 \rightarrow Me_2NSbEt_2 + Et^{\cdot} \quad (157)$$

Substitution by Halogen Atoms

Trialkylarsines and triarylarsines react readily with halogens, usually to form the pentavalent derivative by addition rather than to bring about As-C cleavage.

$$R_3As + X_2 \rightarrow R_3AsX_2 \quad (158)$$

[166] See footnote 101, Chapter III.
[167] See footnote 51, Chapter IV.

A similar reaction occurs between the antimony analogues and halogens or cyanogen halides; even triallylantimony and tribenzylantimony, which have weak C–Sb bonds, yielding the dibromides when treated with bromine.[124,168] These reactions and the cleavage of the As–As bond in tetraalkyldiarsines are probably polar processes, but the alternative radical chain addition, which could occur in favorable circumstances, does not appear to have been eliminated by experiment.

$$X^{\cdot} + MR_3 \rightarrow X\dot{M}R_3 \tag{159}$$

$$R_3\dot{M}X + X_2 \rightarrow R_3MX_2 + X^{\cdot} \tag{160}$$

$$M = As, Sb.$$

Halogens react with pentamethyl-arsenic and -antimony to bring about C–As or C–Sb cleavage, but the mechanism is again probably heterolytic.[169] In contrast to the alkyls of arsenic and antimony, those of bismuth react vigorously with halogens by substitution with C–Bi bond cleavage.[159]

$$Me_3Bi + Br_2 \rightarrow Me_2BiBr + MeBr \tag{161}$$

This difference could simply be due to the instability of the pentavalent dihalides in the case of bismuth, or homolytic substitution at bismuth by halogen atoms might occur, perhaps via an intermediate as in reaction 159.

$$Br^{\cdot} + BiMe_3 \rightarrow BrBiMe_2 + Me^{\cdot} \tag{162}$$

An interesting reaction of pentavalent antimony has been observed by Razuvaev and co-workers.[170] If tetraphenyl-*t*-butylperoxyantimony in chloroform or carbon tetrachloride solution is irradiated with UV light, tetraphenylstibonium chloride and triphenylstibine oxide are formed.

$$Ph_4SbOOBu^t \xrightarrow{h\nu} \begin{cases} \xrightarrow{CCl_4} Ph_4Sb^+Cl^-(76\%) + Ph_3SbO(20\%) \\ \xrightarrow{CHCl_3} Ph_4Sb^+Cl^-(61\%) + Ph_3SbO(36\%) \end{cases} \tag{163}$$

The tetraphenylstibonium chloride was thought to arise from reactions

[168] A. E. Borisov, N. V. Novikova, and A. N. Nesmeyanov, *Izv. Akad. Nauk SSSR, Ser. Khim.*, 1566 (1963).

[169] G. Wittig and K. Torssell, *Acta Chem. Scand.*, **7**, 1293 (1953).

[170] G. A. Razuvaev, T. L. Finov'eva, and T. G. Brilkina, *Izv. Akad. Nauk SSSR, Ser. Khim.*, 2007 (1969); *Dokl. Akad. Nauk SSSR*, **193**, 353 (1970).

164–166, the dichlorocarbene produced in reaction 165 was identified by trapping with cyclohexene.

$$Cl_3C^{\cdot} + Ph_4SbOOBu^t \rightarrow [Ph_4SbCCl_3] + {}^{\cdot}OOBu^t \quad (164)$$

$$[Ph_4SbCCl_3] \rightarrow Ph_4Sb^+Cl^- + :CCl_2 \quad (165)$$

$$Cl^{\cdot} + Ph_4SbOOBu^t \rightarrow Ph_4Sb^+Cl^- + {}^{\cdot}OOBu^t \quad (166)$$

It would appear that reactions 164 and 166 represent the first examples of homolytic substitution at quinquevalent antimony, by a carbon radical and a chlorine atom, respectively.

VII S_H2 Reactions at Group VIA Elements

A. OXYGEN

Although a host of S_H2 reactions at oxygen have been identified, they are almost entirely confined to processes which lead to the scission of the O–O bond of peroxides. There are two reasons for this. In the first place, the long-standing and widespread use of peroxides as initiators of free-radical processes has led to innumerable studies of the reactions between free radicals and peroxides and hence, to the identification of many S_H2 reactions on peroxides. Secondly, peroxides are useful free-radical chain initiators because the O–O bond is rather weak and this means it can be cleaved by an attacking free-radical much more readily than the strong O–C or O–H bonds which are encountered in most non-peroxidic oxygen-containing compounds.[1]

The following discussion of S_H2 reactions at oxygen has been subdivided into eight sections: (a) aroyl peroxides, (b) acyl peroxides, (c) acyl aroyl peroxides, (d) peresters, (e) peracids, (f) alkyl peroxides, (g) hydroperoxides, and (h) nonperoxidic compounds.

(a) AROYL PEROXIDES

The thermal decomposition of benzoyl peroxide is a rather complex reaction in most solvents.[4] In general, decomposition roughly follows a first-order

[1] For example,[2,3] $D(CH_3O–OCH_3) = 37$ kcal/mole; $D(CH_3C(O)O–OC(O)CH_3) = 30$ kcal/mole; $D(CH_3O–CH_3) = 81$ kcal/mole; $D(CH_3–OH) = 91$ kcal/mole; and $D(CH_3O–H) = 104$ kcal/mole.

[2] See footnote 46, Chapter V.

[3] S. W. Benson and R. Shaw, *Adv. Chem. Series*, **75**, 288 (1968).

[4] Ch. 10 in footnote 11, Chapter III.

rate law but the calculated unimolecular rate constants increase with concentration.[5–12] The results are consistent with the simultaneous occurrence of two decomposition reactions, one of the first order,

$$\mathrm{Ph\overset{\overset{O}{\|}}{C}OO\overset{\overset{O}{\|}}{C}Ph} \longrightarrow \mathrm{2Ph\overset{\overset{O}{\|}}{C}O^{\cdot}} \qquad (1)$$

and one of higher order, as was first suggested by Brown[5] in 1940. The higher order process represents an "induced" decomposition of the peroxide. The relative importance of the induced decomposition increases with decreasing temperature.

In 1946, both Bartlett and Nozaki[7,8] and Cass[9] concluded that the inducing agent must be a free radical since the rate can be increased by added sources of radicals and can be decreased by oxygen and by typical free-radical inhibitors such as quinones and polynitroaromatics.

$$\mathrm{R^{\cdot}} + \mathrm{Ph\overset{\overset{O}{\|}}{C}OO\overset{\overset{O}{\|}}{C}Ph} \longrightarrow \mathrm{Ph\overset{\overset{O}{\|}}{C}OR} + \mathrm{Ph\overset{\overset{O}{\|}}{C}O^{\cdot}} \qquad (2)$$

This conclusion has been amply verified by subsequent work.[4]

The free radical $R^{\cdot}$ which induces the decomposition of the peroxide has its unpaired electron centered on a carbon atom whether it is derived from the peroxide itself or from the solvent. In the former case, $R^{\cdot}$ is a phenyl-radical which is formed by the decarboxylation of a benzoyloxy radical,

$$\mathrm{Ph\overset{\overset{O}{\|}}{C}O^{\cdot}} \longrightarrow \mathrm{Ph^{\cdot}} + \mathrm{CO_2} \qquad (3)$$

In the latter case, $R^{\cdot}$ may be produced by a terminal S_H2 process, as must be the case in solvents such as CCl_4 or cyclohexane,

$$\mathrm{Ph^{\cdot}} + \mathrm{CCl_4} \longrightarrow \mathrm{PhCl} + \mathrm{CCl_3^{\cdot}} \qquad (4)$$

$$\mathrm{Ph^{\cdot}(Ph\overset{\overset{O}{\|}}{C}O^{\cdot})} + \mathrm{C_6H_{12}} \longrightarrow \mathrm{Ph(Ph\overset{\overset{O}{\|}}{C}O)H} + \mathrm{C_6H_{11}^{\cdot}} \qquad (5)$$

[5] D. J. Brown, *J. Amer. Chem. Soc.*, **62,** 2657 (1940).
[6] P. D. Bartlett and R. Altschul, *ibid.*, **67,** 816 (1945).
[7] P. D. Bartlett and K. Nozaki, *ibid.*, **68,** 1495 (1946).
[8] K. Nozaki and P. D. Bartlett, *ibid.*, **68,** 1686 (1946).
[9] W. E. Cass, *ibid.*, **68,** 1976 (1946).
[10] B. Barnett and W. E. Vaughan, *J. Phys. Colloid Chem.*, **51,** 926, 942 (1947).
[11] W. E. Cass, *J. Amer. Chem. Soc.*, **69,** 500 (1947).
[12] P. D. Bartlett and K. Nozaki, *ibid.*, **69,** 2299 (1947).

Alternatively, in polymerizable monomers such as styrene or maleic anhydride the radical R· will be produced by addition of the primary radical to the double bond, for example,

$$\mathrm{Ph\overset{\overset{\displaystyle O}{\|}}{C}O\cdot + PhCH{=}CH_2 \longrightarrow Ph\overset{\overset{\displaystyle O}{\|}}{C}OCH_2\dot{C}HPh} \qquad (6)$$

If we make the simplest possible assumptions regarding the mechanism, then the overall rate of benzoyl peroxide decomposition is given by,[8]

$$\frac{-d[(\mathrm{Ph\overset{\overset{\displaystyle O}{\|}}{C}O})_2]}{dt} = k_1[(\mathrm{Ph\overset{\overset{\displaystyle O}{\|}}{C}O})_2] + k_i[(\mathrm{Ph\overset{\overset{\displaystyle O}{\|}}{C}O})_2]^{3/2} \qquad (7)$$

where $k_i = k_2(k_1/k_8)^{1/2}$ and k_8 is the rate constant for the reaction

$$\mathrm{R}^{\cdot} + \mathrm{R}^{\cdot} \rightarrow \text{nonradical products} \qquad (8)$$

Nozaki and Bartlett[8] have listed values of k_1 and k_i for benzoyl peroxide at 80° in a variety of solvents. Their data is reproduced in Table I together with values of k_2 calculated on the assumption that $k_8 = 2 \times 10^9\ M^{-1}$ sec^{-1} which is true of most simple alkyl radicals in solvents of normal viscosity.[13,14] The derived values of k_2 are all in the range $3 \times 10^2 - 3 \times 10^3\ M^{-1}$ sec^{-1} which indicates that the S_H2 reaction on benzoyl peroxide by simple alkyl radicals is quite fast. It must, however, be emphasized that these rate constants have no more than qualitative significance. In the first place, the kinetic assumptions which are required to separate k_1 and k_i involve a gross oversimplification of the real situation and, moreover, in non-polymerizable solvents reaction 2 probably involves a variety of radicals. The kinetics indicate that k_1 is itself subject to solvent effects, a result which has been confirmed by other workers who used inhibitors to eliminate the induced decomposition.[15–18] There is an appreciable variation in different workers' estimates of k_1 even in the same solvents. More recent values[15,19,20] of k_1 are lower than those which were reported by Nozaki and Bartlett and which are given in Table I.

A much more serious criticism of the k_2 values listed in Table I from the present point of view, is that they may not represent S_H2 reactions at oxygen.

[13] See footnote 18, Chapter III.

[14] D. J. Carlsson, K. U. Ingold, and L. C. Bray, *Int. J. Chem. Kin.*, **1,** 315 (1969).

[15] G. S. Hammond, *J. Amer. Chem. Soc.*, **72,** 3737 (1950).

[16] G. S. Hammond and L. M. Soffer, *ibid.*, **72,** 4711 (1950).

[17] C. G. Swain, W. H. Stockmayer, and J. T. Clarke, *ibid.*, **72,** 5426 (1950).

[18] C. E. H. Bawn and S. F. Mellish, *Trans. Faraday Soc.*, **47,** 1216 (1951).

[19] G. A. Russell, *J. Amer. Chem. Soc.*, **78,** 1044 (1956).

[20] K. F. O'Driscoll and P. J. White, *J. Polymer Sci.*, **3A,** 283 (1965).

Table I The Decomposition of 0.2 M Benzoyl Peroxide at 80° [8]

Solvent	$k_1 \times 10^5$ (sec^{-1})	$k_i \times 10^5$ ($M^{-1/2}$ sec^{-1})	$k_2 \times 10^{-2}$ (M^{-1} sec^{-1})[a]	% Decomposition in 1 hr
Cyclohexene	1.9	4.6	4.6	14.0
Carbon tetrachloride	2.1	3.7	3.7	13.5
Benzene	3.3	4.3	3.3	15.5
Toluene	3.3	4.3	3.3	17.4
Nitrobenzene	3.3	4.3	3.3	15.5
t-Butylbenzene	3.3	15	12	28.5
Ethyl iodide	4.0	6.8	4.8	23.4
Cyclohexane	6.4	33	18	51.0
Ethyl acetate	9.0	37	17	53.5
Acetic acid	8.1	51	25	59.3
Acetic anhydride	7.5	31	16	48.5

a. Calculated on the assumption that $k_8 = 2 \times 10^9\ M^{-1}\ \text{sec}^{-1}$.

In 1960 Walling and Savas[21] pointed out the significance of the already known fact that the induced decomposition of benzoyl peroxide in aliphatic hydrocarbons, in CCl_4, and in benzene at high concentrations gave appreciable yields of *ortho*-substituted and *para*-substituted benzoic acids. This led them to propose that at least one path for the induced chain is radical addition to the aromatic system with a concerted cleavage of the O–O bond and the formation of an α-lactone structure, for example

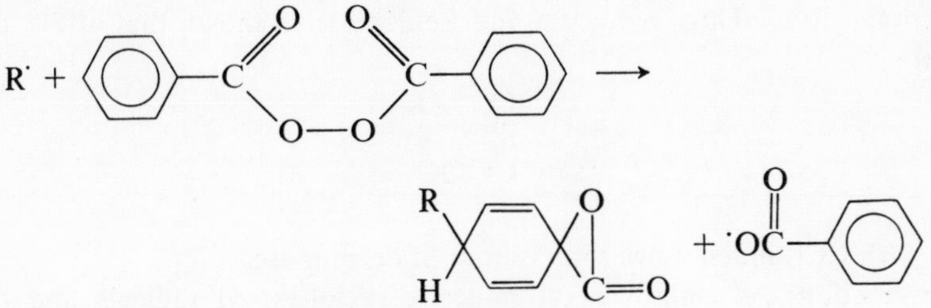

$$R^\cdot + C_6H_5C(=O)O{-}OC(=O)C_6H_5 \rightarrow \text{R(H)C}_6\text{H}_5\text{-spiro-}\alpha\text{-lactone} + {}^\cdot OC(=O)C_6H_5 \quad (9)$$

which rearranges to the observed product.

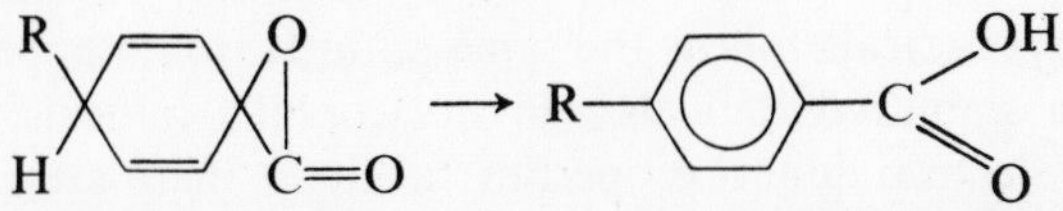

$$\text{R(H)C}_6\text{H}_5\text{-spiro-}\alpha\text{-lactone} \rightarrow R{-}C_6H_4{-}C(=O)OH \quad (10)$$

In addition to the formation of *ortho*-substituted and *para*-substituted benzoic acids, the principal argument for this scheme was the high reactivity of

[21] C. Walling and E. S. Savas, *J. Amer. Chem. Soc.*, **82**, 1738 (1960).

benzoyl peroxide towards attack by polystyryl radicals compared with other aromatic compounds such as, for example, benzoic anhydride. However, this argument is not very compelling because diacyl peroxides are at least as reactive towards polymeric radicals as benzoyl peroxide (see Table II below) which implies that the high reactivity of aroyl and of acyl peroxides is principally a function of the

$$-\overset{\overset{\displaystyle O}{\|}}{C}O O\overset{\overset{\displaystyle O}{\|}}{C}-$$

group.

More recently, Walling and Čekovič have re-examined this question by studying the induced decomposition of neat acetyl benzoyl peroxide.[22] Major products of the induced process (up to 20% each) are *ortho*-toluic and *para*-toluic acids. Acetyl 4-chlorobenzoyl peroxide and acetyl 2,6-dichlorobenzoyl peroxide similarly yield 4-chloro-*o*-toluic acid and 2,6-dichloro-*p*-toluic acid, respectively. These results certainly support a reaction mechanism in which the methyl radical attacks the aromatic ring with a concerted cleavage of the peroxide link. The α-lactone can, apparently, also eliminate CO_2 since toluene (or the appropriate chlorotoluenes) are significant products (~20%) of the reaction. In this connection, it is worth

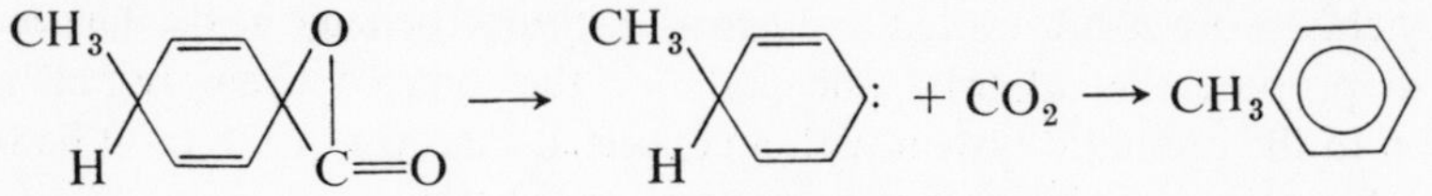

$$\text{(CH}_3\text{)(H)C}_6\text{H}_4\text{(O)C=O} \longrightarrow \text{(CH}_3\text{)(H)C}_6\text{H}_4\!: + CO_2 \longrightarrow CH_3C_6H_5 \quad (11)$$

noting that the formation of a σ radical from the peroxide does *not* induce its decomposition. Thus, Schwartz and Leffler[23] have shown that at 80° the radical

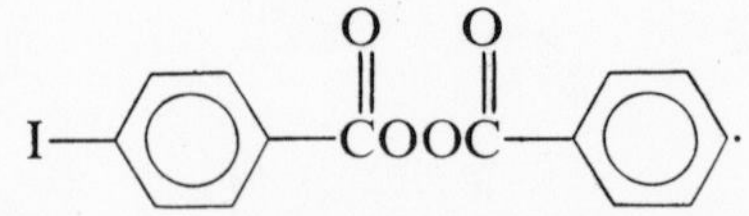

$$I-C_6H_4-\overset{\overset{\displaystyle O}{\|}}{C}OO\overset{\overset{\displaystyle O}{\|}}{C}-C_6H_4\cdot$$

reacts with CCl_4 much more rapidly than it decomposes.

The reactions of α-alkoxyalkyl radicals, cyclohexenyl radicals and triphenylmethyl radicals with benzoyl peroxide are undoubtedly S_H2 reactions at the peroxidic oxygen (see below). It is also likely that α-hydroxyalkyl radicals and acyl radicals follow the same path. It would appear, therefore, that resonance stabilized radicals and nucleophilic radicals (i.e., radicals whose strong electron-donor properties facilitate their attack on electronegative peroxidic oxygen[21,24]) enter into S_H2 reactions at oxygen. However,

[22] C. Walling and Z. Čekovič, *ibid.*, **89,** 6681 (1967).
[23] M. M. Schwartz and J. E. Leffler, *ibid.*, **90,** 1368 (1968).
[24] F. R. Mayo and C. Walling, *Chem. Rev.*, **46,** 269 (1950).

product studies[21,22,25,25a] in solvents such as CCl_4, cyclohexane and acetic acid indicate that the attack of nonstabilized radicals and of electrophilic radicals occurs predominantly, if not exclusively, at the *ortho*- and *para*-positions of the aromatic ring. This means that many of the k_2 values listed in Table II overestimate the importance of the S_H2 process indicated in reaction 2.

DeTar[26] has used a mathematical model for the decomposition of benzoyl peroxide in benzene at 80° to assist in the estimation of the rate constants of the more important reactions that are involved. The induced decomposition reactions for which he estimated rate constants are given below.

	k (M^{-1} sec^{-1})	
$C_6H_5(H)^\cdot$ (phenylcyclohexadienyl radical) + PhC(O)OOC(O)Ph → 1-Ph-4-(OC(O)Ph)-cyclohexa-2,5-diene (H, H) + PhC(O)O$^\cdot$	1	(12)
phenylcyclohexadienyl radical + PhC(O)OOC(O)Ph → Ph—Ph + PhC(O)OH + PhC(O)O$^\cdot$	10	(13)
Ph$^\cdot$ + PhC(O)OOC(O)Ph → PhOC(O)Ph + PhC(O)O$^\cdot$	10	(14)
Ph$^\cdot$ + PhC(O)OOC(O)Ph → Ph—C$_6$H$_4$—C(O)OH + PhC(O)O$^\cdot$	500	(15)
cyclohexenyl$^\cdot$ + PhC(O)OOC(O)Ph → cyclohexenyl—OC(O)Ph + PhC(O)O$^\cdot$	100	(16)
cyclohexenyl$^\cdot$ + PhC(O)OOC(O)Ph → cyclohexadiene + PhC(O)OH + PhC(O)O$^\cdot$	4000	(17)

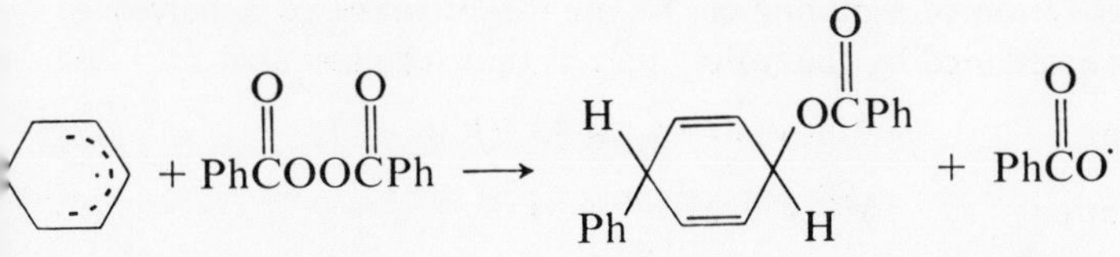

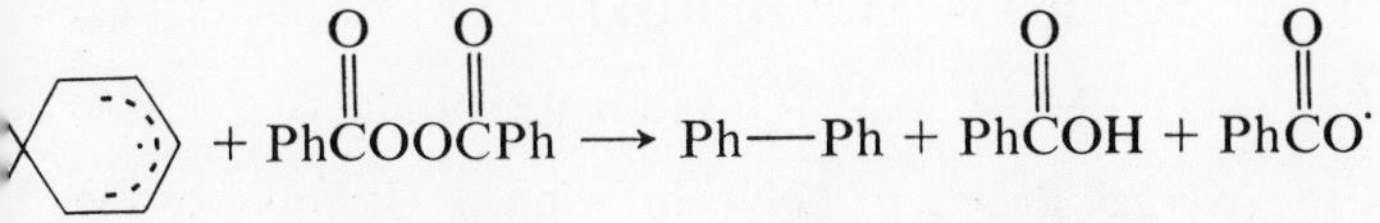

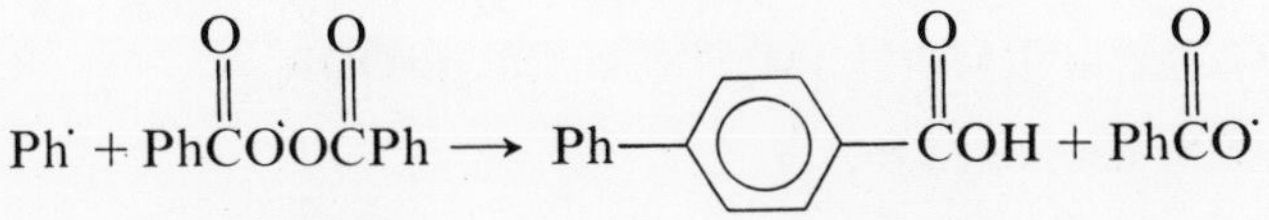

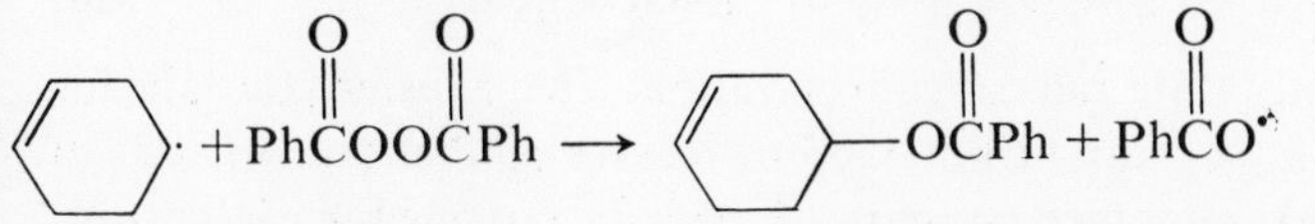

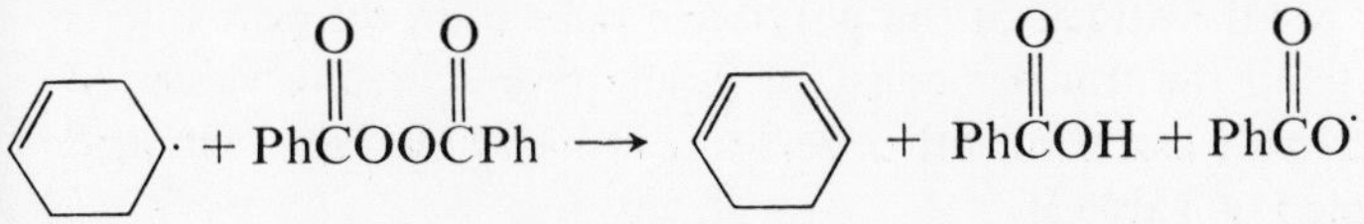

[25] See A. V. Tobolsky and R. B. Mesrobian, *Organic Peroxides*, Interscience, New York, 1954, p. 85, for a summary and pertinent references.

[25a] For a recent product and kinetic study of bis(α-naphthoyl) peroxide and α-naphthoyl benzoyl peroxide see, J. E. Leffler and R. G. Zepp, *J. Amer. Chem. Soc.*, **92**, 3713 (1970).

[26] D. F. DeTar, *ibid.*, **89**, 4058 (1969).

By analogy with the behavior of cyclohexenyl and triphenylmethyl radicals (see below) it seems probable that the resonance stabilized polymeric radicals present in polymerizing vinyl monomers such as styrene, methyl methacrylate and vinyl acetate, will attack the peroxidic oxygen in preference to the aromatic ring. The copolymerization behavior of these three monomers indicates that the corresponding radicals are slightly nucleophilic which also suggests that they will attack the peroxidic oxygen. In these monomers, the rate constant for reaction 2 can be obtained by a competitive procedure which is likely to be much more accurate than rate constants estimated by a kinetic separation of k_1 and k_i. In a polymerizing monomer the peroxide can act both as an initiator and as a chain transfer agent. The overall mechanism of polymerization of a monomer, M, in the presence of a peroxide, ROOR, can be represented by the following reaction scheme.

$$\text{Initiation:} \quad \text{Initiator} \longrightarrow \text{M}^{\cdot} \ (\text{Rate} = \text{R}_i) \tag{18}$$

$$\text{Propagation:} \quad \text{M}^{\cdot}_{n-1} + \text{M} \xrightarrow{k_p} \text{M}^{\cdot}_{n} \tag{19}$$

$$\text{Transfer:} \quad \text{M}^{\cdot}_{n} + \text{ROOR} \xrightarrow{k_{tr}} \text{M}_n\text{OR} + \text{RO}^{\cdot} \tag{20}$$

$$\text{RO}^{\cdot} + \text{M} \xrightarrow{\text{fast}} \text{ROM}^{\cdot} \tag{21}$$

$$\text{Termination:} \quad \text{M}^{\cdot}_{n} + \text{M}^{\cdot}_{m} \xrightarrow{2k_t} \text{molecular products} \tag{22}$$

In such a system, the average degree of polymerization, $\overline{\text{P}}$, is equal to the number of monomers incorporated into the average polymer molecule and this is equal to the ratio of the rate of propagation to the sum of the rates of all processes which prevent the further growth of a polymer molecule,[27] that is

$$\overline{\text{P}} = k_p[\text{M}^{\cdot}_n][\text{M}]/(2k_t[\text{M}^{\cdot}_n]^2 + k_{\text{tr}}[\text{M}^{\cdot}_n][\text{ROOR}]) \tag{23}$$

This equation can be simplified to,

$$1/\overline{\text{P}} - (k_t R_i)^{1/2}/k_p[\text{M}] = \text{C}[\text{ROOR}]/[\text{M}] \tag{24}$$

by assuming a steady state radical concentration. The transfer constant for the peroxide, $\text{C} = k_{tr}/k_p$, can be obtained by measuring $\overline{\text{P}}$ at known rates of chain initiation and monomer concentrations.[27–31] Provided k_p is known, the rate constant for the attack of the polymeric radical on the peroxide is readily obtained from the transfer constant. Some representative values of k_{tr} for benzoyl peroxide and some other peroxides which have been obtained in this way are listed in Table II.

[27] Ch. 4 in footnote 11, Chapter III.
[28] F. R. Mayo, *J. Amer. Chem. Soc.*, **65,** 2324 (1943); **75,** 6133 (1953).
[29] R. A. Gregg and F. R. Mayo, *Disc. Faraday Soc.*, **2,** 328 (1947).
[30] R. A. Gregg and F. R. Mayo, *J. Amer. Chem. Soc.*, **70,** 2373 (1948).
[31] G. Henrici-Olivé and S. Olivé, *Fortshritte Hochpolymeren Forschung*, **2,** 496 (1960–1961).

Table II Rate Constants for Free-Radical Attack on Some Aroyl and Acyl Peroxides[a]

Peroxide	Temperature (°C)	Radical: Polystyryl[b]	Radical: Poly (Methyl Methacrylate)	Radical: Poly (Vinyl Acetate)
Benzoyl	60	3.9[c]	14	210
2-Methylbenzoyl	70	13	—	—
2-Chlorobenzoyl	70	145	570[d]	400[d]
2-Bromobenzoyl	70	165	—	8200[d]
3-Bromobenzoyl	70	35	—	1400[d]
Octanoyl	70	8	—	—
Hexadecanoyl	70	11	110[d]	400[d]
Cinnamoyl	70	84	—	—

a. Calculated from the chain transfer constants listed in Table 18 of footnote 31. Propagation rate constants were calculated from the activation parameters for chain propagation (E_p and PZ) given in footnote 11 (Chapter III). p. 95. Rate constants are given in M^{-1} sec^{-1} units.
b. O'Driscoll and White[20] have reported k_i values for several substituted benzoyl peroxides in styrene at 90°. Taking k_t for polystyryl radicals $= 4 \times 10^6$ M^{-1} sec^{-1} at this temperature gives $k_{tr} \sim 10$ M^{-1} sec^{-1} for benzoyl peroxide, $\sim$200 M^{-1} sec^{-1} for 3-bromo, 4-chloro-, 3,4-dichloro-, and 4-cyano-benzoyl peroxides and $\sim$2000 M^{-1} sec^{-1} for 4-nitrobenzoyl peroxide.
c. At 70°, $k_{tr} = 5.7$ M^{-1} sec^{-1}.
d. At 60°.

The value of k_{tr} increases with increasing electronegativity of the substituents attached to the aromatic ring and, in fact, the data for *meta* and *para* substituted benzoyl peroxides (footnote b in Table II) can be roughly correlated with the σ constants of the substituents by means of the Hammett equation.[20] The rate constants for the two saturated aliphatic diacyl peroxides are not very different from those of benzoyl peroxide but the value for cinnamoyl peroxide, $(PhCH{=}CHCO_2)_2$, is significantly larger. This must be due to the attack of the polymeric radicals on the double bond since Muramoto et al.[32] have recently shown that induced decomposition of this acyl peroxide can be initiated in toluene at temperatures as low as 35° if *t*-butyl peroxyoxalate is added to the system to provide a source of free radicals to start the reaction. A similar induced decomposition can be initiated in diphenylpropioloyl peroxide, $(PhC{\equiv}CCO_2)_2$, and in the *t*-butyl peresters of both acids. However, the decomposition of diacetyl peroxide and dibenzoyl peroxide was not initiated under the same experimental conditions. The products formed in the decompositions of the unsaturated peroxides support

[32] N. Muramoto, T. Ochiai, O. Simamura, and M. Yoshida, *Chem. Comm.*, 717 (1968).

the idea of a radical attack at the multiple bond. The reaction sequence for cinnamoyl peroxide can be represented as,

$$Bu^tO^\cdot + PhCH_3 \longrightarrow Bu^tOH + PhCH_2^\cdot \tag{25}$$

$$PhCH_2^\cdot + (PhCH{=}CHC(=O)O)_2 \longrightarrow \left[\begin{array}{c} Ph\dot{C}HCHC(=O)OOC(=O)CH{=}CHPh \\ | \\ CH_2Ph \end{array}\right] \tag{26}$$

$$\left[\begin{array}{c} Ph\dot{C}HCHC(=O)OOC(=O)CH{=}CHPh \\ | \\ CH_2Ph \end{array}\right] \longrightarrow PhCH{=}CHCH_2Ph + CO_2 + PhCH{=}CHC(=O)O^\cdot \tag{27}$$

$$PhCH{=}CHC(=O)O^\cdot + PhCH_3 \longrightarrow PhCH{=}CHC(=O)OH + PhCH_2^\cdot \tag{28}$$

The 1,3-diphenylpropene was formed in 60–90% yield based on the starting peroxide.

The free-radical induced decomposition of benzoyl peroxide is very much faster in dialkyl ethers and in primary and secondary alcohols than in any of the solvents listed in Table I.[8–12] Thus a 0.2 *M* solution of the peroxide at 80° has a half life of less than 5 minutes in diethyl ether, ethanol and isopropanol, and less than 15 minutes in diisopropyl ether, di-*n*-butyl ether and methanol.[12] These values may be contrasted with half-lives of 4 hr or over for the first five solvents listed in Table I. The decomposition in ethers is retarded by oxygen, by monomers (methyl methacrylate and styrene) and by many typical free-radical inhibitors (quinone, hydroquinone, nitrobenzene, etc.) and the reaction must therefore be a free-radical chain process. In the presence of very efficient inhibitors (stilbene, acrylonitrile, 3,4-dichlorostyrene, styrene, 1,4-diphenylbutadiene, iodine, methyl methacrylate, and trinitrobenzene[17]) the decomposition rate in dioxane is reduced to a constant value which presumably corresponds to the unimolecular decomposition, reaction 1, unaffected by any induced decomposition.[17] This unimolecular decomposition rate constant has a value similar to that found in inactive solvents[17] and therefore there is not a rapid initiation process between the peroxide and the ether. The kinetics of the decomposition are not the same from one ether to another, presumably because of changes in the relative importance of the various possible chain terminating steps. Thus, the order in peroxide at 30° has been reported[9] to be 1.5 in diethyl ether and ethylene glycol diethyl ether and 2.0 in dioxan, while at 80° the reaction is close to first order in di-*n*-butyl ether.[12] The products of the reaction at ~40° for the first three ethers have been determined by Cass.[11] In diethyl ether the products (in mole/mole of peroxide decomposed) are CO_2, 0.20–0.25; benzoic acid, 0.8 and 1-ethoxyethyl benzoate, **1,** 0.84–0.95. The propagation steps in the chain are therefore,

$$\text{Ph}\overset{\overset{\text{O}}{\|}}{\text{C}}\text{O}^{\cdot} + \text{CH}_3\text{CH}_2\text{OCH}_2\text{CH}_3 \longrightarrow \text{Ph}\overset{\overset{\text{O}}{\|}}{\text{C}}\text{OH} + \text{CH}_3\dot{\text{C}}\text{HOCH}_2\text{CH}_3 \qquad (29)$$

$$\text{CH}_3\dot{\text{C}}\text{HOCH}_2\text{CH}_3 + \text{Ph}\overset{\overset{\text{O}}{\|}}{\text{C}}\text{OO}\overset{\overset{\text{O}}{\|}}{\text{C}}\text{Ph} \longrightarrow \underset{\mathbf{1}}{\text{Ph}\overset{\overset{\text{O}}{\|}}{\text{C}}\text{O}\underset{\underset{\text{OCH}_2\text{CH}_3}{|}}{\text{C}}\text{HCH}_3} + \text{Ph}\overset{\overset{\text{O}}{\|}}{\text{C}}\text{O}^{\cdot} \qquad (30)$$

The CO_2 which is formed may arise from decarboxylation of the benzoyloxy radicals, reaction 3, or perhaps reaction 30 gives some phenyl radicals and CO_2 directly. The product composition from ethylene glycol diethyl ether is similar but in this case the acylol product of reaction 30 consists of 1-(2-ethoxyethoxy)-ethyl benzoate, **2,** and 1,2-diethoxyethyl benzoate, **3,** in about a 3:1 ratio.

$$\underset{\mathbf{2}}{\text{Ph}\overset{\overset{\text{O}}{\|}}{\text{C}}\text{O}\overset{\overset{\text{CH}_3}{|}}{\text{C}}\text{HOCH}_2\text{CH}_2\text{OCH}_2\text{CH}_3} \qquad\qquad \underset{\mathbf{3}}{\text{Ph}\overset{\overset{\text{O}}{\|}}{\text{C}}\text{O}\overset{\overset{\text{OCH}_2\text{CH}_3}{|}}{\text{C}}\text{HCH}_2\text{OCH}_2\text{CH}_3}$$

Similar products were obtained with dioxan but the yield of the acylol was lower than in the other two ethers. Rate constants for reaction 30 and for the analogous reactions with the other α-alkoxyalkyl radicals have not been determined because of the complex nature of the kinetics. It seems likely that they will be larger than for simple unsubstituted alkyl radicals.

The bimolecular homolytic substitution, reaction 30, might in principal occur by attack of the radical on either the peroxide oxygen or on the carbonyl oxygen with a subsequent β-scission of the O–O bond to give **1**. Although the latter possibility looks attractive, reaction 30 actually involves predominant, and probably exclusive, attack on the peroxide oxygen. This was proved independently by Drew and Martin[33] and by Denney and Feig[34] in 1959. Both pairs of workers labeled benzoyl peroxide with ^{18}O in the carbonyl positions and allowed it to decompose in boiling diethyl ether. Reduction of the 1-ethoxyethyl benzoate with lithium aluminum hydride gave benzyl alcohol with 71%[33] and 80%[34] of the excess ^{18}O originally incorporated into

[33] E. H. Drew and J. C. Martin, *Chem. Ind. (London)*, 925 (1959).
[34] D. B. Denny and G. Feig, *J. Amer. Chem. Soc.*, **81**, 5322 (1959).

the carbonyl position of the peroxide. Since unlabeled **1** incorporates ^{18}O from ^{18}O labeled benzoic acid these figures indicate only the minimum percentage of attack on the peroxidic oxygen. It is likely that all attack actually occurs at his position.

$$CH_3\dot{C}HOCH_2CH_3 + PhC(=^{18}O)OOC(=^{18}O)Ph \longrightarrow PhC(=^{18}O)OCH(CH_3)OCH_2CH_3 + PhC(\cdots{}^{18}O)(\cdots O^{18})\cdot \quad (31)$$

$$PhC(=^{18}O)OCH(CH_3)OCH_2CH_3 \xrightarrow{LiAlH_4} PhCH_2{}^{18}OH$$

The conclusion that α-ethoxyethyl radicals attack benzoyl peroxide predominantly, and probably exclusively, at the peroxidic oxygen has been confirmed by Doering, Okamoto, and Krauch.[35] These workers obtained a similar result with cyclohexenyl radicals, while with the triphenylmethyl radical they were able to show that attack occurs exclusively on the peroxidic oxygen. It was suggested that the site of attack is determined by the fact that there are fewer large changes in bond angles and distances and in π-electronic interactions for attack at the peroxidic oxygen than for attack on the carbonyl oxygen.

Suehiro et al.[36] have reported the rate constants and activation parameters for the reaction of triphenylmethyl radicals with symmetrically substituted benzoyl peroxides which are given in Table III. The reported variations in ΔH and ΔS have probably little or no significance as they appear to be well within the limits of the experimental errors. The reaction is accelerated by electron withdrawing substituents which indicates that dipolar structures make an important contribution to the transition state.

$$Ph_3C\cdot + PhC(=O)OOC(=O)Ph \longrightarrow \left[\overset{\delta+}{Ph_3C}\cdots\cdots OC(=O)Ph \atop |\ OC(=\overset{\delta-}{O})Ph\right] \longrightarrow Ph_3COC(=O)Ph + PhC(=O)O\cdot \quad (32)$$

The rate constants can be correlated with the σ constants of the substituents by means of the Hammett equation with $\rho = 1.45$. The importance of dipolar

[35] W. von E. Doering, K. Okamoto, and H. Krauch, *ibid.*, **82**, 3579 (1960).

[36] T. Suehiro, A. Kanoya, H. Hara, T. Nakahama, M. Omori, and T. Komori, *Bull. Chem. Soc. Japan*, **40**, 668 (1967).

Table III Rate Constants for the Reaction of Triphenylmethyl Radicals with Substituted Benzoyl Peroxides in Benzene[36]

Substituents	k_{32} at 25° (M^{-1} sec^{-1})	ΔH_{32} (kcal/mole)	ΔS_{32} (e.u.)
$4,4'(MeO)_2$	0.2	9.4	−30
$4,4'(Me)_2$	0.4	10	−27
$3,3'(Me)_2$	0.7	6.5	−38
None	1.2	10	−24
$3,3'(MeO)_2$	1.5	6.9	−35
$4,4'(F)_2$	2.3	8.0	−30
$4,4'(Cl)_2$	5.9	9.2	−24
$3,3'(Cl)_2$	15	6.7	−31

contributions to the transition state is further emphasized by the fact that the rate constants increase with increasing solvent polarity. For example, at 25° the rate constant for reaction 32 with benzoyl peroxide is 1.2 M^{-1} sec^{-1} in benzene, 2.3 in anisole, 2.4 in chlorobenzene and 6.6 in nitrobenzene.

The fast induced decomposition of benzoyl peroxide in ethanol[12] and other primary and secondary alcohols[37] at 80° also appears to be a free radical chain process with a rate which is roughly proportional to the peroxide concentration. The obvious reaction sequence is

$$\mathrm{Ph\overset{\overset{O}{\|}}{C}O^{\cdot} + CH_3CH_2OH \longrightarrow Ph\overset{\overset{O}{\|}}{C}OH + CH_3\dot{C}HOH} \tag{33}$$

$$\mathrm{CH_3\dot{C}HOH + Ph\overset{\overset{O}{\|}}{C}O\overset{\overset{O}{\|}}{O}CPh \longrightarrow Ph\overset{\overset{O}{\|}}{C}O\overset{\overset{OH}{|}}{C}HCH_3 + Ph\overset{\overset{O}{\|}}{C}O^{\cdot}} \tag{34}$$

$$\mathrm{Ph\overset{\overset{O}{\|}}{C}O\overset{\overset{OH}{|}}{C}HCH_3 \longrightarrow Ph\overset{\overset{O}{\|}}{C}OH + CH_3CHO} \tag{35}$$

The first-order dependence on peroxide concentration implies that the chains are ended by cross termination between the two types of radicals carrying the chain,

$$\mathrm{Ph\overset{\overset{O}{\|}}{C}O^{\cdot} + CH_3\dot{C}HOH \longrightarrow Ph\overset{\overset{O}{\|}}{C}OH + CH_3CHO} \tag{36}$$

This overall reaction scheme accounts for such facts as the formation of aldehyde and the much lower rate of decomposition of benzoyl peroxide in *t*-butanol which has no α-hydrogens. However, the products from the decomposition in isobutyl alcohol (70% CO_2, 24% aldehyde, 24% benzene, 23%

[37] S. Kato and F. Mashio, *Kogyo Kagaku Zasshi*, **59**, 380 (1956).

isobutyl benzoate, and a mixture of benzoic acid, isobutoxy benzoic acid and phenyl benzoic acid)[38] indicate that both phenyl radicals and alkoxy radicals may also play some role in the chain processes.

Reaction 34 may, in fact, occur by a hydrogen transfer to give acid and aldehyde directly, that is,

$$R_1R_2CO{-}{-}{-}\dot{H}{-}{-}{-}O{-}\overset{\overset{O}{\|}}{C}Ph \;(\text{with } O{-}\underset{\underset{O}{\|}}{C}Ph \text{ on the lower O}) \longrightarrow R_1R_2C{=}O + HO\overset{\overset{O}{\|}}{C}Ph + {}^{\cdot}O\underset{\underset{O}{\|}}{C}Ph \quad (37)$$

Support for such a transition state has recently been adduced by Smith and Rossiter[39] from their study of the hydroxydiphenylmethyl radical, $Ph_2\dot{C}OH$, induced decomposition of benzoyl peroxide in benzene containing benzhydrol at 25°. Chain lengths were short and the rate controlling propagation step was hydrogen atom abstraction from the alcohol by benzoyloxy radicals. The induced decomposition did not occur when the alcohol was replaced by its methyl ether which indicates that methoxydiphenylmethyl radicals do not attack the peroxide as readily as hydroxydiphenylmethyl radicals. An analogous transition state had been proposed previously for the attack of α-hydroxyalkyl radicals on dialkyl peroxides (see below).

The decomposition of benzoyl peroxide in benzene in the presence of high pressures of carbon monoxide reduces the extent of phenyl radical attack on the solvent and yields benzoic anhydride, presumably as a result of an induced decomposition of the peroxide by benzoyl radicals.[21]

$$Ph^{\cdot} + CO \longrightarrow Ph\dot{C}{=}O \quad (38)$$

$$Ph\dot{C}{=}O + Ph\overset{\overset{O}{\|}}{C}OO\overset{\overset{O}{\|}}{C}Ph \longrightarrow Ph\overset{\overset{O}{\|}}{C}O\overset{\overset{O}{\|}}{C}Ph + Ph\overset{\overset{O}{\|}}{C}O^{\cdot} \quad (39)$$

Benzoic anhydride is also produced when the decomposition is carried out in the presence of benzaldehyde as the source of benzoyl radicals.[21] In cyclohexane under high pressures of carbon monoxide the mixed anhydride is formed.[21]

$$C_6H_{11}^{\cdot} + CO \longrightarrow C_6H_{11}\dot{C}{=}O \quad (40)$$

$$C_6H_{11}\dot{C}{=}O + Ph\overset{\overset{O}{\|}}{C}OO\overset{\overset{O}{\|}}{C}Ph \longrightarrow C_6H_{11}\overset{\overset{O}{\|}}{C}O\overset{\overset{O}{\|}}{C}Ph + Ph\overset{\overset{O}{\|}}{C}O^{\cdot} \quad (41)$$

It seems most likely that these acyl radicals attack the peroxidic oxygen.

O'Driscoll and White[20] have shown that the relative rate constants for the

[38] H. Gelissen and P. H. Hermans, *Chem. Ber.*, **58B**, 765 (1925).
[39] W. F. Smith, Jr., and B. W. Rossiter, *Tetrahedron*, **25**, 2059 (1969).

polystyryl radical-induced decomposition of symmetrically substituted peroxides at 90° can be roughly correlated by means of the Hammett equation with $\rho = 1.6$. (Triphenylmethyl gives $\rho = 1.45$).[36] O'Driscoll, Lyons, and Patsiga[40] subsequently found that the relative rate constants for the N,N dimethyl aniline induced decompositions of the substituted benzoyl peroxides were correlated by the same ρ value. This latter reaction probably involves an S_N2 attack on the peroxide oxygen by the amine nitrogen.[41] Since the rate constants for both the S_N2 and the S_H2 reactions follow the same free-energy relation it was concluded that the transition state requirements for the two reactions must be quite similar (see however footnote 34, Chapter I). That is, the polystyryl radical probably attacks the peroxidic oxygen rather than the aromatic ring.

The decomposition of benzoyl peroxide can also be induced by certain nucleophilic organometallic free-radicals which have the unpaired electron centered on the metal atom. Neumann and co-workers have studied the decomposition induced by trialkyltin hydrides and dialkyltin dihydrides.[42,43,43a] The reaction is a free radical chain of considerable length, the decomposition of the peroxide occurring about a hundred times faster in triethyltin hydride than in dibutyl ether under the same conditions. The principal reaction products are consistent with a chain initiated by the decomposition of the peroxide and propagated by

$$PhC(=O)O^{\cdot} + R_3SnH \longrightarrow PhC(=O)OH + R_3Sn^{\cdot} \qquad (42)$$

$$R_3Sn^{\cdot} + PhC(=O)OOC(=O)Ph \longrightarrow R_3SnOC(=O)Ph + PhC(=O)O^{\cdot} \qquad (43)$$

A rate controlling electron transfer reaction is also a possibility,

$$R_3Sn^{\cdot} + PhC(=O)OOC(=O)Ph \longrightarrow [R_3Sn^{+} + PhC(=O)O^{\cdot} + PhC(=O)O^{-}] \longrightarrow R_3SnOC(=O)Ph + PhC(=O)O^{\cdot} \qquad (44)$$

The benzoic acid is not isolated as such because it reacts with the tin hydride to form the trialkyltin benzoate,

$$PhC(=O)OH + R_3SnH \longrightarrow R_3SnOC(=O)Ph + H_2 \qquad (45)$$

[40] K. F. O'Driscoll, P. F. Lyons and R. Patsiga, *J. Polymer Sci.*, **A, 3,** 1567 (1965).
[41] C. Walling and N. Indictor, *J. Amer. Chem. Soc.*, **80,** 5814 (1958).
[42] W. P. Neumann, K. Rubsamen and R. Sommer, *Angew. Chem.*, **77,** 733 (1965).
[43] W. P. Neumann and K. Rubsamen, *Chem. Ber.*, **100,** 1621 (1967).
[43a] See footnote 131, Chapter V.

Some CO_2 is formed by decarboxylation of the benzoyloxy radicals and there is, therefore, some benzene also. More surprisingly, there is also some RH (e.g. ethane from triethyltin hydride) and some dialkyltin dibenzoate. Both of these products are, presumably, the result of an S_H2 reaction at tin, (see Chapter V (D)), that is,

$$\mathrm{Ph\overset{\overset{O}{\|}}{C}O^{\cdot} + R_3SnO\overset{\overset{O}{\|}}{C}Ph \longrightarrow R_2Sn(O\overset{\overset{O}{\|}}{C}Ph)_2 + R^{\cdot}} \tag{46}$$

$$\mathrm{R^{\cdot} + R_3SnH \longrightarrow RH + R_3Sn^{\cdot}} \tag{47}$$

Experiments with benzoyl peroxide labeled with ^{18}O in the carbonyl group indicate that reaction 43 proceeds almost entirely by an attack of the trialkyltin radicals on peroxidic oxygen.[43]

Competition techniques have been used to estimate a rate constant of $4 \times 10^4\ M^{-1}\ \mathrm{sec}^{-1}$ for reaction 43 (R = *n*-butyl) at 25° in benzene.[43b]

Razuvaev and co-workers[44–49] have examined the reactions of many peroxides (aroyl, acyl, alkyl, etc.) with a large variety of organometallic compounds, including many compounds with metal–metal bonds (R_nMMR_n and $R_nMM'MR_n$). The reactions were generally carried out at rather high temperatures (typically 80–150°) and interest was centered more on the products of reaction than on the mechanism of their formation. In a number of the reactions the products certainly imply the intermediacy of free radicals and it seems quite likely that S_H2 reactions at oxygen by free radicals with the radical center at the metal atom are involved. It is probable that many of these reactions also involve S_H2 processes in which oxy radicals attack the metal atom.

(b) ACYL PEROXIDES

The free-radical induced decomposition of acyl peroxides has received rather less attention than the induced decomposition of aroyl peroxides and per-esters. The chief reason is that acyl peroxides are less frequently used as initiators of vinyl polymerization. Induced decomposition must generally involve an S_H2 reaction on the peroxide oxygen since there can be no question

[43b] J. L. Brokenshire and K. U. Ingold. Unpublished results.
[44] See footnote 117, Chapter V.
[45] See footnote 118, Chapter V.
[46] See footnote 124, Chapter V.
[47] See footnote 129, Chapter V.
[48] N. S. Vyazankin, G. A. Razuvaev, and O. A. Kruglaya, *Organometal. Chem. Rev. A*, **3**, 323 (1968).
[49] G. A. Razuvaev and N. S. Vyazankin, *Khim. Perekisnykh Soedin., Akad. Nauk SSSR, Inst. Obshch. i Neorgan Khim.*, 283 (1963) (*Chem. Abstr.*, **60**, 15708g (1964)).

of an attack on an aromatic ring. However, in certain cases decomposition may also be induced by the abstraction of hydrogen from the β-position, that is,

$$R^{\cdot} + H\text{—}\overset{\overset{\displaystyle R'}{|}}{C}HCH_2\overset{\overset{\displaystyle O}{\|}}{C}OO\overset{\overset{\displaystyle O}{\|}}{C}R'' \longrightarrow RH + \overset{\overset{\displaystyle R'}{|}}{C}H{=}CH_2 + CO_2 + {}^{\cdot}O\overset{\overset{\displaystyle O}{\|}}{C}R'' \quad (48)$$

Because of the concerted nature of such a reaction the β-hydrogen of the peroxide should be much more readily abstracted than the β-hydrogen from an acid or ester. A somewhat analogous concerted elimination has been proposed as an explanation for the ease with which the hydrogen atoms of di-*t*-butyl sulfide are abstracted by free radicals,[50] that is,

$$Ph^{\cdot} + (CH_3)_3CSC(CH_3)_3 \rightarrow PhH + (CH_3)_2C{=}CH_2 + (CH_3)_3CS^{\cdot} \quad (49)$$

In an "inert" solvent the rates of decomposition of acyl peroxides such as acetyl peroxide,[4,51] propionyl peroxide,[52] octanoyl peroxide, etc.[53] are rather similar to the rate of decomposition of benzoyl peroxide. Decomposition is approximately first order but the rate constants tend to increase, particularly at peroxide concentrations above $\sim$0.1 *M*. Although the rate constants for the true unimolecular decomposition of acyl peroxides,

$$R\overset{\overset{\displaystyle O}{\|}}{C}OO\overset{\overset{\displaystyle O}{\|}}{C}R \longrightarrow 2R\overset{\overset{\displaystyle O}{\|}}{C}O^{\cdot} \quad (50)$$

are very similar, the sensitivity to induced decomposition varies. Thus, for example, Smid, Rembaum, and Szwarc[52] have reported that propionyl peroxide is 2.5 times as susceptible as acetyl peroxide to induced decomposition in isooctane. It was suggested[52] that this was due to an intrinsic difference in the reactivity of ethyl and methyl radicals. However, it seems more likely that the higher sensitivity of propionyl peroxide to induced decomposition is due to the incursion of a decomposition according to reaction 48. This point could be cleared up by inducing the decomposition of the two peroxides with a free-radical which would not abstract hydrogen such as a trialkyltin radical. Incidentally, reaction 50 does not give "free" acyloxy radicals in significant quantities. These radicals undergo β-scission to give CO_2 and an

[50] J. A. Kampmeier, R. P. Geer, A. J. Meskin, and R. M. D'Silva, *J. Amer. Chem. Soc.*, **88,** 1257 (1966).
[51] W. M. Thomas and M. T. O'Shaughnessy, *J. Polymer Sci.*, **11,** 455 (1953).
[52] J. Smid, A. Rembaum, and M. Szwarc, *J. Amer. Chem. Soc.*, **78,** 3315 (1956).
[53] J. E. Guillett and J. C. Gilmer, *Can. J. Chem.*, **47,** 4405 (1969).

Table IV Apparent First-Order Rate Constants k_d, for the Decomposition of Some Acyl Peroxides at 65° (Peroxide Concentration $\sim$ 0.05 M)

Peroxide	Acetyl[51]	Propionyl[52]	Butyryl[52]	Benzoyl[a]
Solvent	$k_d \times 10^5$ (sec^{-1})	$k_d \times 10^5$ (sec^{-1})	$k_d \times 10^5$ (sec^{-1})	$k_d \times 10^5$ (sec^{-1})
t-Butanol	0.6	—	—	0.2
Benzene	1.0	2.0	2.1	0.3
Toluene	1.0	1.9	2.0	0.3
Dioxan	—	5.5	5.4	4.2
n-Butanol	7	—	—	3.8
s-Butanol	6	—	—	—
Ethanol	20	—	—	14

a. Estimated from data in footnotes 8, 10, and 12.

alkyl radical at a rate comparable to their rate of diffusion from the solvent cage.[54–56]

$$\mathrm{R\overset{\overset{\displaystyle O}{\|}}{C}O^{\cdot}} \xrightarrow{\text{fast}} \mathrm{R^{\cdot} + CO_2} \tag{51}$$

The first order decomposition rate constants for $\sim$0.05 M acetyl, propionyl, and butyryl peroxides at 65° which are listed in Table IV show that the rate of decomposition of these simple acyl peroxides is accelerated in primary and secondary alcohols and in ethers. Similarly,[57] 0.2 M lauroyl peroxide at 30° has a first order decomposition rate constant of 2.5×10^{-7} sec^{-1} in benzene but 1.9×10^{-6} sec^{-1} in diethyl ether (for benzoyl peroxide the comparable rate constants are 4.8×10^{-8} and 1×10^{-5} sec^{-1} respectively). The rates in "fast" solvents are retarded by oxygen and by inhibitors. Thus, for example, the rate of decomposition of 0.2 M lauroyl peroxide in ether is reduced to the same value as in benzene by the addition of 5×10^{-3} M quinone.[57] The induced decomposition must therefore be a chain reaction.

The products isolated from the decomposition of acetyl peroxide in alcohols[58] and ethers[59] prove that α-hydroxyalkyl and α-alkoxyalkyl radicals attack the

$$\mathrm{-\overset{\overset{\displaystyle O}{\|}}{C}O O\overset{\overset{\displaystyle O}{\|}}{C}-}$$

[54] W. J. Braun, L. Rajbenbach, and E. R. Eirich, *J. Phys. Chem.*, **66,** 1591 (1962).
[55] J. W. Taylor and J. C. Martin, *J. Amer. Chem. Soc.*, **88,** 3650 (1966).
[56] J. C. Martin and S. A. Dombchik, *Adv. Chem. Series*, **75,** 269 (1968).
[57] W. E. Cass, *J. Amer. Chem. Soc.*, **72,** 4915 (1950).
[58] M. S. Kharasch, J. L. Rowe, and W. H. Urry, *J. Org. Chem.*, **16,** 905 (1951).
[59] M. S. Kharasch, H. N. Friedlander, and W. H. Urry, *ibid.*, **16,** 533 (1951).

Table V Principal Decomposition Products of Diacetyl Peroxide in Alcohols[58] and Ethers[59]

Products	Product Yields (moles/mole of Diacetyl Peroxide Decomposed)					
	Bu^nOH	Bu^iOH	Pr^iOH	Bu^sOH	Bu^tOH	Pr^iOPr^i
CH_4	0.92	1.03	1.05	1.01	1.01	0.76
CO_2	1.10	1.28	1.05	1.01	1.45	0.85
CO	0.36	0.22	—	—	—	—
$C_3H_8 + C_3H_6$ (9:1)	0.36	0.22	—	—	—	—
R_1R_2CO	0.14[a]	0.12[b]	0.89[c]	0.75[d]	0.03[c]	0.14[c]
CH_3COOR	0.66[e]	0.56[f]	0.33[g]	0.45[h]	0.03[i]	0.23[g]
CH_3COOCH_3	0.08	0.12	0.02	0.03	0.25	—
CH_3COOH	0.01	0.03	0.55	0.41	0.23	0.92
$(CH_3)_2CHOH$	—	—	—	—	—	0.16

a. *n*-Butyraldehyde.
b. Isobutyraldehyde.
c. Acetone.
d. Methyl ethyl ketone.
e. *n*-Butyl acetate.
f. Isobutyl acetate.
g. Isopropyl acetate.
h. *s*-Butyl acetate.
i. *t*-Butyl acetate.

group (see Table V). It seems a reasonable presumption that such attack occurs at the peroxidic oxygen. The products are similar to those formed with benzoyl peroxide except for the much higher yield of carbon dioxide which arises from the rapid decarboxylation of the acetoxy radical. In a secondary alcohol, for example, isopropanol, the chain propagating steps can be represented as,

$$CH_3\overset{\overset{\displaystyle O}{\|}}{C}O^{\cdot} \longrightarrow CH_3^{\cdot} + CO_2 \tag{52}$$

$$CH_3^{\cdot} + CH_3\underset{\underset{\displaystyle H}{|}}{\overset{\overset{\displaystyle \overset{\displaystyle H}{O}}{|}}{C}}CH_3 \longrightarrow CH_4 + CH_3\underset{\cdot}{\overset{\overset{\displaystyle \overset{\displaystyle H}{O}}{|}}{C}}CH_3 \tag{53}$$

$$CH_3\underset{\cdot}{\overset{\overset{\displaystyle \overset{\displaystyle H}{O}}{|}}{C}}CH_3 + CH_3\overset{\overset{\displaystyle O}{\|}}{C}O\overset{\overset{\displaystyle O}{\|}}{C}CH_3 \longrightarrow \left[CH_3\overset{\overset{\displaystyle O}{\|}}{C}O\underset{\underset{\displaystyle CH_3}{|}}{\overset{\overset{\displaystyle \overset{\displaystyle H}{O}}{|}}{C}}CH_3\right] + CH_3\overset{\overset{\displaystyle O}{\|}}{C}O^{\cdot} \tag{54}$$

$$\left[CH_3\overset{\overset{\displaystyle O}{\|}}{C}O\underset{\underset{\displaystyle CH_3}{|}}{\overset{\overset{\displaystyle \overset{\displaystyle H}{O}}{|}}{C}}CH_3\right] \longrightarrow CH_3\overset{\overset{\displaystyle O}{\|}}{C}OH + CH_3COCH_3$$

Reaction 54 may not involve a discrete intermediate, that is,

$$CH_3\overset{\overset{\displaystyle H}{\displaystyle O}}{\underset{\cdot}{C}}CH_3 + CH_3\overset{O}{\overset{\|}{C}}OO\overset{O}{\overset{\|}{C}}CH_3 \longrightarrow CH_3\overset{O}{\overset{\|}{C}}OH + CH_3COCH_3 + CH_3\overset{O}{\overset{\|}{C}}O^{\cdot} \quad (55)$$

The methyl radical attacks the C–H bond in the alcohol, reaction 53, since deuteration at this position yields CH_3D (partial deuteration gives $k_H/k_D = 6.3$). In a primary alcohol, for example, *n*-butanol, an aldehyde is formed in the analogue of reaction 54 or 55,

$$\left[CH_3\overset{O}{\overset{\|}{C}}O\underset{H}{\overset{\overset{\displaystyle H}{\displaystyle O}}{C}}CH_2CH_2CH_3\right] \longrightarrow CH_3\overset{O}{\overset{\|}{C}}OH + CH_3CH_2CH_2CHO \quad (56)$$

The *n*-butyraldehyde will be much more reactive towards the methyl radical than the starting alcohol. The acyl radical will lose carbon monoxide and the resultant propyl radical will act as the source of the C_3 hydrocarbons, propane and propylene, for example,

$$CH_3 + CH_3CH_2CH_2CHO \rightarrow CH_4 + CH_3CH_2CH_2\dot{C}O \quad (57)$$

$$CH_3CH_2CH_2\dot{C}O \rightarrow CH_3CH_2CH_2 + CO \quad (58)$$

$$CH_3CH_2CH_2 \xrightarrow{RH} CH_3CH_2CH_3 \quad (59)$$

The products of decomposition in *t*-butanol indicate relatively little attack on the solvent and hence the absence of a solvent induced reaction. This is consistent with the low decomposition rate constant found in this alcohol (Table IV). The products in vicinyl glycols are consistent with a solvent induced decomposition of acetyl peroxide.[60]

The products identified in the decomposition of acetyl peroxide in diisopropyl ether[59] are somewhat surprising. The rather small amounts of acetone and isopropyl acetate can be explained by a β-scission of the isopropoxyisopropyl radical followed by attack of the isopropyl radical on the peroxide.

$$(CH_3)_2\dot{C}OCH(CH_3)_2 \rightarrow CH_3COCH_3 + (CH_3)_2\dot{C}H \quad (60)$$

However, it is difficult to account for the high yield of acetic acid and for the absence of the expected isopropoxyisopropyl acetate, **4,** unless this ester decomposes under the reaction conditions.[61]

[60] *Idem, ibid.*, **14,** 91 (1949).

[61] Unspecified, but probably reflux; b.p. isopropyl ether = 68°.

$$CH_3-\overset{\overset{O}{\|}}{C}-O-\underset{\underset{CH_3}{|}}{\overset{\overset{CH_3}{|}}{C}}-O-\underset{\underset{CH_3}{|}}{\overset{\overset{CH_3}{|}}{CH}} \longrightarrow CH_3\overset{\overset{O}{\|}}{C}OH + CH_2{=}\underset{\underset{CH_3}{|}}{C}-O-\underset{\underset{CH_3}{|}}{\overset{\overset{CH_3}{|}}{CH}} \tag{61}$$

4

Rather high yields of "residue" were obtained in this solvent, perhaps because of the polymerization of the isopropyl isopropenyl ether.

The products from the decomposition of lauroyl peroxide in diethyl ether at 37° are[57] $CO_2 = 0.92$, *n*-undecane = 0.84, 1-ethoxyethyl laurate = 0.88, an unknown ester = 0.16, and a trace of acid (lauric?) = 0.04. These products indicate a conventional induced chain decomposition with extensive CO_2 loss from the acyloxy radical so that it is the undecyl radical that attacks the ether.

The reaction of acetyl peroxide and lauroyl peroxide with trialkyltin hydrides occurs in an analogous manner to the benzoyl peroxide reaction.[42,62] Some of the acyloxy radicals liberated in the S_H2 reaction,

$$R_3Sn^{\cdot} + R'\overset{\overset{O}{\|}}{C}O O\overset{\overset{O}{\|}}{C}R' \longrightarrow R_3SnO\overset{\overset{O}{\|}}{C}R' + R'\overset{\overset{O}{\|}}{C}O^{\cdot} \tag{62}$$

react with the tin hydride before decarboxylating, the lauroyloxy radical being somewhat less stable or reacting more slowly with tin hydride, than the acetoxy radical.

$$R'\overset{\overset{O}{\|}}{C}O^{\cdot} \nearrow CO_2 + R'^{\cdot} \xrightarrow{R_3SnH} R'H + R_3Sn^{\cdot}$$
$$R'\overset{\overset{O}{\|}}{C}O^{\cdot} \xrightarrow{R_3SnH} R'\overset{\overset{O}{\|}}{C}OH + R_3Sn^{\cdot} \tag{63}$$

The presence of some RH in the products again indicates the occurrence of an S_H2 reaction at tin. Experiments with ^{18}O labeled acetyl peroxide showed that the trialkyltin radical attacks the peroxidic oxygen.[43] The rate constant for reaction 62 (R = *n*-butyl, R′ = methyl) is $1.4 \times 10^5\ M^{-1}\ sec^{-1}$ at 25° in benzene.[43b]

The decomposition of acetyl peroxide is also induced by phenyl hydrazine.[63] The following reaction sequence seems reasonable.

$$CH_3^{\cdot} + PhNHNH_2 \rightarrow CH_4 + Ph\dot{N}NH_2 \tag{64}$$

$$Ph\dot{N}NH_2 + CH_3\overset{\overset{O}{\|}}{C}O O\overset{\overset{O}{\|}}{C}CH_3 \longrightarrow PhN{=}NH + CH_3\overset{\overset{O}{\|}}{C}OH + CO_2 + CH_3^{\cdot} \tag{65}$$

[62] See footnote 131, Chapter V.
[63] E. S. Huyser and R. H. S. Wang, *J. Org. Chem.*, **33**, 3901 (1968).

Potassium persulfate has sometimes been regarded as an analog of the diacyl peroxides. Its decomposition in aqueous solutions which have been buffered against the accumulation of hydrogen ion follows the overall course,

$$2\ {}^{-}O\overset{\overset{O}{\|}}{\underset{\underset{O}{\|}}{S}}OO\overset{\overset{O}{\|}}{\underset{\underset{O}{\|}}{S}}O^{-} + 2H_2O \longrightarrow 4HSO_4^- + O_2 \quad (66)$$

In contrast to the decomposition of diacyl peroxides in organic solvents the reaction is first order and is not retarded by inhibitors and it is therefore unlikely that the decomposition is a chain reaction. The rate of decomposition of persulfate can, however, be greatly increased by the addition of ethyl acetate,[64] methanol,[65,66] ethanol,[66] isopropanol[67] and allyl alcohol.[66] These fast reactions can be inhibited by allyl acetate[66] and so must be free radical chain processes. In methanol at 60–80° and at a pH of 7–8 the rate of reaction is proportional to the square root of the methanol concentration and the 3/2 power of the persulfate concentration, with a small deviation from these kinetics at the very beginning of the reaction.[65,66] The chain propagating reactions in the presence of methanol can probably be represented by

$$SO_4^{\cdot-} + CH_3OH \rightarrow HSO_4^- + \dot{C}H_2OH \quad (67)$$

$$\dot{C}H_2OH + S_2O_8^{--} \rightarrow HCHO + HSO_4^- + SO_4^{\cdot-} \quad (68)$$

In isopropanol the rate is proportional to the first power of the persulfate concentration and is independent of the alcohol concentration,[67] suggesting that the alcohol is oxidized by a hydroxyl radical, that is,

$$SO_4^{\cdot-} + H_2O \rightarrow HSO_4^- + {}^{\cdot}OH \quad (69)$$

$${}^{\cdot}OH + (CH_3)_2CHOH \rightarrow H_2O + (CH_3)_2\dot{C}OH \quad (70)$$

(c) ACYL AROYL PEROXIDES

Razuvaev and Latiaeva[68] have reported that the decomposition of acetyl benzoyl peroxide in methanol, ethanol, isopropanol, and *n*-butanol is a rapid first order reaction yielding benzoic acid, CO_2, methane and oxidized alcohol (e.g., acetone from isopropanol) as the major products. This result has been confirmed by Walling and Azar[69] who found similar products in

[64] P. D. Bartlett and K. Nozaki, *J. Polymer Sci.*, **3**, 216 (1948).
[65] P. D. Bartlett and J. D. Cotman, Jr., *J. Amer. Chem. Soc.*, **71**, 1419 (1949).
[66] I. M. Kolthoff, E. J. Meehan, and E. M. Carr, *ibid.*, **75**, 1439 (1953).
[67] K. B. Wiberg, *ibid.*, **81**, 252 (1959).
[68] G. A. Razuvaev and V. N. Latiaeva, *Zhur. Obshch. Khim.*, **26**, 1986 (1956).
[69] C. Walling and J. C. Azar, *J. Org. Chem.*, **33**, 3888 (1968).

sec-butanol and in isopropyl ether. Acetyl *p*-chlorobenzoyl peroxide behaved similarly in alcohols and in isopropyl ether. The substantial yields of methane and CO_2 and the low or zero yields of benzene (or chlorobenzene) indicate that most of the chain is propagated via acetoxy and methyl radicals, that is,

$$CH_3\overset{\overset{O}{\|}}{C}O^{\cdot} \longrightarrow CO_2 + CH_3^{\cdot} \tag{52}$$

$$CH_3^{\cdot} + RCHOHR' \longrightarrow CH_4 + R\dot{C}OHR' \tag{71}$$

$$R\dot{C}OHR' + CH_3\overset{\overset{O}{\|}}{C}OO\overset{\overset{O}{\|}}{C}Ph \longrightarrow CH_3\overset{\overset{O}{\|}}{C}O^{\cdot} + HO\overset{\overset{O}{\|}}{C}Ph + RCOR' \tag{72}$$

Neumann and co-workers[70] have examined the reaction of triethyl tin hydride with a large number of acyl aroyl peroxides. The triethyltin radical which is formed induces a very rapid decomposition of the peroxides. This organometallic radical is nucleophilic and so its rate of attack on a peroxidic oxygen atom is greater the more positive the oxygen. Thus, for a series of ring-substituted acetyl benzoyl peroxides the rate of the induced decomposition tends to increase as the electron withdrawing inductive effect of the substituent increases. This occurs in spite of the fact that the stability of the peroxide itself (which in part reflects changes in the rate of its decomposition to free radicals) increases in the same direction. At 60° the rate of the triethyltin radical induced decomposition increases along the series *p*-methoxy < *p*-ethoxy < *p*-cyano < unsubstituted < *p*-chloro = *m*-chloro. Only the *p*-cyano compound—which by itself is the most stable—appears out of place in this series. The composition of the products clearly indicates that electron attracting substituents on the aromatic ring increase the extent of attack on the aroyl peroxidic oxygen, while electron releasing substituents decrease it. The overall reaction and the % product yields are given on p. 170. The products from lauroyl benzoyl peroxide, propionyl benzoyl peroxide, chloroacetyl benzoyl peroxide and chloroacetyl *p*-methoxybenzoyl peroxide further serve to confirm that the triethyltin radical preferentially attacks the most positive oxygen in the peroxide bridge. (That is, the triethyltin radical is nucleophilic like the methyl and cyclohexyl radicals, whereas halocarbon radicals (e.g., $CF_3^{\cdot}$) and most radicals with the unpaired electron centered on oxygen or sulfur are electrophilic.)

(d) PERESTERS

The thermal decomposition behavior of peresters, aroyl peroxides and acyl peroxides are all rather similar. For example, in 1951 Blomquist and

[70] See footnote 134, Chapter V.

$$Et_3Sn^{\cdot} + CH_3C(=O)-O_1-O_2-C(=O)C_6H_4X$$

$$\xrightarrow{S_H2 \text{ on } O_2} CH_3^{\cdot} + CO_2 + Et_3SnOC(=O)C_6H_4X$$

$$\xrightarrow{S_H2 \text{ on } O_1} {}^{\cdot}OC(=O)C_6H_4X + Et_3SnOC(=O)CH_3$$

$$\longrightarrow CH_4 \quad CO_2 \quad Et_3SnOC(=O)C_6H_4X \quad + H_2 \quad Et_3SnOC(=O)CH_3 \qquad (73)$$

X	CH_4	CO_2	$Et_3SnOC(=O)C_6H_4X$	H_2	$Et_3SnOC(=O)CH_3$
p-CH_3O	24	25	95	59	74
—	29	37	96	54	71
p-Cl	34	38	98	53	66
m-Cl	45	50	95	40	52
p-CN	54	54	95	35	42

Ferris[71] found that the first order rate constant for the decomposition of *t*-butyl perbenzoate in aromatic solvents showed a small increase with increasing initial concentration of peroxide. The rate of decomposition of this perester in a number of "fast" solvents paralleled the decomposition rate found for benzoyl peroxide in these same solvents.[72] In particular,[72] the decomposition was markedly accelerated in di-*n*-butyl ether and in *n*-butanol. Moreover, in *n*-butyl acetate, another fast solvent, the rate was reduced to the same value as in aromatic solvents by the addition of 0.2 *M* styrene as a free-radical trap. It was concluded that the decomposition of the perester can be induced by free radicals.

More recently, Walling and Azar[69] have examined the rates and products of the decomposition of *t*-butyl perbenzoate and *t*-butyl peracetate in alcohol and ether solvents and have confirmed the occurrence of an induced free radical chain. For 0.2–0.4 *M* *t*-butyl peracetate at 75° the chain length was estimated as ~250–300 in secondary alcohols and ~60 in *n*-butanol. The products (in moles/mole of perester) from the decomposition of perbenzoate and peracetate in *sec*-butanol at 100° were, respectively: *t*-butanol, 0.95 and 0.61; acetone, 0.05 and 0.03; benzene, 0.05 and 0.00; 2-butanone, 1.41 and 0.81; CO_2, 0.04 and 0.07. The high yield of *t*-butanol and low yield of CO_2 are highly significant because the acetoxy radical, in particular, would be certain to undergo extensive decarboxylation if it had been formed. The products therefore indicate that the *t*-butoxy radical is the chain carrier and, by analogy with benzoyl peroxide, that the α-hydroxyalkyl radical attacks the acylperoxy oxygen, for example,

$$(CH_3)_3CO^{\cdot} + CH_3CH_2CHOHCH_3 \longrightarrow (CH_3)_3COH + CH_3CH_2\dot{C}OHCH_3 \qquad (74)$$

$$CH_3CH_2\dot{C}OHCH_3 + R\overset{\overset{\displaystyle O}{\|}}{C}OOC(CH_3)_3 \longrightarrow CH_3CH_2COCH_3 + R\overset{\overset{\displaystyle O}{\|}}{C}OH + (CH_3)_3CO^{\cdot} \qquad (75)$$

The minor reaction products are consistent with this direction of cleavage. Secondary alcohols give the expected ketone in high yield, but aldehyde yields from *n*-butanol were low, presumably because of the further oxidation of the expected *n*-butyraldehyde.

Decomposition of the peresters in ethers also gives almost quantitative yields of *t*-butanol and very little CO_2.[69] The rates of decomposition were slowest in isopropyl ether, next in *n*-butyl ether and were comparable to the rates in alcohols only in cyclic ethers such as tetrahydropyran. This difference

[71] A. T. Blomquist and A. F. Ferris, *J. Amer. Chem. Soc.*, **73**, 3408 (1951).
[72] A. T. Blomquist and A. F. Ferris, *ibid.*, **73**, 3412 (1951).

was attributed to the varied importance of β-scission in the reaction of the α-alkoxyalkyl radicals involved in the ether reactions and the resulting competition for example, with isopropyl ether,

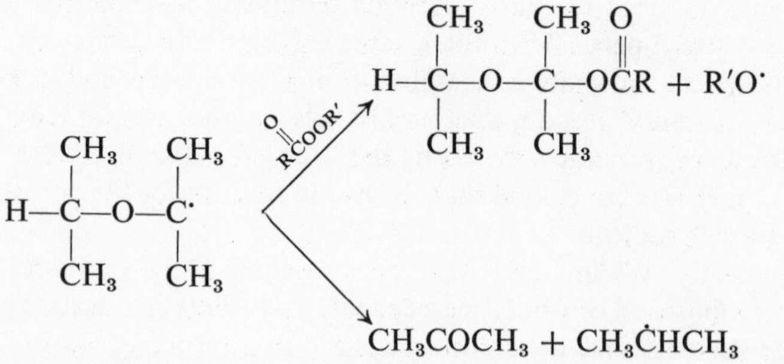

$$(CH_3)_2CH{-}O{-}\dot{C}(CH_3)_2 \xrightarrow{RC(=O)OOR'} (CH_3)_2CH{-}O{-}C(CH_3)_2{-}OC(=O)R + R'O\cdot$$

$$(CH_3)_2CH{-}O{-}\dot{C}(CH_3)_2 \longrightarrow CH_3COCH_3 + CH_3\dot{C}HCH_3 \tag{76}$$

The alkyl radical formed in the β-scission shows less tendency to propagate the induced chain than the α-alkoxyalkyl radical it was derived from. In the case of the cyclic ethers reversibility of the β-scission is important because the fragments cannot separate after scission has occurred.

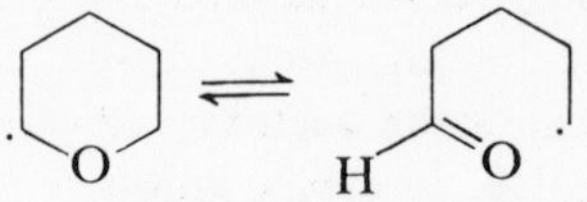

$$\text{(tetrahydropyran-2-yl radical)} \rightleftharpoons \text{(H–C(=O)–CH}_2\text{CH}_2\text{CH}_2\dot{C}\text{H}_2) \tag{77}$$

Walling and Azar[69] point out that the most obvious explanation for the attack occurring at the acylperoxy oxygen is a steric one, since substitution on the other peroxide oxygen would involve the same crowding as in a neopentyl system. Energetic considerations actually predict that the resonance stabilized acyloxy radical would be liberated by the S_H2 process. Since fast induced decompositions are limited to radicals with strong electron donor properties it is evident that polar contributions of the type

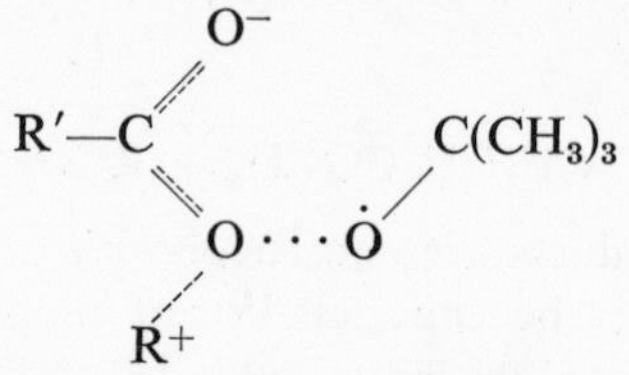

$$R'{-}C(\cdots O^-)(\cdots O\cdots\dot{O}{-}C(CH_3)_3)\cdots R^+$$

play an important role in the transition state.

A few other examples of S_H2 reactions involving peresters and carbon radicals are worth noting. Bartlett, Benzing and Pincock[73] have shown that the decomposition of di-*t*-butylperoxyoxalate in isopropyl ether occurs by a fast induced chain reaction which can be inhibited by oxygen or styrene. The products of the induced decomposition are, surprisingly, isopropyl isopropenyl ether, 0.89 (rather than the expected acylal), together with *t*-butanol,

[73] P. D. Bartlett, E. P. Benzing, and R. E. Pincock, *ibid.*, **82**, 1762 (1960).

2.02, and CO_2, 1.96 moles/mole of peroxide. The most probable mechanism would seem to involve the intermediate formation of mono-*t*-butyl peroxalate,

$$(CH_3)_3COOC(=O)C(=O)OOC(CH_3)_3 + (CH_3)_2\dot{C}OCH(CH_3)_2 \longrightarrow (CH_3)_3CO^{\cdot}$$

$$+ HOC(=O)C(=O)OOC(CH_3)_3 + CH_3C(=CH_2)OCH(CH_3)_2 \qquad (78)$$

$$\downarrow$$

$$2CO_2 + (CH_3)_3COH$$

As was mentioned above, the decomposition of acetyl peroxide in isopropyl ether also does not give the anticipated acylal.[59] In this case, isopropyl isopropenyl ether was not detected but there was a high yield of "residue" which may have been largely the polymerized ether.

Van Sickle[74] has reported that dicyclohexyl peroxydicarbonate decomposes in benzene at 50° both by simple cleavage and by a radical induced reaction, the two processes being of comparable importance at 0.1 *M* concentration of peroxide. The decomposition products are CO_2, cyclohexanol and cyclohexanone. It is suggested that the induced decomposition occurs via abstraction of the α-hydrogen from the cyclohexane ring.

$$C_6H_{11}{-}O{-}C(=O){-}O^{\cdot} + C_6H_{11}{-}O{-}C(=O){-}O{-}O{-}C(=O){-}O{-}C_6H_{11} \longrightarrow C_6H_{11}{-}O{-}C(=O){-}OH + \dot{C}_6H_{10}{-}O{-}C(=O){-}O{-}O{-}C(=O){-}O{-}C_6H_{11}$$

$$C_6H_{11}{-}O{-}C(=O){-}OH \longrightarrow C_6H_{11}OH + CO_2 \qquad (79)$$

$$\dot{C}_6H_{10}{-}O{-}C(=O){-}O{-}O{-}C(=O){-}O{-}C_6H_{11} \longrightarrow C_6H_{10}{=}O + CO_2 + {}^{\cdot}O{-}C(=O){-}O{-}C_6H_{11}$$

[74] D. E. Van Sickle, *J. Org. Chem.*, **34,** 3446 (1969).

Intramolecular homolytic substitutions of peresters by carbon radicals are also known. In 1958, Milas and Golubovic[75] reported the formation of a polyester of α-hydroxyisobutyric acid from the decomposition of *t*-butyl peroxyisobutyrate in the absence of solvent. Based on their studies of the decomposition of di-*t*-butyl peroxydiphenylmalonate in cumene, Bartlett and Gortler[76] proposed the following mechanism to account for Milas' result.

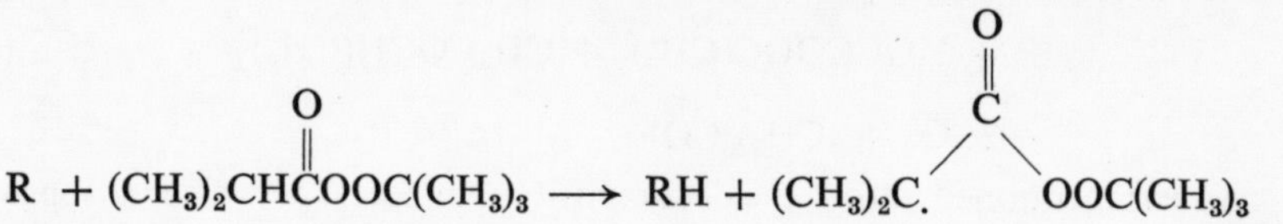

$$R\cdot + (CH_3)_2CH\overset{O}{\overset{\|}{C}}OOC(CH_3)_3 \longrightarrow RH + (CH_3)_2\dot{C}\overset{O}{\overset{\|}{C}}OOC(CH_3)_3 \quad (80)$$

$$(CH_3)_2\dot{C}\overset{O}{\overset{\|}{C}}OOC(CH_3)_3 \longrightarrow (CH_3)_2C\underset{\text{---}O\text{---}}{\overset{O}{\overset{\|}{C}}} + (CH_3)_3CO\cdot \quad (81)$$

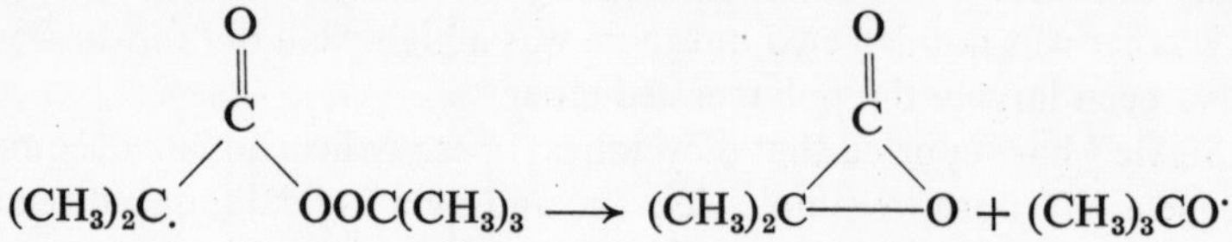

$$n(CH_3)_2C\underset{\text{---}O\text{---}}{\overset{O}{\overset{\|}{C}}} \longrightarrow \left[\begin{array}{c} CH_3\ \ O \\ |\ \ \ \ // \\ -C-C-O- \\ | \\ CH_3 \end{array}\right]_n \quad (82)$$

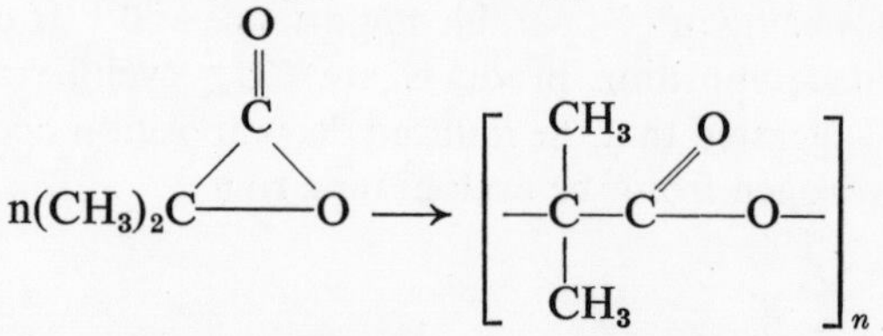

The expulsion of a *t*-butoxy radical by the *t*-butylperoxy isobutyryl radical, reaction 81, is an S_Hi process which leads to the formation of an unstable α-lactone which, in turn, polymerizes to the polyester. The occurrence of this reaction has been confirmed by Gortler and Saltzman[77] who generated the *t*-butylperoxy isobutyryl radical by the decomposition of di-*t*-butyl peroxy-dimethylmalonate.

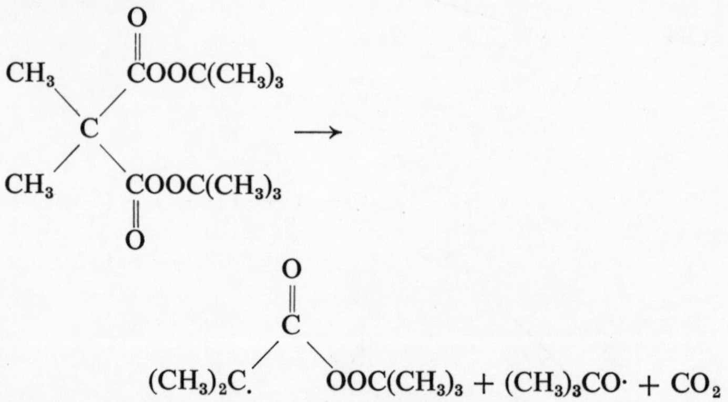

$$(CH_3)_2C[\overset{O}{\overset{\|}{C}}OOC(CH_3)_3]_2 \longrightarrow (CH_3)_2\dot{C}\overset{O}{\overset{\|}{C}}OOC(CH_3)_3 + (CH_3)_3CO\cdot + CO_2 \quad (83)$$

[75] N. A. Milas and A. Golubovic, *J. Amer. Chem. Soc.*, **80,** 5994 (1958).
[76] P. D. Bartlett and L. B. Gortler, *ibid.*, **85,** 1864 (1963).
[77] L. B. Gortler and M. D. Saltzman, *J. Org. Chem.*, **31,** 3821 (1966).

The polyester of α-hydroxyisobutyric acid is again produced. The chief product from the decomposition of di-*t*-butyl peroxydiphenylmalonate in cumene is a polyester of benzilic acid produced by an analogous S_Hi reaction,[76] that is,

$$Ph_2C\begin{cases} \overset{O}{\overset{\|}{C}}-OOC(CH_3)_3 \\ \underset{\underset{O}{\|}}{C}-OOC(CH_3)_3 \end{cases} \longrightarrow Ph_2\dot{C}-\overset{O}{\overset{\|}{C}}-OOC(CH_3)_3 + (CH_3)_3CO^{\cdot} + CO_2 \quad (84)$$

$$Ph_2\dot{C}-\overset{O}{\overset{\|}{C}}-OOC(CH_3)_3 \longrightarrow Ph_2\underset{\text{(ring)}}{C-\overset{O}{\overset{\|}{C}}-O} + (CH_3)_3CO^{\cdot} \quad (85)$$

$$nPh_2\underset{\text{(ring)}}{C-\overset{O}{\overset{\|}{C}}-O} \longrightarrow \left[Ph_2C\overset{O}{\overset{\|}{C}}O-\right]_n \quad (86)$$

This material is identical with the polyester obtained by the oxidation of diphenylketene,[76] that is,

$$Ph_2C{=}C{=}O + ROO^{\cdot} \longrightarrow Ph_2\dot{C}-\overset{O}{\overset{\|}{C}}-OOR \longrightarrow Ph_2\underset{\text{(ring)}}{C-\overset{O}{\overset{\|}{C}}-O} + RO^{\cdot} \quad (87)$$

An S_Hi reaction also appears to be involved in the vapor phase decomposition at 225° of 4-iodobutyryl peroxide.[78] 1,3-Diiodopropane and γ-butyrolactone were isolated in 66% yield in exactly a 1:1 ratio, and the gaseous products were CO_2 (92%) and cyclopropane (2–3%). Two reaction mechanisms were suggested. The first involves an S_Hi at oxygen by a carbon radical.

[78] See footnote 64, Chapter V.

$$ICH_2CH_2\dot{C}H_2 + I(CH_2)_3\overset{O}{\overset{\|}{C}}OO\overset{O}{\overset{\|}{C}}(CH_2)_3I \longrightarrow I(CH_2)_3I + \cdot CH_2CH_2CH_2\overset{O}{\overset{\|}{C}}OO\overset{O}{\overset{\|}{C}}(CH_2)_3I \quad (88)$$

$$\cdot CH_2CH_2CH_2\overset{O}{\overset{\|}{C}}OO\overset{O}{\overset{\|}{C}}(CH_2)_3I \longrightarrow \text{(γ-butyrolactone)} + I(CH_2)_3\overset{O}{\overset{\|}{C}}O\cdot \quad (89)$$

$$I(CH_2)_3\overset{O}{\overset{\|}{C}}O\cdot \longrightarrow ICH_2CH_2\dot{C}H_2 + CO_2 \quad (90)$$

The second suggested mechanism seems much less probable since it involves an S_Hi reaction at carbon.

$$I(CH_2)_3\overset{O}{\overset{\|}{C}}O\cdot \longrightarrow \text{(γ-butyrolactone)} + I\cdot \quad (91)$$

The rate of thermal decomposition of *t*-butyl perbenzoate (and of benzoyl peroxide) is enormously accelerated by the introduction into the *ortho* position of methylthio,[79] phenylthio,[79–82] iodo[79,83,84,85] or vinyl[86,87] groups. Although these reactions have occasionally been labeled as intramolecular S_H2 reactions they are actually anchimerically assisted homolyses in which bond formation occurs simultaneously with bond cleavage,[87a] for example,[87]

$$o\text{-}(Ph_2C{=}CH)C_6H_4C(O)OOC(CH_3)_3 \longrightarrow \text{(3-(diphenylmethyl radical)phthalide: } Ph\text{—}\dot{C}\text{—}Ph \text{ on CH of lactone)} + (CH_3)_3CO\cdot \quad (92)$$

[79] W. G. Bentrude and J. C. Martin, *J. Amer. Chem. Soc.*, **84**, 1561 (1962).
[80] J. C. Martin, D. L. Tuleen, and W. G. Bentrude, *Tetrahedron Lett.*, 229 (1962).
[81] D. L. Tuleen, W. G. Bentrude, and J. C. Martin, *J. Amer. Chem. Soc.*, **85**, 1938 (1963).
[82] T. H. Fisher and J. C. Martin, *ibid.*, **88**, 3382 (1966).
[83] W. Cooper, *J. Chem. Soc.*, 3106 (1951).
[84] J. E. Leffler, R. D. Faulkner, and C. Petropoulos, *J. Amer. Chem. Soc.*, **80**, 5435 (1958).
[85] W. Honsberg and J. E. Leffler, *J. Org. Chem.*, **26**, 733 (1961).
[86] T. W. Koenig and J. C. Martin, *ibid.*, **29**, 1520 (1964).
[87] J. C. Martin and T. W. Koenig, *J. Amer. Chem. Soc.*, **86**, 1771 (1964).
[87a] For recent comments on the nature of these and of other, related, transition states, see: C. Walling, H. P. Waits, J. Milovanovic, and C. G. Pappiaonnou, *ibid.*, **92**, 4927 (1970).

The intermolecular analogue of this reaction has also been observed. Thus, by means of ^{18}O labeling experiments, Martin and Drew[88] have shown that the cyclohexyl acetate which is produced when acetyl peroxide is decomposed in cyclohexene is formed by attack of the cyclohexene on the peroxidic oxygen and not by the addition of acetoxy radicals to the double bond.

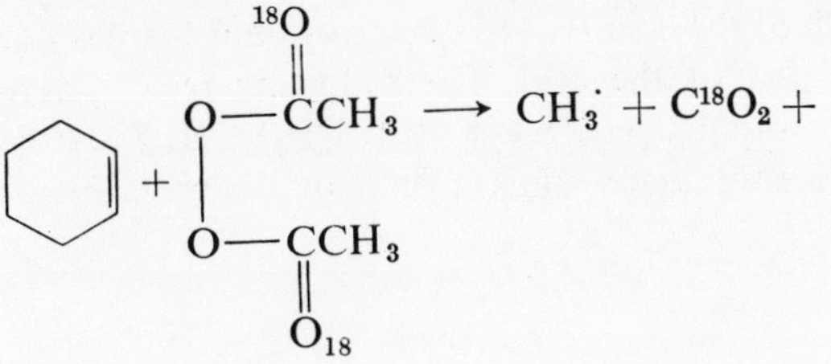

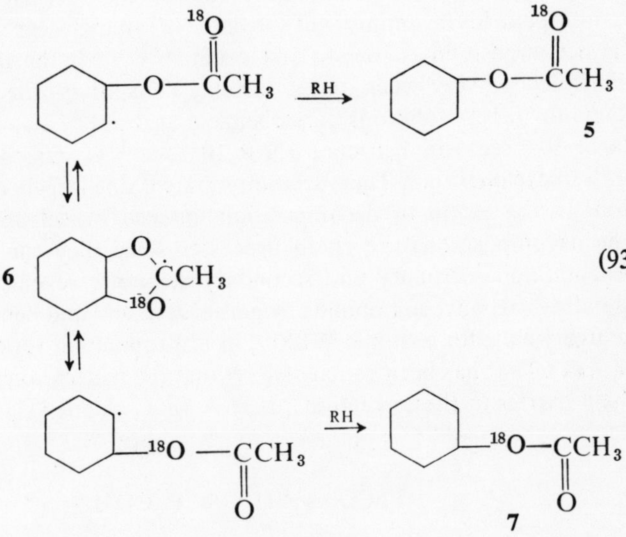

$$\text{C}_6\text{H}_{10} + (\text{CH}_3\text{C}(=^{18}\text{O})\text{O})_2 \longrightarrow \text{CH}_3^{\cdot} + \text{C}^{18}\text{O}_2 +$$

$$\text{RH} \rightarrow \mathbf{5}; \quad \rightleftarrows \mathbf{6} \rightleftarrows; \quad \text{RH} \rightarrow \mathbf{7} \tag{93}$$

Scrambling of the ^{18}O label via the intermediate, **6**, is slightly slower in cyclohexene than is hydrogen atom abstraction. That is, the product **5** with ^{18}O labeled carbonyl oxygen is formed to an extent of 58% and thus predominates over the product, **7**, (42%) which has the ^{18}O labeled oxygen in the saturated position.

The induced decomposition of a number of *t*-butyl peresters by tri-*n* butyltin hydride have been studied.[43b] The production of *t*-butanol is virtually quantitative and so the propagating steps of the chain reaction can be

[88] J. C. Martin and E. H. Drew, *ibid.*, **83,** 1232 (1961).

formulated as,

$$Bu_3^nSn\cdot + Bu^tOOC(O)R \rightarrow Bu_3^nSnOC(O)R + Bu^tO\cdot \qquad (94)$$

$$Bu^tO\cdot + Bu_3^nSnH \rightarrow Bu^tOH + Bu_3^nSn\cdot \qquad (95)$$

Rate constants for reaction 94 in benzene at 10° were estimated by means of competitive experiments with an alkyl halide. The rate constants tend to decrease as the strength of the acid RCO_2H decreases but there is not a linear correlation with the pK_a of the acid. The following results were obtained (R, k_{94} in M^{-1} sec^{-1} units): C_6H_5, 1.4×10^5; $(CH_3)_2CH$, 7×10^4; CH_3, 4×10^4; di-*t*-butylperterephthalate, 56×10^6; and di-*t*-butylperisophthalate, 3×10^7.

(e) PERACIDS

Several groups of workers have studied the thermal decomposition of various peracids in a number of solvents.[89–97] In hydrocarbons and in alcohols the decomposition is pseudo-first order,[91,94,95] but the measured rate constants depend very much on the solvent. Thus, at 79° the apparent rate constants for 0.1 *M* solutions of perbenzoic acid are[95]· 2×10^{-6} sec^{-1} in CCl_4, 2.6×10^{-5} sec^{-1} in benzene, 1.9×10^{-4} sec^{-1} in cumene and $\sim 4 \times 10^{-3}$ sec^{-1} in cyclohexanol. This variation in rate is due largely or entirely to variations in the extent of decomposition induced by solvent derived radicals. The decompositions are chain processes since they can be inhibited with benzoquinone. Primary and secondary alcohols are oxidized to the corresponding carbonyl compounds by perbenzoic acid and benzoic acid is formed in high yields, for example,[95] 100% in isopropanol at room temperature and 92% at 60°. It has been concluded[92,95] that the benzoyloxy radical is the chief chain carrier in these reactions, that is, in a solvent SH,

$$Ph\overset{\overset{O}{\|}}{C}O\cdot + SH \longrightarrow Ph\overset{\overset{O}{\|}}{C}OH + S\cdot \qquad (96)$$

$$S\cdot + Ph\overset{\overset{O}{\|}}{C}OOH \longrightarrow Ph\overset{\overset{O}{\|}}{C}O\cdot + SOH \qquad (97)$$

[89] D. Lefort, C. Paquot, and J. Sorba, *Bull. Soc. Chim. France*, 1385 (1959).
[90] D. Lefort, J. Sorba, and D. Rouillard, *ibid.*, 2219 (1961).
[91] D. Lefort and J. Sorba, *ibid.*, 2373 (1961).
[92] T. M. Luong and D. Lefort, *ibid.*, 827 (1962).
[93] D. L. Heywood, B. Phillips, and H. A. Starbury, Jr., *J. Org. Chem.*, **26,** 281 (1961).
[94] K. Tokumaru, O. Simamura, and M. Fukuyama, *Bull. Chem. Soc. Japan*, **35,** 1673 (1962).
[95] K. Tokumaru and O. Simamura, *ibid.*, **35,** 1678 (1962).
[96] K. Tokumaru and O. Simamura, *ibid.*, **35,** 1955 (1962).
[97] S. R. Cohen and J. O. Edwards, *J. Phys. Chem.*, **64,** 1086 (1960).

Further evidence in favor of an S_H2 reaction at the outer oxygen comes from Lefort, Paquot, and Sorba's report[89] that the decomposition of peroxylauric acid in petroleum ether at 65–70° gives good yields (~80%) of *n*-undecanol.

$$C_{11}H_{23}^{\cdot} + C_{11}H_{23}\overset{\overset{\displaystyle O}{\|}}{C}OOH \longrightarrow$$

$$C_{11}H_{23}OH + (C_{11}H_{23}\overset{\overset{\displaystyle O}{\|}}{C}O^{\cdot}) \longrightarrow C_{11}H_{23}^{\cdot} + CO_2 \quad (98)$$

In addition, a number of hydrocarbon solvents can be quite efficiently hydroxylated by decomposing a peracid in them, for example, cyclohexanol is formed by the decomposition of perlauric acid[90] or peracetic acid[93] in cyclohexane.

The decomposition of perbenzoic acid in ethers is fast but appears to be autoretarding and the rate does not conform to simple first order kinetics.[96] The ethers are converted to carbonyl compounds and alcohols, presumably via the intermediate α-hydroxyalkyl ether, for example,

$$(CH_3)_2\overset{\overset{\displaystyle H}{|}}{C}O\dot{C}(CH_3)_2 + Ph\overset{\overset{\displaystyle O}{\|}}{C}OOH \longrightarrow Ph\overset{\overset{\displaystyle O}{\|}}{C}O^{\cdot} + \left((CH_3)_2\overset{\overset{\displaystyle H}{|}}{C}O\overset{\overset{\displaystyle OH}{|}}{C}(CH_3)_2\right) \quad (99)$$

$$\left((CH_3)_2\overset{\overset{\displaystyle H}{|}}{C}O\overset{\overset{\displaystyle OH}{|}}{C}(CH_3)_2\right) \longrightarrow (CH_3)_2\overset{\overset{\displaystyle H}{|}}{C}OH + (CH_3)_2C{=}O \quad (100)$$

Walling and Azar[69] have questioned whether the benzoyloxy radical can be the principal chain carrier in the induced decomposition of perbenzoic acid because reported yields of benzoic acid are higher in these reactions than in the decomposition of benzoyl peroxide in the same solvents. Although the data of different workers is difficult to compare directly this does appear to be generally the case. For example, Cass[11] reported 80% benzoic acid formed in the decomposition of benzoyl peroxide in diethyl ether at reflux whereas Tokumaru and Simamura[96] found 90% from perbenzoic acid under the same conditions. Even larger differences in benzoic acid yields appear in the alcohol solvents. The gaseous products from the decomposition of perbenzoic acid afford additional evidence that its decomposition is more complex than the decomposition of benzoyl peroxide. In diisopropyl ether at 65° the products include 15% oxygen[96] (together with 10% CO_2 and 80% benzoic acid). Small, but approximately equal quantities of O_2 and CO_2 are formed in the decomposition at 60° in isopropanol[94] (2% O_2, 1.5% CO_2 and 92% benzoic acid). The most likely source of oxygen is the benzoylperoxy radical. The oxygen yields imply that these peroxy radicals must play an important role in the overall decomposition. The simplest reaction sequence would be,

$$\mathrm{Ph\overset{O}{\overset{\|}{C}}O^{\cdot} + Ph\overset{O}{\overset{\|}{C}}OOH \longrightarrow Ph\overset{O}{\overset{\|}{C}}OH + Ph\overset{O}{\overset{\|}{C}}OO^{\cdot}} \quad (101)$$

$$\mathrm{2Ph\overset{O}{\overset{\|}{C}}OO^{\cdot} \longrightarrow 2Ph\overset{O}{\overset{\|}{C}}O^{\cdot} + O_2} \quad (102)$$

An alternative, but rather unlikely, possibility is that the oxygen is formed by an S_H2 reaction of the benzoylperoxy radical on the peracid.

$$\mathrm{Ph\overset{O}{\overset{\|}{C}}O{-}O^{\cdot} \quad \underset{O-O}{\overset{H}{|}}\!\!\diagup\overset{O}{\overset{\diagdown\!\!\diagdown}{C}}Ph \longrightarrow Ph\overset{O}{\overset{\|}{C}}O^{\cdot} + O_2 + \overset{H-O}{\diagdown}CPh\underset{O}{\diagup\!\!\diagup}} \quad (103)$$

(f) ALKYL PEROXIDES

The thermal decomposition of di-*t*-butyl peroxide yields two *t*-butoxy radicals in a first order process.

$$\mathrm{(CH_3)_3COOC(CH_3)_3 \rightarrow 2(CH_3)_3CO^{\cdot}} \quad (104)$$

Raley, Rust, and Vaughan[98] found that the decomposition in such diverse solvents as cumene, *t*-butylbenzene and tri-*n*-butylamine occurred at essentially the same rate as in the gas phase. It was concluded that in these solvents the decompositions were not complicated by any induced chain involving a solvent derived radical. However, the neat liquid was later found to decompose 2–5 times as fast at 110° as a dilute solution of the peroxide in *t*-butylbenzene.[99] Isobutylene oxide, which is not formed in dilute solution in significant amounts, became a major product in the neat liquid (35% based on *t*-butoxy units at 110° and 20% for a photochemically initiated decomposition at 17°). It was suggested, therefore, that the *t*-butoxy radicals abstract hydrogen from the peroxide and the resultant radical then undergoes an S_Hi reaction, that is,

$$\mathrm{(CH_3)_3CO^{\cdot} + (CH_3)_3COOC(CH_3)_3 \longrightarrow (CH_3)_3COH + {}^{\cdot}CH_2(CH_3)_2COOC(CH_3)_3} \quad (105)$$

$$\mathrm{{}^{\cdot}CH_2(CH_3)_2COOC(CH_3)_3 \longrightarrow \underset{\diagdown O \diagup}{CH_2{-}C(CH_3)_2} + (CH_3)_3CO^{\cdot}} \quad (106)$$

[98] J. H. Raley, F. F. Rust, and W. E. Vaughan, *J. Amer. Chem. Soc.*, **70,** 1336 (1948).
[99] E. R. Bell, F. F. Rust, and W. E. Vaughan, *ibid.*, **72,** 337 (1950).

The decomposition of neat di-*t*-amyl peroxide in the liquid phase yields rather less epoxide (5% 1,1,2-trimethylethylene oxide and 1% 1-ethyl-1-methylethylene oxide at 115°).[99a] Up to 1% of *t*-butanol and 1–2% of isobutylene oxide can also be formed in the gas phase decomposition of di-*t*-butylpe roxide which implies that reactions 105 and 106 are not confined to solution.[100] Similarly, an S_Hi reaction must be invoked to account for the formation of furans and other cyclic ethers in the decomposition at 120° of neat bis-(2-methyl-2-hexyl)-peroxide,[101] that is,

$$\text{cyclic: } (H_3C)(CH_3)C(-CH_2-CH_2-\dot{C}H(CH_3))-O-O-C(CH_3)_2-C_4H_9 \longrightarrow (H_3C)(CH_3)C(-CH_2-CH_2-CH(CH_3)-)O + {}^{\cdot}O-C(CH_3)_2-C_4H_9 \quad (107)$$

The formation of a certain amount of epoxide during the liquid phase autoxidation of most olefins occurs in a completely analogous manner.[102,103]

$$ROO^{\cdot} + -\underset{H}{C}-C{=}C- \xrightarrow{b} -\underset{H}{C}-\underset{O-OR}{C}-\dot{C}- \xrightarrow{S_Hi} -\underset{H}{C}-C\overset{O}{—}C- + RO^{\cdot}$$

$$\downarrow a \qquad\qquad \downarrow O_2 \qquad (108)$$

$$ROOH + -\dot{C}-C{=}C- \qquad -\underset{H}{C}-\underset{OOR}{C}-COO^{\cdot} \longrightarrow \text{polyperoxide}$$

The relative rates of hydrogen abstraction, reaction 108a, and peroxy radical addition to the double bond, reaction 108b determine the relative yield of hydroperoxide and addition products (polyperoxide and epoxide). The yield of epoxide relative to polyperoxide will be reduced at high oxygen pressures. At a given oxygen pressure, the greater the stability of the β-peroxyalkyl radical the more likely it is to react with oxygen rather than undergo the S_Hi reaction.

The decomposition of methyl *t*-amyl peroxide at 195° in the gas phase[104]

[99a] E. S. Huyser and K. J. Jankauskas, *J. Org. Chem.*, **35,** 3196 (1970).
[100] L. Batt and S. W. Benson, *J. Chem. Phys.*, **36,** 895 (1962).
[101] H. E. De La Mare and F. F. Rust, *J. Amer. Chem. Soc.*, **81,** 2691 (1959).
[102] F. R. Mayo, *Accounts Chem. Res.*, **1,** 193 (1968).
[103] K. U. Ingold, *ibid.*, **2,** 1 (1969).
[104] J. H. Raley and D. O. Collamer, *J. Amer. Chem. Soc.*, **74,** 1606 (1952).

gives an 8% yield of methyl ethyl ether. This might be formed either in the radical combination reaction,

$$CH_3O^{\cdot} + {}^{\cdot}C_2H_5 \rightarrow CH_3OC_2H_5 \tag{109}$$

or, in an attack of the ethyl radical on the peroxide,

$${}^{\cdot}C_2H_5 + CH_3OOC(CH_3)_2C_2H_5 \rightarrow CH_3OC_2H_5 + {}^{\cdot}OC(CH_3)_2C_2H_5 \tag{110}$$

By way of contrast, the decomposition of di-*t*-butyl peroxide in the gas phase gives only 0.02% *t*-butyl methyl ether.[100] This difference between the two peroxides might be due to the different degrees of steric protection of the O-O bond.

Pryor and co-workers[105–108] have shown that dialkyl peroxides (ethyl, propyl, isopropyl, butyl, *sec*-butyl, and *t*-butyl) are effective chain transfer agents in polymerizing styrene. It was concluded that the polystyryl radical abstracts hydrogen from the peroxide rather than attacking the peroxidic linkage. Similarly, hydrogen atom abstraction from di-*t*-butylperoxide is involved in the induced decomposition of this peroxide by chlorine atoms.[109]

Although the rate of decomposition of di-*t*-butyl peroxide was initially thought to be independent of the solvent, more recent studies by Huyser and co-workers have shown that this peroxide is subject to induced decomposition by radicals which have an hydroxy group or a primary or secondary amino group attached directly to the radical center.[110–116] That is, decomposition is induced by the radicals $>\dot{C}OH$ and $>\dot{C}NHR$. The rates of

[105] See footnote 32, Chapter I.
[106] W. A. Pryor and E. P. Pultinas, *J. Amer. Chem. Soc.*, **85,** 133 (1963).
[107] W. A. Pryor, *Tetrahedron Lett.*, 1201 (1963).
[108] W. A. Pryor, D. M. Huston, T. R. Fiske, T. L. Pickering, and E. Ciuffarin, *J. Amer. Chem. Soc.*, **86,** 4237 (1964).
[109] G. Archer and C. Hinshelwood, *Proc. Roy. Soc.* (*London*), **261A,** 293 (1961).
[110] E. S. Huyser and C. J. Bredeweg, *J. Amer. Chem. Soc.*, **86,** 2401 (1964).
[111] E. S. Huyser, C. J. Bredeweg, and R. M. Van Scoy, *ibid.*, **86,** 4148 (1964).
[112] E. S. Huyser and B. Amini, *J. Org. Chem.*, **33,** 576 (1968).
[113] E. S. Huyser and R. H. C. Feng, *ibid.*, **34,** 1727 (1969).
[114] E. S. Huyser and A. A. Kahl, *Chem. Comm.*, 1238 (1969).
[115] The rate of decomposition, even in "inert" solvents is not really independent of the solvent.[116] The activation parameters for the unimolecular decomposition vary considerably but in a manner such as to nearly compensate one another at the temperature of most experiments. Thus, although the activation energy for the decomposition decreases from 41 kcal/mole in cyclohexane to 31 kcal/mole in acetonitrile the decomposition rate constants at 125° differ by a factor of only 2.3 ($k_{cyclohexane} = 1.5 \times 10^{-5}$ sec^{-1}, $k_{acetonitrile} = 3.5 \times 10^{-5}$ sec^{-1}).[116]
[116] E. S. Huyser and R. M. Van Scoy, *J. Org. Chem.*, **33,** 3524 (1968).

peroxide decomposition are appreciably faster in primary and secondary alcohols[110] (which are oxidized to the corresponding carbonyl compounds) and in primary and secondary amines[111] (which are oxidized to imines) than in various "inert" solvents including hydrocarbons, ethers and *t*-amines (see Table VI). In all the solvents the decomposition is first order in peroxide which implies that in the "fast" solvents the chains are terminated by reaction between the two chain carrying radicals. The induced decompositions can be inhibited by *t*-butoxy radical scavengers such as toluene, 1-octene and methyl methacrylate. The chain propagation steps are

$$(CH_3)_3CO^{\cdot} + \left\{ \begin{array}{c} \text{HCOH} \\ \\ \text{HCNHR} \end{array} \right. \longrightarrow \left. \begin{array}{c} {}^{\cdot}\text{COH} \\ \\ {}^{\cdot}\text{CNHR} \end{array} \right\} + (CH_3)_3COH \qquad (111)$$

$$\left. \begin{array}{c} {}^{\cdot}\text{COH} \\ \\ {}^{\cdot}\text{CNHR} \end{array} \right\} + (CH_3)_3COOC(CH_3)_3 \longrightarrow$$

$$\left. \begin{array}{c} \text{C=O} \\ \\ \text{C=NR} \end{array} \right\} + (CH_3)_3COH + (CH_3)_3CO^{\cdot} \qquad (112)$$

Since decomposition is not induced by tetrahydropyran or by *t*-amines it was concluded that the attack on the peroxide occurs through the hydroxy or amino hydrogen. The transition state for the α-hydroxy alkyl radical, for example, can be represented as,

$$\left[\begin{array}{c} \qquad\qquad\qquad \overset{\delta-}{O}C(CH_3)_3 \\ \qquad\qquad\qquad \vdots \\ >\overset{\delta+}{C}\text{=\!=}O\text{---}H\text{---}OC(CH_3)_3 \end{array} \right]$$

This mechanism receives strong support from Huyser and Kahl's[114] discovery that the decomposition rate is markedly slower in $CH_3CHODC_2H_5$ than in $CH_3CHOHC_2H_5$ (see Table VI). However, these workers observed no difference in the rates of the induced decompositions of either acetyl

Table VI First-Order Rate Constant for the Decomposition at 125° of ~10% Di-*t*-Butyl Peroxide in Various Solvents[110,111,114]

Solvent	$k \times 10^5$ (sec^{-1})	Half Life $\times 10^5$ (sec)
"Inert"[a]	1.4–1.7	4.0–4.9
Cyclohexanol	2.4	2.8
1-Propanol	2.8	2.4
Norborneol	4.4	1.6
2-Butanol[b]	4.5	1.4
2-Butanol-OD[b]	2.8	2.3
2-Octanol	5.5	1.3
Benzhydrol	8.7	0.8
Piperidine	3.5	2.0
N-Ethylcyclohexylamine	4.0	1.7
Cyclohexylamine	5.5	1.3

a. Cyclohexane, benzene, toluene, cumene, *t*-butylbenzene, *n*-butyl mercaptan, tetrahydropyran, tri-*n*-butylamine, N-methyl piperidine and γ-collidine.
b. 20% peroxide.

peroxide or *t*-butyl peracetate in 2-butanol and in 2-butanol-OD. They conclude that in these two cases the α-hydroxyalkyl radical reacts with the peroxide linkage either via a direct S_H2 attack by the carbon radical center on oxygen as is normally assumed or else, perhaps, by a rate controlling electron-transfer reaction of the type,

$$\cdot\overset{|}{\underset{|}{C}}OH + CH_3\overset{O}{\overset{\|}{C}}OOR \longrightarrow \underbrace{\overset{+}{>C}OH + CH_3\overset{O}{\overset{\|}{C}}O^-}_{} + RO\cdot \tag{113}$$

$$\downarrow$$

$$>C{=}O + CH_3\overset{O}{\overset{\|}{C}}OH$$

Although the decomposition of di-*t*-butyl peroxide is not induced in tetrahydropyran its decomposition is induced by benzyl methyl ethers.[117,118] At 110°, the decomposition of the peroxide in substituted benzyl methyl ethers, $ArCH_2OMe$, gave, besides small quantities of the aldehyde ArCHO and varying amounts of the dimer $(Ar\overset{|}{C}HOMe)_2$, the acetals $ArCH(OMe)(OBu^t)$

[117] R. L. Huang, T. W. Lee, and S. H. Ong, *Chem. Comm.*, 1251 (1968).
[118] *Idem*, *J. Chem. Soc.* (*C*), 2522 (1969).

in yields of 85% (Ar = p-$NO_2C_6H_4$), 62% (3, 4-$Cl_2C_6H_3$), 36% (p-ClC_6H_4), 17% (m-ClC_6H_4), 8.5% (p-$Bu^tC_6H_4$), 2.8% (C_6H_5) and 0% (p-$C_6H_5C_6H_4$ and p-$MeOC_6H_4$). The half life in p-chlorobenzyl methyl ether was reduced by a factor of 2.5 compared with the value found in t-butylbenzene which certainly implies an induced decomposition. Further evidence in support of an induced decomposition was obtained by carrying out the reaction at 60° at which temperature di-t-butyl peroxide is thermally stable. However, in the presence of p-chlorobenzyl methyl ether (10 m moles) and t-butyl hyponitrite (1 m moles) to initiate the reaction, 25% of the peroxide (2 m moles initially) was consumed in 280 min. The S_H2 reaction must involve the attack of the α-alkoxyalkyl radical through the carbon radical center, for example,

$$p\text{-}ClC_6H_4\dot{C}H(OCH_3) + (CH_3)_3CO\text{—}OC(CH_3)_3 \longrightarrow p\text{-}ClC_6H_4CH(OCH_3)\text{—}O\text{—}C(CH_3)_3 + (CH_3)_3CO \tag{114}$$

The substituent effect indicates that this reaction is facilitated by an increase in the electrophilicity of the attacking ether radical. It is suggested that the reaction may be a stepwise rather than a synchronous process.

Additional support for the S_H2 reaction 114 comes from Goh and Ong's[119] report that the rates of decomposition of 1 M di-t-butyl peroxide in p-chlorobenzyl, m-chlorobenzyl, and 3,4-dichlorobenzyl methyl ether at 110° are reduced some 20–30% by the addition of 1 M α-methylstyrene as a radical trap. Relative rates of decomposition of the peroxide suggest that induced decomposition may occur in solvents other than benzyl ethers and alcohols. After 72 hours at 110° the percentage decompositions in some of the solvents studied were, decalin 43%, t-butylbenzene 49%, dibenzyl ether 50%, benzyl methyl ether 56%, p-chlorotoluene 59%, benzyl chloride 60%, benzyl ethyl ether 62%, benzylamine 62%, benzaldehyde 68%, chlorobenzene 69%, carbon tetrachloride 71%, p-chlorobenzyl methyl ether 85%, benzyl alcohol 85%, α-methylbenzyl alcohol 88%, 3,4-dichlorobenzyl methyl ether 99%. Some of these differences are probably due to solvent effects on the rate of the initial homolytic scission of the peroxide rather than to the occurrence of an induced decomposition. For instance, it seems improbable that the induced decomposition could be more important in chlorobenzene than in decalin or t-butylbenzene. On the other hand, an induced decomposition must be

[119] S. H. Goh and S. H. Ong, *J. Chem. Soc.* (*B*), 870 (1970).

involved in CCl_4 as the extent of decomposition is reduced from 71% to 58% by the addition of iodine as a free radical trap.

The results of Huang, Lee, and Ong[117] that show the decomposition of di-*t*-butyl peroxide is not induced in tetrahydropyran stand in interesting contrast to Walling and Azar's[69] finding that the decomposition of *t*-butyl perbenzoate is faster in cyclic ethers than in acyclic ethers (isopropyl and *n*-butyl). It would be interesting to know whether or not benzyl methyl ethers are also especially effective in inducing the decomposition of peresters and acyl peroxides.

The decomposition of di-*t*-butyl peroxide in secondary alcohols is catalyzed by small amounts of 4-pyridone.[113] The catalysis arises from the ability of 4-pyridone to act as a hydrogen atom carrier by reacting with the α-hydroxyalkyl radical derived from the alcohol. The resultant 4-hydroxymonohydropyridyl radical reacts rapidly with the peroxide. The overall propagation sequence is

$$(CH_3)_3CO^{\cdot} + H\overset{|}{\underset{|}{C}}OH \longrightarrow (CH_3)_3COH + {}^{\cdot}\overset{|}{\underset{|}{C}}OH \qquad (115)$$

$${}^{\cdot}\overset{|}{\underset{|}{C}}OH + O\!=\!C_5H_4NH \longrightarrow \;>\!C\!=\!O + HO\!-\!C_5H_4NH \qquad (116)$$

$$HO\!-\!\dot{C}_5H_4NH + (CH_3)_3COOC(CH_3)_3 \longrightarrow$$

$$(CH_3)_3CO^{\cdot} + (CH_3)_3COH + HO\!-\!C_5H_4N \;\left(\longrightarrow O\!=\!C_5H_4NH\right) \qquad (117)$$

The decomposition of di-*t*-butyl peroxide is also induced by 1,4-dihydropyridines,[111] for example,

$$\text{(2,6-dimethyl-3,5-bis(EtO}_2\text{C)-1,4-dihydropyridine, 4-H)} + (CH)_3COOC(CH_3)_3 \longrightarrow$$

$$(CH_3)_3CO^{\cdot} + (CH_3)_3COH + \text{(2,6-dimethyl-3,5-bis(EtO}_2\text{C)pyridine)} \qquad (118)$$

Both reactions 117 and 118 involve an S_H2 reaction at the peroxide linkage in which the radical attacks the oxygen with the *formal* site of the unpaired electron being oriented away from the reaction center.

The S_H2 reaction between 2,6-di-*t*-butyl-4-substituted phenoxy radicals and di-*t*-butyl peroxide provides another example of a free radical attacking the peroxide with the *formal* site of its unpaired electron oriented away from the reaction center.[120]

$$\cdot O\text{—}C_6H_2(Bu^t)_2\text{—}R + (CH_3)_3COOC(CH_3)_3 \longrightarrow O{=}C_6H_2(Bu^t)_2(R)(OC(CH_3)_3) + (CH_3)_3CO\cdot \qquad (119)$$

The reaction is first order in phenoxy radical and in peroxide. From 10° to 50° in a large excess of peroxide the pseudo first order rate constants for the different substituents can be represented by:

R— in Reaction (119)	k_{119} (sec^{-1})
$(CH_3)_3C$—	$4 \times 10^6 \exp(-9900/RT)$
$CH_3C(=O)\text{—}C_6H_4$— (para)	$7 \times 10^6 \exp(-10200/RT)$
1 naphthyl	$1.2 \times 10^7 \exp(-10400/RT)$
$Cl\text{—}C_6H_4$— (para)	$9 \times 10^6 \exp(-10500/RT)$
C_6H_5—	$1.3 \times 10^7 \exp(-11000/RT)$
$CH_3\text{—}C_6H_4$— (para)	$1.2 \times 10^7 \exp(-11200/RT)$
$C_6H_4(OCH_3)$— (ortho)	$2 \times 10^7 \exp(-11400/RT)$
$CH_3O\text{—}C_6H_4$— (para)	$2.6 \times 10^7 \exp(-11900/RT)$

[120] A. I. Prokof'ev, S. P. Solodovnikov, G. N. Bogdanov, G. A. Nikiforov, and V. V. Ershov, *Teor. Eksp. Khim.*, 3, 416 (1967).

The increase in activation energy down this list of substituents is paralleled by a decrease in the spin density at the *para* position of the phenoxy radical, that is, a decrease in spin density at the reaction center. Further work on this reaction[120a] has yielded the following bimolecular rate constants for 4-(X-substituted) phenyl-2,6-di-*t*-butylphenoxy radicals at 20° (X given and $k_{119} \times 10^5$ in M^{-1} sec^{-1} units): MeO, 0.63; *t*-Bu, 1.04; *iso*-Pr, 1.02; Et, 1.08; Me, 1.04; H, 1.37; Cl, 2.75; $COCH_3$, 3.80; and PhO, 1.10.

Huyser and Amini[112] have utilized the α-hydroxyalkyl radical induced decomposition of di-*t*-butyl peroxide to alkylate 2-hydroxy-1,4-naphthoquinone in the 3-position. In the presence of a hydrocarbon such as toluene the following chain reaction is set up,

$$(CH_3)_3CO^{\cdot} + RH \longrightarrow (CH_3)_3COH + R^{\cdot} \qquad (120)$$

R˙ + [2-hydroxy-1,4-naphthoquinone] ⟶ [3-R-2-hydroxy-1,4-dioxo-naphthyl radical] (121)

[3-R-2-hydroxy-1,4-dioxo-naphthyl radical] + $(CH_3)_3COOC(CH_3)_3$

⟶ [3-R-1,2,4-trioxo-tetrahydronaphthalene] + $(CH_3)_3CO^{\cdot}$ + $(CH_3)_3COH$ (122)

[3-R-1,2,4-trioxo-tetrahydronaphthalene] ⟶ [2-hydroxy-3-R-1,4-naphthoquinone] (123)

[120a] A. I. Prokof'ev, S. P. Solodovnikov, G. A. Nikiforov, and V. V. Ershov, *Izv. Akad. Nauk, SSSR, Ser. Khim.*, 558 (1970).

This hydroxy naphthoquinone can also be alkylated in the 3-position by acyl peroxides.[121,122] Presumably in this case reaction proceeds via an S_H2 attack by carbon to form an intermediate hemiacylol, that is,

$$\text{(2-hydroxy-3-R-1,4-naphthoquinone radical)} + RC(=O)OOC(=O)R \longrightarrow \text{(2-hydroxy-2-acyloxy-3-R-1,4-naphthalenedione)} + R^{\cdot} + CO_2^{\cdot} \quad (124)$$

$$\text{(2-hydroxy-2-acyloxy-3-R-1,4-naphthalenedione)} \longrightarrow \text{(2-hydroxy-3-R-1,4-naphthoquinone)} + RC(=O)OH \quad (125)$$

The powerful reducing radical $^{\cdot}CO_2H$ (which is the conjugate acid of the carbon dioxide radical-anion, $^{\cdot}CO_2^-$) has been reported to induce the decomposition of hydrogen peroxide, *t*-butyl hydroperoxide and di-*t*-butyl peroxide.[123] The reaction is formulated as a one electron transfer process, for example,

$$^{\cdot}CO_2H + (CH_3)_3COOR \rightarrow CO_2 + H^+ + (CH_3)_3CO^{\cdot} + RO^- \quad (126)$$

$$(R = H \quad \text{or} \quad C(CH_3)_3)$$

which seems not unlikely for H_2O_2 or the hydroperoxide. However, such a reaction would be very slow with the peroxide since this compound is relatively unreactive even towards such powerful one electron reducing agents as cobalt (II) and iron (II).

The decomposition of di-*t*-butyl peroxide and *t*-butyl cumyl peroxide is not induced by trialkyl tin hydrides.[42] This result is surprising in view of the high reactivity of trialkyltin radicals towards acyl peroxides and even esters. Presumably these tertiary peroxides are too hindered for attack by trialkyltin radicals since reaction occurs readily with diethyl peroxide.[43b] The rate constant for the attack of tributyltin radicals on diethyl peroxide in benzene at 10° is $7.5 \times 10^4 \ M^{-1} \text{ sec}^{-1}$.[43b]

[121] L. F. Fieser and A. E. Oxford, *J. Amer. Chem. Soc.*, **64,** 2060 (1942).
[122] L. F. Fieser et al., *ibid.*, **70,** 3174 (1948).
[123] R. O. C. Norman and P. R. West, *J. Chem. Soc. (B)*, 389 (1969).

(g) HYDROPEROXIDES

The decomposition of cumene hydroperoxide in cumene at 110–160° involves an induced chain, the propagation steps for which were originally proposed to be,[124]

$$RO^{\cdot} + RH \rightarrow ROH + R^{\cdot} \qquad (127)$$

$$R^{\cdot} + ROOH \rightarrow ROH + RO^{\cdot} \qquad (128)$$

More recent studies on the free radical induced decomposition of a number of tertiary hydroperoxides have shown that the principal reaction sequence actually involves attack of the alkoxy radical on the hydroperoxide to yield an alkylperoxy radical which then regenerates the alkoxy radical,[125–129] that is,

$$RO^{\cdot} + ROOH \longrightarrow ROH + ROO^{\cdot} \qquad (129)$$

$$ROO^{\cdot} + ROO^{\cdot} \begin{cases} \nearrow ROOR + O_2 \quad (10\text{–}20\%) \\ \searrow 2RO^{\cdot} + O_2 \quad (90\text{–}80\%) \end{cases} \qquad (130)$$

However there is abundant evidence that carbon radicals can induce the decomposition of hydroperoxides, though attack does not necessarily occur at oxygen. For instance, Berezin, Kazanskaya, and Ugarova[130] have used tritium labeling to show that methyl radicals can abstract hydrogen from *t*-butyl hydroperoxide both from the hydroperoxidic group (k_{OH}) and from the butyl group (k_{CH}). In heptane the relative rates of attack on the two positions is given by $k_{OH}/k_{CH} = 0.03 \exp(-7000/RT) = 640$ at 80°. The high reactivity of the hydroperoxidic hydrogen is reduced in water because of hydrogen bonding. In this solvent $k_{OH}/k_{CH} = 87 \exp(-1100/RT) = 18$ at 80°.

The thermal (150°) and photo (2537 Å) initiated reaction of *t*-butyl hydroperoxide with methyl isobutyrate yields methyl 2-hydroxyisobutyrate in 14–34% yield.[131,132] The same product is formed in 5% yield when 2-carbomethoxy-2-propyl radicals were independently generated by the thermolysis

[124] G. H. Twigg, G. W. Godin, H. C. Bailey, and J. Holden, *Erdoel Kohle*, **15**, 74 (1962).
[125] R. Hiatt, J. Clipsham, and T. Visser, *Can. J. Chem.*, **42**, 2754 (1964).
[126] A. Factor, C. A. Russell, and T. G. Traylor, *J. Amer. Chem. Soc.*, **87**, 3692 (1965).
[127] J. R. Thomas, *ibid.*, **87**, 3935 (1965).
[128] *Idem. ibid.*, **89**, 4872 (1967).
[129] J. A. Howard and K. U. Ingold, *Can. J. Chem.*, **47**, 3797 (1969).
[130] I. V. Berezin, N. F. Kazanskaya, and N. N. Ugarova, *Zhur. Fiz. Khim.*, **40**, 766 (1966).
[131] D. J. Trecker and R. S. Foote, *Chem. Comm.*, 841 (1967).
[132] *Idem.*, *J. Org. Chem.*, **33**, 3527 (1968).

(79–92°) of dimethyl 2,2′-azobis(isobutyrate) under deoxygenated conditions. It was suggested that the methyl 2-hydroxyisobutyrate was formed by the following S_H2 reaction at oxygen.

$$(CH_3)_2\dot{C}COOCH_3 + (CH_3)_3COOH \longrightarrow HOC(CH_3)_2COOCH_3 + (CH_3)_3CO^{\cdot} \quad (131)$$

However, even under "deoxygenated conditions" the solution containing the hydroperoxide and a source of free radicals would rapidly become saturated with oxygen.[125,129] It seems unlikely that the hydroperoxide could compete with oxygen for the alkyl radical and the following alternate mode of formation of the hydroxyisobutyrate is suggested.

$$(CH_3)_2\dot{C}COOCH_3 \xrightarrow{O_2} {}^{\cdot}OOC(CH_3)_2COOCH_3 \xrightarrow{t\text{-BuOO}^{\cdot}}$$

$$\underset{(+O_2)}{{}^{\cdot}OC(CH_3)_2COOCH_3} \xrightarrow{t\text{-BuOOH}} \underset{(+t\text{-BuOO}^{\cdot})}{HOC(CH_3)_2COOCH_3} \quad (132)$$

More than 20 years ago Merz and Waters[133] showed that the oxidation of many organic compounds by Fentons reagent (H_2O_2 + Fe(II)) involved a short chain reaction. The propagation steps can be represented as,

$$RH + {}^{\cdot}OH \rightarrow R^{\cdot} + H_2O \quad (133)$$

$$R^{\cdot} + H_2O_2 \rightarrow ROH + {}^{\cdot}OH \quad (134)$$

Compounds which underwent this chain reaction included primary and secondary alcohols, aldehydes, α-hydroxy-acids, amino-acids, and ethers. These compounds yield α-hydroxyalkyl radicals, α-aminoalkyl radicals and α-alkoxyalkyl radicals all of which can induce the decomposition of diacyl peroxides and the first two of which induce the decomposition of dialkyl peroxides. Although reaction 134 may occur in the straightforward manner indicated above, Merz and Waters were careful to point out that the radicals produced from the inducing solvents readily reduced mercuric ion or iodine, for example,

$$Hg^{++} + CH_3\dot{C}HOH \rightarrow Hg^{+} + CH_3\overset{+}{C}HOH \rightarrow CH_3CHO + H^{+} \quad (135)$$

[133] J. H. Merz and W. A. Waters, *J. Chem. Soc.*, S, 15 (1949).

This raises the interesting question, still unanswered, as to whether the induced decomposition involves a rate controlling electron transfer.

$$R^{\cdot} + H_2O_2 \rightarrow R^+ + OH^- + {}^{\cdot}OH \rightarrow ROH + {}^{\cdot}OH \qquad (136)$$

Norman and West[123] have recently reinvestigated the reduction of hydrogen peroxide by hydroxyalkyl radicals using e.s.r. spectroscopy to follow the reactions. The hydroxyalkyl radicals were prepared by oxidation of alcohols with hydroxyl radicals produced in the titanium (III)–hydrogen peroxide reaction. The order of reactivity towards hydrogen peroxide increased along the series ${}^{\cdot}CH_2CH(OH)Me < {}^{\cdot}CH_2CMe_2OH < {}^{\cdot}CH_2OH < {}^{\cdot}CMe_2OH < {}^{\cdot}CO_2H$. This order is that of increasing electron donating ability (i.e., increasing stability of the resulting carbonium ion). It is also the same order as that obtained for the one electron reduction of nitro-compounds and biacetyl. This might seem to favor the electron transfer reaction. However, the radicals exhibit exactly the same order of reactivity towards oxygen. This might indicate that in the reaction with oxygen an initial electron transfer is followed by the union of the organic cation and the oxygen radical anion,

$$R^{\cdot} + O_2 \rightarrow R^+ + O_2^{-\cdot} \rightarrow ROO^{\cdot} \qquad (137)$$

but, on the other hand, it might only imply that the direct formation of the peroxy radical occurs via a transition state which is significantly stabilized by contributions from dipolar canonical structures, $R^{+\cdot}\text{---}O_2^-$.[134] By analogy, the reduction of hydrogen peroxide might also involve only the partial transfer of an electron in the transition state

$$R^{\cdot} + H_2O_2 \longrightarrow \left[\overset{\delta+}{R}\text{---}\overset{H\delta-}{OOH}\right]^{\cdot} \longrightarrow ROH + {}^{\cdot}OH \qquad (138)$$

Alternatively, the reduction of the hydrogen peroxide might involve the hydroxy hydrogen of the radical, that is,

$$H_2\dot{C}OH + H_2O_2 \longrightarrow \left[H_2CO\text{---}H\text{---}\overset{H}{OOH}\right]^{\cdot} \longrightarrow H_2C{=}O + H_2O + {}^{\cdot}OH \qquad (139)$$

It is clear that the detailed mechanisms of these reactions have still to be determined.

Seddon and Allen[135] have estimated a rate constant of $1.5 \times 10^5\ M^{-1}\ \text{sec}^{-1}$ for the room-temperature reaction between the 1-hydroxyethyl radical and hydrogen peroxide. For the hydroxymethyl radical a rate constant of $4.0 \times 10^4\ M^{-1}\ \text{sec}^{-1}$ has been obtained.[136] The rate constants reported by James and Sicilio[137] for the induced decomposition of hydrogen peroxide in isopropanol, namely, for $(CH_3)_2\dot{C}OH$, $k = 4 \times 10^2\ M^{-1}\ \text{sec}^{-1}$ at room temperature with $E = 11.5$ kcal/mole and for ${}^{\cdot}CH_2CH(CH_3)OH$, $k = 2.5 \times 10^2\ M^{-1}\ \text{sec}^{-1}$,

[134] cf. G. A. Russell and R. F. Bridger, *J. Amer. Chem. Soc.*, **85**, 3765 (1963).
[135] W. A. Seddon and A. O. Allen, *J. Phys. Chem.*, **71**, 1914 (1967).
[136] I. S. Ginns, Ph.D. thesis, University of Manitoba (1970).
[137] R. E. James and F. Sicilio, *J. Phys. Chem.*, **74**, 1166 (1970).

$E = 11.6$ kcal/mole, appear to be very small. It is also surprising that the α-hydroxyalkyl radical should not be very much more efficient than the β-hydroxyalkyl radical at promoting this decomposition.

Although intermolecular homolytic substitutions involving the attack of carbon radicals on the peroxide link of hydroperoxides are not very well established there appears to be little doubt that intramolecular substitutions play an important role in the gas phase oxidation of hydrocarbons. These S_Hi reactions are probably responsible for the formation of cyclic ethers in relatively high yields in the cool-flame region.[138–141] Thus, for example,[139] 2-methylpentane in the cool-flame temperature range (250°–300°, [RH] = 50–150 mm, $[O_2]/[RH] = 2$) yields substantial quantities of the O-heterocycles, 2,2-dimethyltetrahydrofuran and 2,4-dimethyltetrahydrofuran.

$$CH_3C(CH_3)(OO^{\cdot})CH_2CH_2CH_3 \longrightarrow CH_3C(CH_3)(O\text{—}OH)CH_2CH_2\dot{C}H_2 \longrightarrow CH_3C(CH_3)CH_2CH_2(O\text{—}CH_2) + {}^{\cdot}OH \qquad (140)$$

$$CH_3C(CH_3)(H)CH_2CH(OO^{\cdot})CH_3 \longrightarrow \dot{C}H_2C(CH_3)(H)CH_2CH(O\text{—}OH)CH_3 \longrightarrow HC(CH_3)CH_2CHCH_3(CH_2\text{—}O) + {}^{\cdot}OH \qquad (141)$$

Although the S_Hi reactions outlined above are attractive these O-heterocycles may, in fact, be formed via the biradicals

$$(CH_3)_2C(\dot{O})CH_2CH_2\dot{C}H_2 \quad \text{and} \quad \dot{C}H_2(CH_3)CCH_2CH(\dot{O})CH_3$$

(h) NONPEROXIDIC COMPOUNDS

Examples of S_H2 reactions at nonperoxidic oxygen are relatively rare except for carbonyl oxygen, that is, except for oxygen in a terminal position. For nonterminal oxygen the two most firmly established S_H2 reactions involve substitutions by trialkyltin radicals.

Neumann and Lind[142,143] have shown that the decomposition of alkyl hyponitrites is induced by trialkyltin hydrides. The overall reaction can be

138 J. H. Jones and M. R. Fenske, *Ind. Eng. Chem.*, **51,** 262 (1959).
139 A. A. Fish, *Proc. Roy. Soc.* (*London*), **A298,** 204 (1967).
140 *Idem., Advances in Chem. Series*, **76,** 69 (1968).
141 *Idem., Angew Chem. Intern. Eng. Ed.*, **7,** 45 (1968).
142 W. P. Neumann and H. Lind, *Angew. Chem. Intern. Eng. Ed.*, **6,** 76 (1967).
143 *Idem., Chem. Ber.*, **101,** 2837 (1968).

represented as,

$$\text{RONNOR} \longrightarrow 2\text{RO}^{\cdot} + \text{N}_2 \tag{142}$$

$$\text{RO}^{\cdot} + \text{R}'_3\text{SnH} \longrightarrow \text{ROH} + \text{R}'_3\text{Sn}^{\cdot} \tag{143}$$

$$\text{R}'_3\text{Sn}^{\cdot} + \text{RONNOR} \longrightarrow \left[\begin{array}{l}\text{R}'_3\text{Sn}\text{---}\text{OR} \\ \quad\quad\quad\;\;\vdots \\ \quad\quad\quad\;\;\text{NNOR}\end{array}\right] \longrightarrow \text{R}'_3\text{SnOR} + \text{N}_2 + {}^{\cdot}\text{OR} \tag{144}$$

The decomposition of alkyl hyponitrites in solution is not induced either by increasing the concentration of hyponitrite or by the addition of galvinoxyl, iodine, chloranil, methanol or 4-*t*-butylthiophenol. However, at 30° a 0.2 *M* solution of dibenzyl hyponitrite has a half life of 165 minutes which is reduced to 84 min by the addition of 0.4 *M* triethyltin hydride and to 22 min with 1.6 *M* tin hydride. Similarly, the rate of decomposition of di-*t*-butyl hyponitrite is increased by tin hydrides. For both hyponitrites the induced reaction can be inhibited by the powerful radical trap, galvinoxyl. The ease of reaction 144 presumably reflects the low strength of the O–N bond in the reactant and the high strength of the Sn–O bond in the product.

Khoo and Lee[144] have shown that the reduction of esters by trialkyltin hydrides at temperatures of 80° and higher is a free-radical chain process. The products are the trialkyltin ester and the hydrocarbon.

$$\text{R}'_3\text{SnH} + \text{RO}\overset{\overset{\text{O}}{\|}}{\text{C}}\text{R}'' \longrightarrow \text{R}'_3\text{SnO}\overset{\overset{\text{O}}{\|}}{\text{C}}\text{R}'' + \text{RH} \tag{145}$$

The reduction of optically active α-phenylethyl benzoate with tri-*n*-butyltin deuteride in benzene at 80° gave racemic α-deuteriophenylethane. The unreacted ester retained its optical activity so racemization could not have occurred prior to reduction. The reaction must be an S_H2 process in which the trialkyltin radical attacks either the ethereal oxygen atom of the ester,

$$\text{Bu}^n_3\text{Sn}^{\cdot} + \underset{\underset{\text{O=C—Ph}}{|}}{\text{O}}\text{—}\overset{\overset{\text{CH}_3}{*|}}{\underset{\underset{\text{H}}{|}}{\text{C}}}\text{—Ph} \longrightarrow \text{Bu}^n_3\text{Sn—}\underset{\underset{\text{O=C—Ph}}{|}}{\text{O}} + {}^{\cdot}\overset{\overset{\text{CH}_3}{|}}{\underset{\underset{\text{H}}{|}}{\text{C}}}\text{—Ph} \tag{146}$$

or the carbonyl oxygen,

$$\text{Bu}^n_3\text{Sn}^{\cdot} + \text{O=}\underset{\underset{\text{Ph}}{|}}{\text{C}}\text{—O—}\overset{\overset{\text{CH}_3}{*|}}{\underset{\underset{\text{H}}{|}}{\text{C}}}\text{—Ph} \longrightarrow \text{Bu}^n_3\text{Sn—O—}\underset{\underset{\text{Ph}}{|}}{\text{C}}\text{=O} + {}^{\cdot}\overset{\overset{\text{CH}_3}{|}}{\underset{\underset{\text{H}}{|}}{\text{C}}}\text{—Ph} \tag{147}$$

[144] L. E. Khoo and H. H. Lee, *Tetrahedron Lett.*, 4351 (1968).

The displaced 1-phenylethyl radical racemizes before picking up deuterium.

$$\text{Ph\dot{C}HCH}_3 + \text{Bu}^n_3\text{SnD} \longrightarrow \text{PhC(D)HCH}_3 + \text{Bu}^n_3\text{Sn}^\cdot \qquad (148)$$

Competitive experiments showed that benzyl benzoate was about $\frac{1}{40}$ as reactive as benzyl chloride at 80°. The rate constant for chlorine atom abstraction from benzyl chloride by the tin radical is $6.4 \times 10^5\ M^{-1}\ \text{sec}^{-1}$ at 25°.[144a] Assuming a normal pre-exponential factor of $10^{8.5}\ M^{-1}\ \text{sec}^{-1}$, the activation energy for chlorine abstraction will be ~3.7 kcal/mole and, hence, the rate constant will be $1.8 \times 10^6\ M^{-1}\ \text{sec}^{-1}$ at 80°. The absolute rate constant for the attack of the tributyltin radical on benzyl benzoate is therefore $\sim 4.5 \times 10^4\ M^{-1}\ \text{sec}^{-1}$ at 80°. Khoo and Lee[144] also used competitive experiments to measure the reactivity of three other benzyl esters relative to benzyl benzoate:

$$\text{Bu}^n_3\text{Sn}^\cdot + \text{RC(=O)OCH}_2\text{Ph} \longrightarrow \text{Bu}^n_3\text{SnOC(=O)R} + \text{PhCH}_2^\cdot \qquad (149)$$

R	Relative k_{149} at 80°	Absolute k_{149} (M^{-1} sec^{-1})
CH_3	1.0	0.7×10^4
Ph	6.3	4.5×10^4
H	7.5	5.3×10^4
CF_3	35	25×10^4

The reaction is clearly facilitated by electron withdrawing R groups as we would expect for an attack by the nucleophilic tributyltin radical.

Hydrogen atoms also seem capable of effecting S_H2 reactions at nonterminal oxygen. Thus, Merak[145] has shown that when powdered alloxan is exposed to a beam of thermal hydrogen atoms at room temperature an α-hydroxy radical is produced.

$$\text{(alloxan hydrate: } O{=}C(NH{-}C(=O))_2C(OH)_2\text{)} + H^\cdot \longrightarrow O{=}C(NH{-}C(=O))_2\dot{C}(OH) + H_2O \qquad (150)$$

The same radical was produced when deuterium atoms were used in place of

[144a] See footnote 18, Chapter III.
[145] J. N. Merak, *J. Amer. Chem. Soc.*, **91**, 5171 (1969).

hydrogen. An analogous reaction is the proposed possible substitution of D for H in water.[146]

$$D^{\cdot} + HOH \longrightarrow \left[\begin{array}{c} D \\ | \\ O \\ / \quad \backslash \\ H \qquad H \end{array} \right]^{\cdot} \longrightarrow DOH + H^{\cdot} \qquad (151)$$

The reaction of a protonated α-aminoalkyl radical with a protonated trialkylamine oxide has been suggested to involve an S_H2 reaction at oxygen,[147] that is,

$$(CH_3)_2\underset{H}{\overset{+}{N}}\dot{C}H_2 + (CH_3)_3\overset{+}{N}OH \rightarrow (CH_3)_3\overset{+}{N}^{\cdot} + (CH_3)_2\underset{H}{\overset{+}{N}}CH_2OH \qquad (152)$$

$$(CH_3)_2\underset{H}{\overset{+}{N}}CH_2OH \rightarrow (CH_3)_2\overset{+}{N}H_2 + CH_2O \qquad (153)$$

No account of S_H2 reactions at oxygen would be complete without some mention of substitutions at the oxygen of carbonyl groups. This reaction is particularly easily effected by α-hydroxyalkyl radicals, the most notable reaction being the photochemical reduction of benzophenone to benzpinacol in isopropanol. Bäckström[148] put forward the following reaction scheme,

$$Ph_2CO \xrightarrow{h\nu} Ph_2CO^* \longrightarrow Ph_2\dot{C}{-}\dot{O} \qquad (154)$$

$$Ph_2\dot{C}{-}\dot{O} + Me_2CHOH \longrightarrow Ph_2\dot{C}OH + Me_2\dot{C}OH \qquad (155)$$

$$2Ph_2\dot{C}OH \longrightarrow \underset{OH}{Ph_2C}{-}\underset{OH}{CPh_2} \qquad (156)$$

$$2Me_2\dot{C}OH \longrightarrow Me_2CO + Me_2CHOH \qquad (157)$$

He reported a quantum yield for the formation of pinacol and acetone of 0.5. Subsequently, Pitts et al.[149] reported that the quantum yield of pinacol formation is approximately unity and explained this by replacing Bäckström's reaction 157 by reaction 158.

$$Me_2\dot{C}OH + Ph_2CO \rightarrow Me_2CO + Ph_2\dot{C}OH \qquad (158)$$

Beckett and Porter[150] have since shown that the quantum yield in isopropanol is concentration-dependent, increasing with increasing benzophenone concentration from a probable lower limit of 0.5 at concentrations $\sim 10^{-6}$ *M*

[146] See p. 107 of Footnote 42, Chapter V.
[147] J. P. Ferris, R. D. Gerwe, and G. R. Gapski, *J. Amer. Chem. Soc.*, **89,** 5270 (1967).
[148] H. L. J. Bäckström, *The Svedburg (1884–1944)—60th Birthday Memorial Volume*, Uppsala University, 1944, p. 45.
[149] J. N. Pitts, R. L. Letsinger, P. P. Taylor, J. M. Patterson, G. Rechtenwald, and R. B. Martin, *J. Amer. Chem. Soc.*, **81,** 1068 (1959).
[150] A. Beckett and G. Porter, *Trans. Faraday Soc.*, **59,** 2038 (1963).

towards an upper limit of 2.0 at concentrations above 0.1 *M*. This change in quantum efficiency reflects the competition between the two routes by which $Me_2\dot{C}OH$ radicals can disappear, that is, reaction 157 at low benzophenone concentrations and reaction 158 at high benzophenone concentrations. The quantum yield also shows a significant variation with light intensity, decreasing as the light intensity is increased.[151] At high light intensity (but not at low) the quantum efficiency decreases on O-deuteration of the isopropanol, that is, reaction 158 has a deuterium isotope effect.[151]

The photoreduction of benzophenone to benzpinacol can also be achieved in secondary alcohols other than isopropanol and in primary alcohols. Reduction with α-hydroxyalkyl radicals can also be achieved by generating these radicals thermally by decomposing di-*t*-butyl peroxide at 130° in the alcohol-ketone mixture.[152]

The reduction of carbonyl groups by α-hydroxyalkyl radicals is not confined to diaryl ketones. Acetophenones, for example, can also be reduced in isopropanol both photochemically[153,154] and thermally.[152] The rate of the thermal reaction is accelerated by electron withdrawing substituents attached to the aromatic nucleus ($\rho = 1.59$).[152] Acetophenone is also photoreduced in suitable amine solvents.[155] The amines show an intrinsically greater reactivity than the corresponding alcohols, thus, the photo-reduction is, for example, 35% faster in *sec*-butylamine than in isopropanol.

The reaction of α-hydroxyalkyl radicals and α-aminoalkyl radicals with ketones probably involves attack on oxygen via hydrogen rather than via the carbon center of the radical for example,

$$Me_2C \overset{\delta+}{=\!=} O \cdots H \cdots \overset{\delta-}{O} =\!= CPh_2$$

Starnes[156] has shown that the oxidation of triarylmethanols with lead tetracetate is a free-radical chain reaction and has suggested as propagation steps,

$$Ar_2\dot{C}OAr + Ar_3COPb(OAc)_3 \rightarrow Ar_2C(OAc)OAr + Ar_3CO\dot{P}b(OAc)_2 \quad (159)$$

$$Ar_3CO\dot{P}b(OAc)_2 \rightarrow Pb(OAc)_2 + Ar_3CO^{\cdot} \quad (160)$$

$$Ar_3CO^{\cdot} \rightarrow Ar_2\dot{C}OAr \quad (161)$$

[151] N. C. Yang and S. Murov, *J. Amer. Chem. Soc.*, **88,** 2852 (1966).
[152] E. S. Huyser and D. C. Neckers, *ibid.*, **85,** 3641 (1963).
[153] A. Beckett, A. D. Osborne, and G. Porter, *Trans. Faraday Soc.*, **60,** 873 (1964).
[154] N. C. Yang, D. S. McClure, S. L. Murov, J. J. Houser, and R. Dusenbury, *J. Amer. Chem. Soc.*, **89,** 5466 (1967).
[155] S. G. Cohen and B. Green, *ibid.*, **91,** 6824 (1964).
[156] W. H. Starnes, Jr., *ibid.*, **89,** 3368 (1967).

The first of these reactions is an S_H2 process in which the α-phenoxydiphenylmethyl radical attacks, presumably, the carbonyl oxygen. The triarylmethoxy lead compound is formed in the reaction,

$$Ar_3COH + Pb(OAc)_4 \rightarrow Ar_3COPb(OAc)_3 + HOAc \qquad (162)$$

Two particularly interesting free-radical chains involving S_H2 reactions at carbonyl oxygen by α-hydroxyalkyl radicals have been reported by Huyser and co-workers. Firstly, ω-methanethioacetophenone is reduced to acetophenone and methyl mercaptan by *sec*-butanol at 125° in the presence of a source of free radicals.[157]

$$CH_3\dot{C}(OH)C_2H_5 + PhC(=O)CH_2SCH_3 \longrightarrow CH_3C(=O)C_2H_5 + Ph\dot{C}(OH)CH_2SCH_3 \qquad (163)$$

$$Ph\dot{C}(OH)CH_2SCH_3 \longrightarrow CH_3S^{\cdot} + PhC(OH){=}CH_2(\rightarrow PhC(=O)CH_3) \qquad (164)$$

$$CH_3S^{\cdot} + CH_3CH(OH)C_2H_5 \longrightarrow CH_3\dot{C}(OH)C_2H_5 \qquad (165)$$

Secondly, 4-methyl-4-trichloromethyl-2,5-cyclohexadienone is reduced to *p*-cresol and chloroform with concurrent oxidation of the alcohol at 80–125° in the presence of a source of free radicals.[158]

$$CH_3\dot{C}(OH)C_2H_5 + \text{4-methyl-4-trichloromethyl-2,5-cyclohexadienone} \longrightarrow CH_3C(=O)C_2H_5 + \text{1-hydroxy-4-methyl-4-trichloromethylcyclohexadienyl radical} \qquad (166)$$

$$\text{1-hydroxy-4-methyl-4-trichloromethylcyclohexadienyl radical} \longrightarrow \text{p-}HOC_6H_4CH_3 + {}^{\cdot}CCl_3 \qquad (167)$$

$${}^{\cdot}CCl_3 + CH_3CH(OH)C_2H_5 \longrightarrow CHCl_3 + CH_3\dot{C}(OH)C_2H_5 \qquad (168)$$

[157] E. S. Huyser and R. M. Kellogg, *J. Org. Chem.*, **31,** 3366 (1966).
[158] E. S. Huyser and K. L. Johnson, *ibid.*, **33,** 3645 (1968).

These reductions can be effected by primary but not by tertiary alcohols. It seems likely that similar reductions could occur with α-aminoalkyl radicals provided that at least one hydrogen was joined to the nitrogen.

The reaction of simple alkyl radicals with carbonyl groups usually involve the carbon end of the C=O bond and yields alkoxy radicals (see Chapter V(A)). One example where attack is on oxygen but an oxy radical is still produced occurs when quinones are used to inhibit vinyl polymerization. Silyl

$$M^{\cdot} + O{=}C_6H_4{=}O \longrightarrow \sim MO{-}C_6H_4{-}O^{\cdot} \tag{169}$$

radicals add to the oxygen of carbonyl groups[159] presumably because of the high strength of Si–O bonds.

An S_Hi reaction involving the attack of a carbon radical on carbonyl oxygen has been reported by Menapace and Kuivila.[160] The reduction of γ-chloro-butyrophenone with tri-*n*-butyltinhydride provided 65% of a product containing 80% 2-phenyltetrahydrofuran and 20% butyrophenone, that is,

$$PhCOCH_2CH_2CH_2Cl + Bu_3^nSn^{\cdot} \longrightarrow PhCOCH_2CH_2CH_2^{\cdot} + Bu_3^nSnCl$$

$$PhCOCH_2CH_2CH_2^{\cdot} \xrightarrow{Bu_3{}^nSnH} PhCOCH_2CH_2CH_3 \ (20\%)$$

$$PhCOCH_2CH_2CH_2^{\cdot} \longrightarrow Ph\dot{C}{-}CH_2{-}CH_2{-}CH_2{-}O \text{ (ring)} \xrightarrow{Bu_3{}^nSnH} PhCH{-}CH_2{-}CH_2{-}CH_2{-}O \text{ (ring)} \ (80\%) \tag{170}$$

The driving force for this intramolecular substitution at oxygen is undoubtedly the formation of a species which is both benzylic and an α-alkoxy radical.

Tanner and Law[161] have reported another interesting intramolecular substitution of a carbon radical at the oxygen of a carbonyl group. The radical generated by decarboxylation of 2-acetoxy-2-methylbutyraldehyde, **8**, in a radical chain process at 75° yields both the unrearranged and rearranged

[159] See footnote 103, Chapter V.

[160] L. W. Menapace and H. G. Kuivila, *J. Amer. Chem. Soc.*, **86**, 3047 (1964).

[161] D. D. Tanner and F. C. P. Law, *ibid.*, **91**, 7535 (1969).

acetate. The acyloxy group undergoes a 1,2 migration which converts the initial primary alkyl radical to the more stable tertiary alkyl radical.

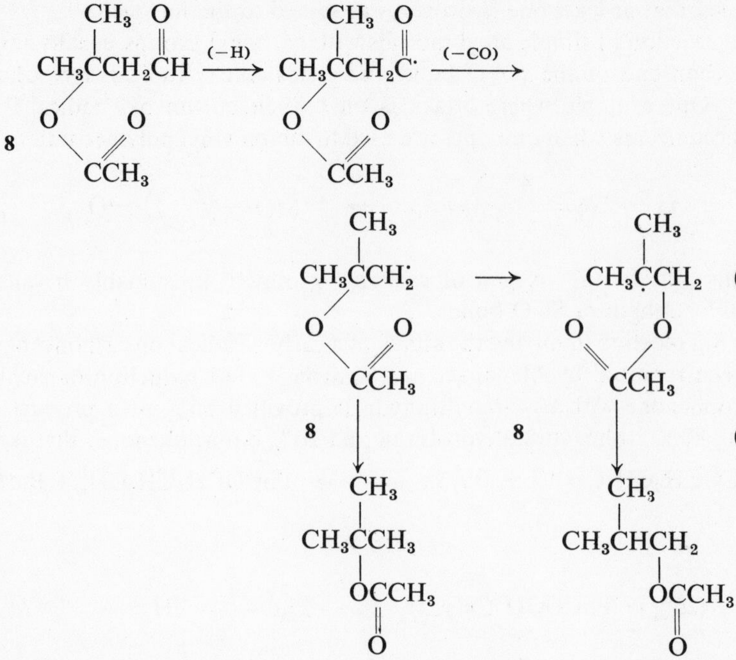

$$\underset{\mathbf{8}}{CH_3C(CH_3)(OCOCH_3)CH_2CHO} \xrightarrow{(-H)} CH_3C(CH_3)(OCOCH_3)CH_2\dot{C}O \xrightarrow{(-CO)}$$

$$CH_3C(CH_3)(OCOCH_3)CH_2^{\cdot} \longrightarrow CH_3\dot{C}(CH_3)CH_2OCOCH_3 \quad (171)$$

$$\downarrow \qquad\qquad\qquad \downarrow \quad \mathbf{8} \qquad (172)$$

$$CH_3C(CH_3)_2OCOCH_3 \qquad CH_3CH(CH_3)CH_2OCOCH_3$$

The acetoxy rearrangement, reaction 172 is analogous to the symmetrical acetoxy rearrangement studied by Drew and Martin[88] and shown in scheme 93. However, Tanner and Law[161] found that a rearrangement analogous to reaction 172 did not take place with 2-acetoxybutyraldehyde. This is presumably because the stability gained by the rearrangement of an acetoxy group from a secondary to a primary position is insufficient to overcome the activation energy necessary for its migration.[161]

B. SULFUR

Extensive studies of homolytic substitutions at sulfur have shown that such reactions occur with particular facility when sulfur is bound to sulfur. The relatively weak S–S bond in disulfides (D(MeS–SMe) = 69 kcal/mole[162]) is readily cleaved both by carbon radicals and by thiyl radicals.[163,164]

$$R'^{\cdot} + RSSR \rightarrow R'SR + RS^{\cdot} \quad (173)$$

$$R'S^{\cdot} + RSSR \rightarrow R'SSR + RS^{\cdot} \quad (174)$$

162 T. F. Palmer and F. P. Lossing, *ibid.*, **84**, 4661 (1962).

163 Chs. 4 and 7 in footnote 11, Chapter III .

164 W. A. Pryor, *Mechanisms of Sulfur Reactions*, McGraw-Hill, New York, 1962, Ch. 3.

The even weaker S–S bond in polysulfides ($D(MeS_2{-}SMe) = 46$ kcal/mole,[165] $D(MeS_2{-}S_2Me) = 37$ kcal/mole[165,166] and $D(\sqsubset S_4{-}S_4 \sqsupset) \approx D(\dot{S}_i{-}\dot{S}_j) \approx$ 33 kcal/mole[165]) can be cleaved by polysulfenyl radicals, for example,

$$R'SS^{\cdot} + RS_4R \rightarrow R'S_3R + RS_3^{\cdot} \tag{175}$$

Polysulfenyl radicals are otherwise almost completely unreactive. Except in special cases, they do not abstract hydrogen or add to double bonds.

Homolytic substitutions on sulfur compounds are discussed in four subsections: (a) elemental sulfur, (b) polysulfides, RS_nR ($n > 2$), (c) disulfides and (d) sulfides.

(a) ELEMENTAL SULFUR

A particularly interesting example of the cleavage of polysulfides by polysulfenyl radicals can be observed in sulfur itself at elevated temperatures. From room temperature to well above its melting point the stable form of sulfur is an eight-membered, puckered, ring, S_8. Up to about 159° liquid sulfur is of normal viscosity, but at this temperature the viscosity starts to increase and, by 187°, has increased by a factor of 10,000. The viscosity then decreases as the temperature is raised further up to the boiling point of sulfur. These unusual properties are believed to be *principally* due to an equilibrium between S_8 rings and long chains of sulfur atoms, $\cdot S_n^{\cdot}$, which reach their maximum length at the temperature of highest viscosity.[163,167–178a]

$$S_8\ (\text{ring}) \rightleftharpoons \cdot S_8^{\cdot} \tag{176}$$

$$\cdot S_8^{\cdot} + S_8 \rightleftharpoons \cdot S_{16}^{\cdot} \xrightarrow{S_8} \cdot S_n^{\cdot} \tag{177}$$

[165] T. L. Pickering, K. J. Saunders, and A. V. Tobolsky, *The Chemistry of Sulfides*, A. V. Tobolsky (Ed.), Interscience, New York, 1968, pp. 61–72.

[166] T. L. Pickering, K. J. Saunders, and A. V. Tobolsky, *J. Amer. Chem. Soc.*, **89,** 2364 (1967).

[167] Ch. 2 of footnote 164.

[168] R. E. Powell and H. Eyring, *J. Amer. Chem. Soc.*, **65,** 648 (1943).

[169] G. Gee, *Trans. Faraday Soc.*, **48,** 515 (1952).

[170] F. Fairbrother, G. Gee, and G. T. Merrall, *J. Polymer Sci.*, **16,** 459 (1955).

[171] A. V. Tobolsky and A. Eisenberg, *J. Amer. Chem. Soc.*, **81** 780 (1959); **82,** 289 (1960).

[172] T. Doi, *Rev. Phys. Chem. Japan*, **33,** 41 (1963); **35,** 1, 11, 18 (1965).

[173] A. V. Tobolsky and W. J. MacKnight, *Polymeric Sulfur and Related Polymers*, Interscience, New York, 1965.

[174] W. J. MacKnight and A. V. Tobolsky, in *Elemental Sulfur*, B. Meyer (Ed.), Interscience, New York, 1965, Ch. 5.

[175] T. K. Wiewiorowski and F. J. Touro, *J. Phys. Chem.*, **70,** 3528 (1966).

[176] A. Eisenberg and L. A. Telter, *ibid.*, **71,** 2332 (1967).

[177] A. T. Ward, *ibid.*, **72,** 4133 (1968).

[178] A. Eisenberg, *Macromolecules*, **2,** 44 (1969).

[178a] R. E. Harris, *J. Phys. Chem.*, **74,** 3102 (1970).

Depolymerization will occur both by random chain scission and by the loss of S_8 rings from the ends of the chain by an intramolecular substitution (S_Hi).

The most detailed picture of the state of liquid sulfur has been obtained from its paramagnetic resonance spectrum.[179,180] The variation in the radical concentration with temperature gives 33.4 kcal/mole for the bond strength of the S_8 ring[179] which is in excellent agreement with the value obtained by other methods. (For example,[171] $k_{176} = 1.1 \times 10^5 \exp(-33{,}000/RT)$ sec^{-1} $\Delta S^\circ_{176} =$ 23 gibbs/mole, $\Delta H^\circ_{176} = 32.8$ kcal/mole.) The variation in the width of the paramagnetic absorption line with temperature gives the rate constant for the propagation of polymerization,[179] $k_{177} = 2.8 \times 10^8 \exp(-3100/RT)\ M^{-1}$ sec^{-1}. The equilibrium constant for this reaction is given by,[171] $K_{177} = 10.4 \exp(-3200/RT)\ M^{-1}$ ($\Delta S^\circ_{177} = 4.6$ gibbs/mole, $\Delta H^\circ_{177} = 3.17$ kcal/mole).

The ring-chain equilibrium of elemental sulfur can be modified by the addition of thiyl or alkyl radicals.[181,182] For example,[181] in mixtures of dimethyl disulfide and sulfur at 120° some 70–80% of the disulfide is destroyed quite rapidly with 50% of the methyl groups appearing in long chain polysulfides (MeS_nMe, $n \geqslant 6$). The long chain compounds break down slowly to give an equilibrium mixture of polysulfides ($n = 2, 3, 4, 5$ etc.). The initial reaction can be represented as

$$S_8 \rightleftharpoons {}^{\cdot}S_8^{\cdot} \tag{176}$$

$$MeS_2Me \rightleftharpoons 2MeS^{\cdot} \tag{178}$$

$$MeS^{\cdot} + S_8 \rightleftharpoons MeS_9^{\cdot} \tag{179}$$

$${}^{\cdot}S_8^{\cdot} + MeS_2Me \rightleftharpoons MeS_9^{\cdot} + MeS^{\cdot} \tag{180}$$

$$MeS^{\cdot} + MeS_9^{\cdot} \rightleftharpoons MeS_{10}Me \quad \text{etc.} \tag{181}$$

The reaction of dimethyl sulfide with elemental sulfur requires temperatures above those for the transition from S_8 rings to long chains.[181] Even at 190°, the reaction is very slow and gives complicated mixtures consisting of dialkyl polysulfides as well as methyl-terminated methylene sulfide chain molecules. Some hydrogen atom abstraction must occur in these reaction mixtures.

[179] D. M. Gardner and G. K. Fraenkel, *J. Amer. Chem. Soc.*, **75**, 5891 (1954); **78**, 3279 (1956).

[180] A. G. Pinkus and L. H. Piette, *J. Phys. Chem.*, **63**, 2086 (1959).

[181] D. Grant and J. R. Van Wazer, *J. Amer. Chem. Soc.*, **86**, 3012 (1964).

[182] H. Shizuka and T. Azami, *Nippon Gomu Kyokaishi*, **39**, 905 (1966) (*Chem. Abstr.*, **67**, 22666g (1967)).

Elemental sulfur is a fairly effective inhibitor of the polymerization of a number of vinyl monomers including vinyl acetate,[183] methyl methacrylate,[184,185] methyl acrylate,[186] and styrene.[187] The resonance stabilized, polymeric, carbon radicals derived from these monomers attack the S_8 ring rather more slowly than the polysulfenyl radicals present in liquid sulfur (see Table VII). The initial reactions may be represented as

$$\sim M^{\cdot} + S_8 \rightarrow \sim MS_8^{\cdot} \quad (182)$$

$$\sim MS_8^{\cdot} + M \rightarrow \sim MS_8M^{\cdot} \quad (183)$$

$$\sim MS_8^{\cdot} + \sim M^{\cdot} \rightarrow \sim MS_8M\sim \quad (184)$$

The linear S_8 units incorporated into the polymer are, of course, themselves liable to attack by polymer radicals. The rate of attack on $\sim MS_8M\sim$ is somewhat less than the rate of attack on the S_8 ring, that is, $k_{182} > k_{185} > \cdots$ etc.[183]

$$\sim M^{\cdot} + \sim MS_8M\sim \rightarrow \sim MS_{7-1}M\sim + \sim MS_{1-7}^{\cdot} \quad (185)$$

$$\sim M^{\cdot} + \sim MS_{7-1}M\sim \rightarrow \sim MS_{6-1}M\sim + \sim MS_{1-6}^{\cdot} \quad (186)$$

The polymer radicals from vinyl acetate and styrene attack sulfur much more rapidly than they attack monomer but the reverse is true for methyl methacrylate. This difference is probably due to polar effects.[163,164] The electron-deficient poly(methyl methacrylate) radical attacks sulfur rather slowly which indicates that the S–S bond is less susceptible to electrophilic attack than to nucleophilic attack.[163,188] Similar polar effects are observed in the reactions of polysulfides and disulfides with free radicals, both being more effective chain transfer agents in polymerizing vinyl acetate or styrene than in polymerizing methyl acrylate or methyl methacrylate.

Cleavage of the S_8 ring by simple alkyl radicals is an even more facile process than the cleavage by polymeric radicals. For example,[189] the rate constant for the homolytic substitution of S_8 by the cyclopentyl radical is $6 \times 10^7\ M^{-1}\ \text{sec}^{-1}$ at $-40°$. Dialkyl and diaryl polysulfides can be formed in quite high yields by reacting carbon centered radicals with elemental

[183] P. D. Bartlett and H. Kwart, *J. Amer. Chem. Soc.*, **74,** 3939 (1952).
[184] J. L. Kice, *ibid.*, **76,** 6274 (1954).
[185] T. Sugimura, Y. Ogata, and Y. Minoura, *J. Polymer Sci.* (*A*), **4,** 2747 (1966),
[186] J. L. Kice. *J. Polymer Sci.*, **19,** 123 (1956).
[187] P. D. Bartlett and D. S. Trifan, *ibid.*, **20,** 457 (1956).
[188] For a recent review of electrophilic and nucleophilic scission of S–S bonds see, J. L. Kice, *Accounts Chem. Res.*, **1,** 58 (1968).
[189] B. Smaller, J. R. Remko, and E. C. Avery, *J. Chem. Phys.*, **48,** 5174 (1968).

Table VII Rate Constants for Free-Radical Attack on Elemental Sulfur

Radical	Temperature (°C)	$C = k_{tr}/k_p^a$	k_p^b (M^{-1} sec^{-1})	k_{tr} (M^{-1} sec^{-1})	Footnote
Polysulfenyl	45	—	—	2×10^6	179
Cyclopentyl	−40	—	—	6×10^7	189
Poly(vinyl acetate)	45	470[c]	1480	7×10^5	183
Polystyryl	81	50–120	138	$0.7–1.7 \times 10^4$	187
Poly(methyl acrylate)	44	0.86	1290	1.1×10^3	186
Poly(methyl methacrylate)	44	0.075	505	38	184
Poly(methyl methacrylate)	50	0.106	575	61[d]	185
Tri-*n*-butyltin	30	—	—	5.5×10^6	198

a. Ratio of rate constants for attack on S_8 to attack on monomer.
b. Calculated from values of the activation parameters for chain propagation (E_p and PZ_p) given on p. 95 of footnote 11, Chapter III.
c. Assuming that each S_8 unit incorporated into the chain causes two further chains to end then $k_{185}/k_p = 57$ ($k_{185} = 8.4 \times 10^4\ M^{-1}$ sec^{-1}). With the same assumption, $k_{186}/k_p = 20$ ($k_{186} = 3 \times 10^4\ M^{-1}$ sec^{-1}).
d. The rate constants in M^{-1} sec^{-1} units for the attack of the poly(methyl methacrylate) radicals on $PhCH_2S_nCH_2Ph$ are:[189] 49 ($n = 4$); 12 ($n = 3$); 9 ($n = 2$); 6 ($n = 1$); 0 ($n = 0$) at 50°.

sulfur,[190–194] for example,

$$CF_3^{\cdot} + S_8 \rightarrow CF_3S_nCF_3 \qquad (187)^{190,191}$$

$$Me_2\dot{C}CN + S_8 \rightarrow Me_2(CN)CS_nC(CN)Me_2 \qquad (188)^{194}$$

The reactions of thiyl, polysulfenyl and carbon radicals with elemental sulfur and polysulfides undoubtedly play an important role in the vulcanization of rubber.

Very few studies have been made of homolytic substitutions at S_8 rings by radicals having the unpaired electron centered on an atom other than carbon or sulfur, and yet many such reactions must be highly exothermic and should proceed readily. One difficulty in studying the reactions of elemental sulfur is that S_8 rings are readily cleaved by electrophilic and nucleophilic species

[190] G. A. R. Brandt, H. J. Emeléus, and R. N. Haszeldine, *J. Chem. Soc.*, 2198, 2549 (1952).
[191] R. N. Haszeldine and J. M. Kidd, *ibid.*, 3219 (1953).
[192] E. I. Tinyakova, B. A. Dolgoplosk, and M. P. Tikhomolova, *Zhur. Obshchei Khim.*, **25,** 1387 (1955).
[193] V. Ya. Andakushkin, B. A. Dolgoplosk, and I. I. Radchenko, *ibid.*, **26,** 2972 (1956).
[194] D. I. Relyea, P. O. Tawney, and A. R. Williams, *J. Org. Chem.*, **27,** 1078 (1962).

as well as by radicals.[167,188] Since the resultant open chain polysulfides probably dissociate into radicals more easily than S_8 rings, an ionic species may catalyze the formation of polysulfenyl radicals by promoting the conversion of the rings to the more readily homolyzed linear polysulfides.[167] For this reason, even the identification of polysulfenyl radicals in a reaction involving elemental sulfur[195–197] does not, of necessity, require a homolytic substitution at the S_8 ring. Species which might be expected to cleave the S_8 ring homolytically include atomic hydrogen and fluorine, and free radicals with the unpaired electron centered on an atom such as phosphorus, tin or lead. The reduction of sulfur by tri-*n*-butyltin hydride is a rapid free-radical chain process that provides one of the few examples of this class of homolytic substitutions.[198]

$$Bu^n_3Sn^\cdot + S_8 \rightarrow Bu^n_3SnS_8^\cdot \tag{189}$$

$$Bu^n_3SnS_8^\cdot + Bu^n_3SnH \rightarrow Bu^n_3SnS_8H + Bu^n_3Sn^\cdot \tag{190}$$

(b) POLYSULFIDES

The reactions of polysulfides are closely analogous to those of elemental sulfur. Pickering, Saunders, and Tobolsky[165,166] have studied the thermal decomposition of dimethyl tetrasulfide at 80° by NMR.[199] The initial decomposition (Figure VII.1) results in the formation of trisulfides, pentasulfides, and hexasulfides, this part of the reaction being essentially complete after about 40 hours. Disulfide is formed very slowly by decomposition of the trisulfide, the first trace appearing after 140 hr and the yield rising to 6% after 1800 hr. No monosulfide or hydrocarbons were formed. The final products would, eventually, be dimethyl disulfide and sulfur. The reaction involves free radicals but has a chain length of less than unity. Because of the absence of disulfide the following scheme, which does not involve thiyl radicals, was suggested.

Initiation:

$$MeS_4Me \rightleftharpoons 2MeS_2^\cdot \tag{191}$$

Propagation:

$$MeS_2^\cdot + MeS_4Me \rightleftharpoons MeS_3Me + MeS_3^\cdot \tag{192}$$

$$MeS_3^\cdot + MeS_4Me \rightleftharpoons MeS_5Me + MeS_2^\cdot \tag{193}$$

$$MeS_3^\cdot + MeS_5Me \rightleftharpoons MeS_6Me + MeS_2^\cdot \tag{194}$$

Termination:

$$MeS_2^\cdot + MeS_3^\cdot \rightleftharpoons MeS_5Me \qquad \text{and so on} \tag{195}$$

[195] W. Hodgson, S. Buckler, and E. Peters, *J. Amer. Chem. Soc.*, **85,** 543 (1963).
[196] E. Poziomek, *Chemist–Analyst*, **55,** 78 (1966).
[197] W. A. Mosher and R. R. Irino, *J. Amer. Chem. Soc.*, **91,** 756 (1969).
[198] J. Spanswick and K. U. Ingold, *Int. J. Chem. Kinetics*, **2,** 157 (1970).
[199] See footnotes 165 and 166 for citations to earlier studies on the decomposition of trisulfides and tetrasulfides.

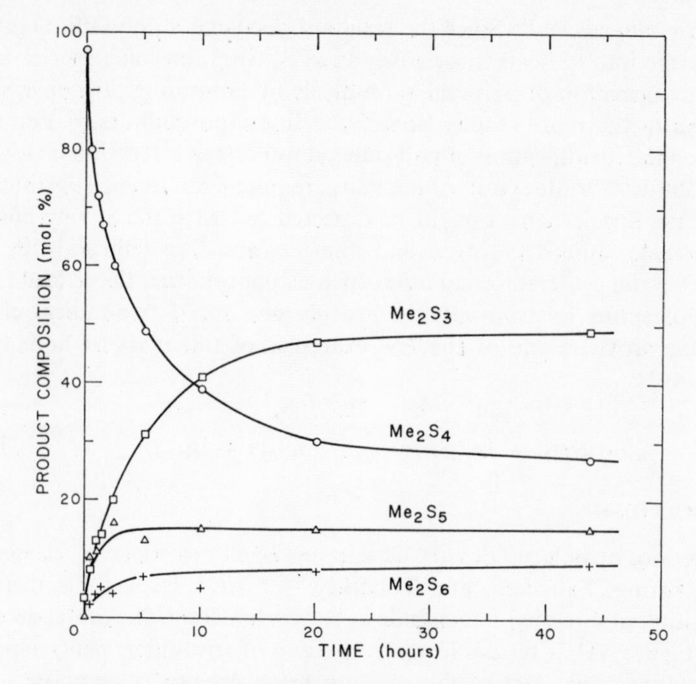

Figure VII.1. The thermal decomposition of dimethyl tetrasulfide at 80°.

Dimethyl trisulfide decomposes much more slowly than the tetrasulfide,[165] an equilibrium product mixture being reached after about 500 hours at 80° (Figure VII.2). Initial decomposition results in the formation of disulfide and tetrasulfide in approximately equimolar quantities, presumably by the reaction sequence,

$$MeS_3Me \rightleftharpoons MeS_2^{\cdot} + MeS^{\cdot} \tag{196}$$

$$MeS^{\cdot} + MeS_3Me \rightarrow MeS_2Me + MeS_2^{\cdot} \tag{197}$$

$$MeS_2^{\cdot} + MeS_2^{\cdot} \rightarrow MeS_4Me \tag{-191}$$

Reaction 197 should be thermodynamically favored over the alternative (identity) reaction which yields $MeS^{\cdot}$ and MeS_3Me (cf, however, footnote 200). In a similar way, the 2-cyano-2-propyl radical induced decomposition of dimethyl trisulfide yields the more resonance stabilized methyldisulfenyl radical,

$$Me_2\dot{C}CN + MeS_3Me \rightarrow Me_2(MeS)CCN + MeS_2^{\cdot} \tag{198}$$

[200] C. D. Trivette, Jr., and A. Y. Coran, *J. Org. Chem.*, **31**, 100 (1966).

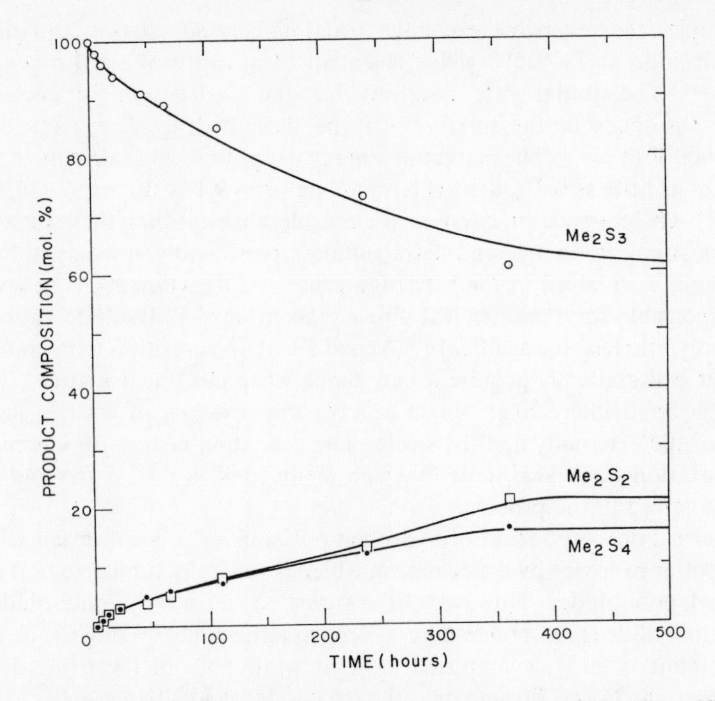

Figure VII.2. The thermal decomposition of dimethyl trisulfide at 80°.

rather than the methylthiyl radical.[165]

$$Me_2\dot{C}CN + MeS_3Me \rightarrow Me_2(MeS_2)CCN + MeS^{\cdot} \qquad (199)$$

The photodecomposition and thermal decomposition of *bis*(trifluoromethyl) polysulfides presumably occurs by a generally similar mechanism.[190,191]

The thermal decomposition of mixtures of two polysulfides will lead to scrambling of the sulfur atoms which are not attached directly to carbon (which has been shown[201–203] by using radioactive sulfur-35) as well as to the scrambling of the alkyl or aryl groups.[200,203]

$$RS\overset{*}{S}SR + R'SSSR' \rightleftharpoons RS\overset{*}{S}SR' + RSSSR' + RSSSR + R'S\overset{*}{S}SR' \qquad (200)$$

[201] E. N. Gur'yanova and V. N. Vasiliyeva, *Zhur. Fiz. Khim.*, **28,** 60 (1954); *Chem. Abstr.*, **48,** 11888a (1954).
[202] E. N. Gur'yanova, *ibid.*, **28,** 67 (1954); *Chem. Abstr.*. **48,** 11888c (1954).
[203] E. N. Gur'yanova, V. N. Vasiliyeva, and L. S. Kuzina, *Rubber Chem. Technol.*, **29,** 534 (1956).

For example, the reversible exchange reaction between diethyl and di-*n*-propyl trisulfide at 130–150° yields a statistical quantity of ethyl *n*-propyl trisulfide.[200] The initial rate constant for the destruction of each of the pure trisulfides in the mixture can be described by $k = 1.3 \times 10^{11} \exp(-29000/RT)\ \text{sec}^{-1}$. The activation energy drops to ~26 kcal/mole in the presence of as little as 0.3% diethyl tetrasulfide ($k = 9.1 \times 10^{9} \exp(-26300/RT)\ \text{sec}^{-1}$). These activation energies are considerably less than the estimated S–S bond strengths in tri- and tetra-sulfides, presumably because a free-radical chain is involved in the exchange processes. Reaction 200 is believed to be responsible for the chemical stress relaxation of polysulfide rubbers such as poly(ethylene tetrasulfide).[173] Above a certain transition temperature the rubber will gradually acquire a new shape when put under a stress. The rate of the bond interchange which relieves the stress is, of course, independent of the externally applied strain. The activation energy for chemical stress relaxation is 26 kcal/mole[173] which again implies that a free-radical chain is involved in the process.

The thermal decomposition of hydrogen polysulfides to yield, eventually, H_2S and sulfur proceeds by a mechanism which is generally similar to that for the dialkyl polysulfides. However, in contrast to the methyl polysulfides, hydrogen trisulfide is less stable than hydrogen tetrasulfide.[204] In CCl_4 at 70° the tetrasulfide is 50% decomposed in 60 hr while 50% of the trisulfide is decomposed in 36 hr. Presumably the trisulfide yields the reactive $HS^{\cdot}$ radical which would not be formed from the tetrasulfide,

$$HS_3H \rightleftharpoons HS_2^{\cdot} + HS^{\cdot} \qquad (201)$$

These radicals would enter into the usual homolytic substitutions at sulfur,

$$HS^{\cdot} + HS_3H \rightarrow HS_2H + HS_2^{\cdot} \qquad (202)$$

and would also be able to abstract hydrogen, something the $HS_2^{\cdot}$ radicals would not do.

$$HS^{\cdot} + HS_3H \rightarrow H_2S + HS_3^{\cdot} \qquad (203)$$

Of the three primary reaction products (H_2S, H_2S_2, and H_2S_4) the H_2S and H_2S_2 will be much less reactive than the starting trisulfide and so cannot help to protect the trisulfide from further reaction. In contrast, the primary products from H_2S_4 are H_2S_3 and H_2S_5 both of which are reactive towards polysulfenyl radicals and so afford their parent some protection from free-radical attack.[204]

The rate constants for homolytic substitution at polysulfides by a particular radical decrease as the number of sulfur atoms in the chain decrease (see for example, the data given in footnotes c and d of Table VII). Presumably, it is

[204] E. Muller and J. B. Hine, *J. Amer. Chem. Soc.*, **91,** 1907 (1969).

for this reason that only 2–3 sulfur atoms out of the entire S_8 ring scavenge the very reactive cyclopentyl radical with a high efficiency.[189]

The low temperature chlorination of certain trisulfides with chlorine can yield a chlorodithio compound (RS_2Cl) together with a sulfenyl chloride (RSCl).[205] This reaction may be an S_H2 process at sulfur by a chlorine atom.

(c) DISULFIDES

Most differences in the behavior of disulfides and polysulfides towards free-radicals can be attributed to the greater S–S bond strength in the former compounds. Disulfides are readily cleaved by thiyl radicals.[164,206–208] Mixtures of organic disulfides exchange mercapto groups under free-radical conditions, that is, heat, light or an added free-radical initiator. (The exchange under ionic conditions is also very rapid.[164])

$$R'SSR' \rightleftharpoons 2R'S^{\cdot} \qquad (204)$$

$$R'S^{\cdot} + RSSR \rightleftharpoons R'SSR + RS^{\cdot} \qquad (174)$$

At elevated temperatures this process may be a chain reaction with a fairly long kinetic chain length.

Exchange of alkyl groups and sulfur also occurs between disulfides and mercaptans[209–215] presumably by a sequence involving reaction 205 followed by reaction 174.

$$RS^{\cdot} + R'SH \rightarrow RSH + R'S^{\cdot} \qquad (205)$$

$$R'S^{\cdot} + RSSR \rightarrow RSSR' + RS^{\cdot} \qquad (174)$$

205 N. Kharasch and Z. S. Ariyan, *Intra-Science Chem. Reports–Special Report on Sulfur Chemistry*, **1,** (4), 337 (1967).
206 E. R. Bertozzi, F. O. Davis, and E. M. Fettes, *J. Polymer Sci.*, **19,** 17 (1956) and references cited therein.
207 A. S. Prashchikina, E. N. Gur'yanova, and A. E. Grinberg, *Vysokomolekul. Soedin*, **6,** 112 (1964), *Chem. Abstr.* **60,** 10895 (1964).
208 L. Haraldson, C. J. Olander, S. Sunner, and E. Vorde, *Acta Chemica Scand.*, **14,** 1509 (1960).
209 G. Gorin, G. Dougherty, and A. V. Tobolsky, *J. Amer. Chem. Soc.*, **71,** 3551 (1949).
210 G. Leandri and A. Tuno, *Ann. Chim.* (*Rome*), **44,** 63 (1954); *Chem. Abstr.*, **49,** 4563d (1955).
211 S. Ikeda, *Nippon Kogaku Zasshi*, **79,** 103 (1958); *Chem. Abstr.*, **52,** 15208h (1958).
212 G. A. Blokh, *Rubber Chem. Technol.*, **31,** 1035 (1958).
213 E. E. Gur'yanova and V. N. Vasil'eva, *Khim. Seraorgan. Soedin. Soderzhashch. V. Neft. i Neftepod. Akad. Nauk SSSR. Bashkirsk, Filial*, **4,** 24 (1961); *Chem. Abstr.* **57,** 5833f (1962).
214 R. D. Obolentsev and Yu. E. Nikitin, *ibid.*, **7,** 104 (1964); *Chem. Abstr.*, **63,** 17205b (1965).
215 A. Fava, G. Reichenbach, and U. Peron, *J. Amer. Chem. Soc.*, **89,** 6696 (1967).

Reaction 174 is rate controlling[215] and its activation energy for $R' = R =$ phenyl has been estimated to be 5 kcal/mole.[215]

The decomposition of hydrogen disulfide probably occurs by a free-radical chain mechanism in which hydrogen atom abstraction by $HS^{\cdot}$ radicals is responsible for the eventual products, H_2S and sulfur.[216] An ionic chain mechanism has also been proposed for this decomposition[217] but this seems much less probable than a free-radical chain.

Disulfides are quite effective chain transfer agents in polymerizing vinyl monomers. That is, they react with and terminate one growing polymer radical and then initiate the growth of a new polymer radical. The formulation of the first of these reactions as an S_H2 process is supported by the fact that styrene thermally polymerized in the presence of dibutyl disulfide[218] or diaryl disulfides[219] incorporates approximately two atoms of sulfur per polymer chain, these being situated, of course, at either end of each chain. The overall propagation process can be formulated as

$$RSM_n^{\cdot} + RSSR \xrightarrow{k_{tr}} RSM_nSR + RS^{\cdot} \tag{206}$$

$$RS^{\cdot} + M \rightarrow RSM^{\cdot} \tag{207}$$

$$RSM^{\cdot} + (n - 1)M \xrightarrow{k_p} RSM_n^{\cdot} \tag{208}$$

As has already been mentioned the rate constant ratio, k_{tr}/k_p, (the chain transfer constant) can be calculated from the degree of polymerization of the monomer. Provided k_p is known, the rate constant for the attack of the polymeric radical on the disulfide is readily obtained. Table VIII includes data for the attack of polystyryl radicals on some representative disulfides.[105,219–222] The agreement between different workers is not particularly good and it is unlikely that the differences can be attributed solely to differences in temperature since transfer constants vary only slightly with temperature.[164,219]

Additional support for the formulation of chain transfer by disulfides as an S_H2 reaction comes from the fact that when a cyclic disulfide is used as the transfer agent the molecular weight of the polymer is not reduced significantly but large quantities of sulfur are incorporated into it. Thus, in the presence

[216] E. I. Tinyakova, E. K. Khrennikova, and B. A. Dolgoplosk, *Zhur. Obshchei Khim.*, **28,** 1632 (1958).

[217] F. Feher and H. Weber, *Z. Elektrochem.*, **61,** 285 (1957).

[218] A. V. Tobolsky and B. Baysal, *J. Amer. Chem. Soc.*, **75,** 1757 (1953).

[219] R. M. Pierson, A. J. Costanza, and A. H. Weinstein, *J. Polymer Sci.*, **17,** 221 (1955).

[220] V. A. Dinaburg and A. A. Vansheidt, *Zhur. Obshchei Khim.*, **24,** 840 (1954).

[221] A. J. Costanza, R. J. Coleman, R. M. Pierson, C. S. Marvel, and C. King, *J. Polymer Sci.*, **17,** 319 (1955).

[222] A more extensive listing of k_{tr} for disulfides can be found on pp. 53–54 of footnote 164.

Table VIII Absolute and Relative Rate Constants for Free-Radical Attack on Some Symmetric Disulfides, RSSR

Radical	°C	Footnote	Absolute Rate Constant (M^{-1} sec^{-1})						
			R = Me	*n*-Bu	*i*-Bu	*sec*-Bu	*t*-Bu	$PhCH_2$	Ph
Polystyryl	50	219, 221	—	—	—	<0.5	<0.5	3	6
Polystyryl	60	105	1.36	0.34	0.30	0.064	0.020	1.4	8.7
Polystyryl	99	220	—	3.4	—	—	—	5	70
Poly(vinyl acetate)	60	223a	—	$<2.3 \times 10^3$	—	—	—	—	—
Poly(methyl methacrylate)	50	185	—	—	—	—	—	9	—
Methyl	100	231	6×10^4	—	—	—	—	—	—
Cyclopentyl	−50	189	—	—	—	—	—	$<1.6 \times 10^6$	—
Tri-*n*-butyltin	30	198	—	1.1×10^6	1.0×10^5	—	7.9×10^4	7×10^5	9×10^6
			Relative Rate Constant						
Polystyryl	60	105	68	17	15	3.2	(1)[a]	70	435
Phenyl	60	234, 232	135	70[b]	—	14[c]	(1)[a]	—	—
Phenyl	60	235	86	—	—	18[c]	(1)[a]	—	—
4-Nitrophenyl	60	235	35	—	—	12[c]	(1)[a]	—	—
Tri-*n*-butyltin	30	198	—	14	13	—	(1)[a]	9	110

a. Assumed.
b. Propyl.
c. Isopropyl.

of 1-oxa-4,5-dithiocycloheptane (diethyl ether disulfide) polymerized styrene contained up to 17 atoms of sulfur per molecule[218] and polymerized vinyl acetate up to 18 atoms per molecule.[223]

$$M_n^{\cdot} + \underline{\lfloor SC_2H_4OC_2H_4S \rfloor} \rightarrow M_nSC_2H_4OC_2H_4S^{\cdot} \xrightarrow{nM} M_nSC_2H_4OC_2H_4SM_n^{\cdot} \rightarrow \text{etc.} \quad (209)$$

Diaryl disulfides react fairly readily with non-polymeric resonance stabilized radicals such as the benzyl,[224] triphenylmethyl[225,226] and 2-cyano-2-propyl[227] radicals, for example,[224]

$$PhCH_2^{\cdot} + ArSSAr \rightarrow PhCH_2SAr + ArS^{\cdot} \quad (210)$$

Cleavage of diaryl disulfides by the 2-cyano-2-propyl radical is facilitated by electron donating *para*-substituents.[227] Dialkyl disulfides are not very reactive towards this radical unless the disulfide link forms a part of a 5 or 6 membered ring.[227] The low reactivity of dialkyl disulfides towards resonance stabilized radicals is a general phenomenon. Thus, for example,[228,229] benzyl radicals react only sluggishly with isobutyl disulfide and 2-cyano-2-propyl radicals failed to effect a substitution on this disulfide. In contrast, dialkyl disulfides are quite reactive towards nonstabilized alkyl,[227–231] acyl,[233] and aryl[230,232,234,235] radicals. For example, Suama and Takezaki[231] have photolyzed azomethane (as a source of methyl radicals) in the presence of dimethyl disulfide in the gas phase at 100°.

$$CH_3^{\cdot} + CH_3SSCH_3 \rightarrow CH_3SCH_3 + CH_3S^{\cdot} \quad (211)$$

The ratio of the rate constants for cleavage of the S–S bond to combination

[223] W. H. Stockmayer, R. O. Howard, and J. T. Clarke, *J. Amer. Chem. Soc.*, **75,** 1756 (1953).
[223a] J. T. Clarke, R. O. Howard, and W. H. Stockmayer, *Makromol. Chem.*, **44–46,** 427 (1961).
[224] J. Degani, M. Tiecco, and A. Tundo, *Gazz. Chim. Ital.*, **92,** 1213 (1962).
[225] H. Lecher, *Chem. Ber.*, **48,** 524 (1915).
[226] A. Schönberg, A. Stephenson, H. Kaltschmitt, E. Petersen, and H. Schulten, *ibid.*, **66,** 237 (1933).
[227] U. Schmidt and A. Müller, *Liebigs Ann. Chem.*, **672,** 90 (1964).
[228] See footnote 41, Chapter VI.
[229] See footnote 43, Chapter VI.
[230] J. Degani and A. Tundo, *Ann. Chim. (Rome)*, **51,** 543 (1961).
[231] M. Suama and Y. Takezaki, *Bull. Inst. Chem. Res. Kyoto Univ.*, **40,** 229 (1962).
[232] W. A. Pryor and P. K. Platt, *J. Amer. Chem. Soc.*, **85,** 1496 (1963).
[233] See footnote 76, Chapter VI.
[234] See footnote 33, Chapter I.
[235] See footnote 34, Chapter I.

of the methyl radicals

$$CH_3^{\cdot} + CH_3^{\cdot} \rightarrow C_2H_6 \tag{212}$$

that is, $k_{211}/k_{212} \approx 3 \times 10^{-6}$. Taking[236,237] $k_{212} = 2 \times 10^{10}\ M^{-1}\ \text{sec}^{-1}$, yields $k_{211} = 6 \times 10^4\ M^{-1}\ \text{sec}^{-1}$ which is remarkably large for a reaction which must be nearly thermoneutral.

Pryor and co-workers[232,234,235] have studied the reactions of phenyl radicals with a number of aliphatic disulfides in solution. The phenyl radicals were generated by the thermal decomposition of phenylazotriphenylmethane at 60°. Pryor and Platt[232] measured the ratio of the yields of phenyl alkyl sulfide to benzene which gave k_{213}/k_{214}, the rate constant ratio for attack on sulfur and hydrogen atom abstraction.

$$Ph^{\cdot} + RSSR \rightarrow PhSR + RS^{\cdot} \tag{213}$$

$$Ph^{\cdot} + RSSR \rightarrow PhH + RSSR(-H)^{\cdot} \tag{214}$$

Subsequently, Pryor, and Guard[234] added CCl_4 to the system and were thus able to measure k_{213} and k_{214} relative to the invariable rate constant k_4.

$$Ph^{\cdot} + CCl_4 \rightarrow PhCl + CCl_3^{\cdot} \tag{4}$$

Relative values of k_{213} obtained in this way are listed in Table VIII. Unfortunately k_4 has not been measured and so the k_{213} values cannot be put on an absolute basis.

Pryor and Smith[235] have repeated Pryor and Guard's work with phenyl radicals and report somewhat different relative reactivities (see Table VIII). However, both sets of data indicate that the highly reactive phenyl radical is more selective than the rather unreactive polystyryl radical, which seems rather surprising. Pryor and Smith have also measured the relative rate of attack of the 4-nitrophenyl radical on three disulfides (Table VIII). This radical is less selective than the phenyl radical.

Pryor and Guard[234] utilized relative k_{213} values (R = Me, Et, Pr^i, and Bu^t) to draw conclusions as to the stereochemistry of this S_H2 reaction. A log–log plot of k_{213} (relative) against the relative rates of S_N2 reactions on disulfides and structurally analogous carbon compounds gave a reasonably good linear correlation. Since the logarithm of a rate constant is proportional to energy it was concluded that similar energetic requirements affect the stereochemistry of both the S_H2 and the S_N2 reactions. The S_N2 reaction was believed to proceed by a Walden inversion, backside attack, mechanism[105,164] and it therefore seemed probable that S_H2 reactions should follow a similar reaction path.

[236] R. Gomer and G. B. Kistiakowsky, *J. Chem. Phys.*, **19,** 85 (1951)
[237] A. Shepp, *ibid.*, **24,** 939 (1956).

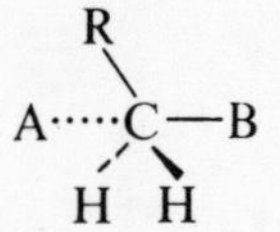

$$\begin{array}{c} R \\ \diagdown \\ A\cdots C{-}B \\ H\ \ H \end{array} \qquad S_N2$$

$$\begin{array}{c} R \\ \diagdown \\ A\cdots S{-}SR \end{array} \qquad S_N2,\ S_H2$$

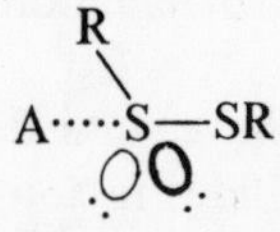

In order to ascertain the significance of a log-log correlation between the rate constants for two reactions in such an aliphatic series (i.e., methyl, isopropyl, *t*-butyl), Pryor and Smith[235] attempted the correlation of all the reactions for which data was available with the S_N2 reaction on carbon. Surprisingly, all of the reactions correlated with the S_N2 reaction at the 1% level of significance. Consequently, a good correlation in a log-log graph between two such reactions is *not* diagnostic of mechanism and even the magnitude of the slope of the graph is not significant. Two reactions which are known to have similar mechanisms may not correlate with a slope any closer to unity than do two reactions with quite different mechanisms. Nevertheless, Pryor and Smith[235] conclude that attack on sulfur, by nucleophiles or by radicals, usually occurs from the backside and that many of these substitutions probably occur in a stepwise, rather than a synchronous, manner.[238] That is, they conclude that S_H2 reactions at sulfur probably involve backside attack with the formation of a metastable species having 9 electrons around the sulfur.

Disulfides are also readily cleaved by vinyl radicals. Thus, Heiba and Dessau[239] have shown that the free-radical addition of disulfides to acetylenes gives high yields of the corresponding 1:1 adducts.

$$RS^{\cdot} + R'C{\equiv}CH \rightarrow R'\dot{C}{=}CHSR \qquad (215)$$

$$R'\dot{C}{=}CHSR + RSSR \rightarrow R'(RS)C{=}CHSR + RS^{\cdot} \qquad (216)$$

The corresponding addition to olefins resulted in poor yields of the 1:1 adducts presumably because the thiyl radical addition is reversible.[239] In addition, the alkyl radical formed from the olefin will be less reactive towards the disulfide than the vinyl radical formed from the acetylene. However, the photo-initiated (3660 Å) addition of di-*n*-butyl disulfide (0.5 mol) to vinyl

[238] It should, however, be noted that other workers have concluded that nucleophilic attack at sulfur is a synchronous process, see for example, L. Senatore, E. Ciuffarin, and A. Fava, *J. Amer. Chem. Soc.*, **92,** 3035 (1970).

[239] E. I. Heiba and R. M. Dessau, *J. Org. Chem.*, **32,** 3837 (1967).

acetate (0.1 mol) at room temperature has been reported to give the adduct in 30% yield.[240]

$$Bu^nSSBu^n + CH_3\overset{\overset{O}{\|}}{C}OCH{=}CH_2 \xrightarrow{h\nu} CH_3\overset{\overset{O}{\|}}{C}O\overset{\overset{SBu^n}{|}}{C}HCH_2SBu^n \quad (217)$$

The cleavage of disulfides by α-hydroxyalkyl radicals has been extensively investigated by Cohen and co-workers.[241–248]

$$>\dot{C}{-}OH + RSSR \rightarrow >C{=}O + RSH + RS^{\cdot} \quad (218)$$

This reaction is formally analogous to the alcohol induced decomposition of peroxides. Diaryl disulfides are more reactive than dialkyl disulfides. Their reactions with the α-hydroxyalkyl radicals produced when benzophenone or acetophenone are photolyzed in the presence of a secondary alcohol[242–245] (or ether[246]) has received detailed attention. The disulfide does not affect the rate of formation of the hydroxydiphenylmethyl radical but it accelerates its rate of disappearance.[247] The α-hydroxy alkyl radical may be oxidized back to the starting ketone by disulfide or it may be reduced to carbinol by the mercaptan formed in this reaction.

$$Ph_2\dot{C}OH + ArSSAr \rightarrow Ph_2C{=}O + ArSH + ArS^{\cdot} \quad (219)$$

$$Ph_2\dot{C}OH + ArSH \rightarrow Ph_2CHOH + ArS^{\cdot} \quad (220)$$

The mercaptan and disulfide are therefore equally effective in retarding the photo-reduction of the ketone since each is converted into the same mixture of the two during irradiation. The sulfur compounds are used repeatedly, each molecule negating the chemical action of many quanta. This work has been reviewed by Cohen[248] and by Kellogg.[249]

[240] K. Yamagishi, K. Araki, T. Suzuki, and T. Hoshino, *Bull. Chem. Soc. Japan*, **33,** 528 (1960).
[241] C. H. Wang and S. G. Cohen, *J. Amer. Chem. Soc.*, **81,** 3005 (1959).
[242] S. G. Cohen, S. Orman, and D. Laufer, *ibid.*, **84,** 1061, 3905 (1962).
[243] S. G. Cohen and W. V. Sherman, *ibid.*, **85,** 1642 (1963).
[244] S. G. Cohen, D. A. Laufer, and W. V. Sherman, *ibid.*, **86,** 3060 (1964).
[245] W. V. Sherman and S. G. Cohen, *ibid.*, **86,** 2390 (1964).
[246] S. G. Cohen and S. Aktipis, *Tetrahedron Lett.*, 579 (1965); *J. Amer. Chem. Soc.*, **88,** 3587 (1966).
[247] W. V. Sherman and S. G. Cohen, *J. Phys. Chem.*, **70,** 178 (1966).
[248] S. G. Cohen, in *Organosulfur Chemistry*, M. J. Janssen (Ed.), Interscience, New York, 1967, Ch. 3.
[249] R. M. Kellogg, *Methods in Free Radical Chem.*, **2,** 1 (1969).

The possibility that the reaction of thiocyanogen with phenyl metallic compounds (Ph_nM, M= Hg, B, etc.) may involve phenyl radical cleavage of the S–S bond, has already been mentioned several times.

$$Ph^{\cdot} + (SCN)_2 \rightarrow PhSCN + S\dot{C}N \quad (221)$$

$$S\dot{C}N + Ph_nM \rightarrow Ph_{n-1}MSCN + Ph^{\cdot} \quad (222)$$

The cleavage of disulfides by radicals with the unpaired electron centered on an atom other than sulfur or carbon has received little attention. Dimethyl disulfide is probably cleaved by hydrogen atoms in the gas phase.[231] Dialkyl and diaryl disulfides are rapidly cleaved in S_H2 processes by the tri-*n*-butyltin radical[198] and the triphenyltin radical.[250] A similar free-radical mechanism may possibly apply to their cleavage in the presence of tetralkyl diarsines ($R'_2AsAsR'_2$)[251,252] and dialkyl arsines[252] (R'_2AsH) to give compounds with arsenic–sulfur bonds (R'_2AsSR). However, an ionic mechanism seems rather more likely for these reactions, particularly at ambient temperatures.

The production of trifluoromethyl-thiodifluoramine by UV irradiation of bis(trifluoromethyl) disulfide and tetrafluorohydrazine,[253]

$$CF_3SSCF_3 + F_2NNF_2 \xrightarrow{h\nu} 2CF_3SNF_2 \quad (223)$$

may be due to the combination of $CF_3S^{\cdot}$ radicals with $F_2N^{\cdot}$ radicals or to an S_H2 reaction of $F_2N^{\cdot}$ radicals on the disulfide. The alternative of an S_H2 reaction of $CF_3S^{\cdot}$ on the hydrazine seems unlikely.

Absolute rate constants for the homolytic scission of some disulfides by the tri-*n*-butyltin radical[198] (k_{224}) are given in Table VIII.

$$Bu^n_3Sn^{\cdot} + RSSR \rightarrow Bu^n_3SnSR + RS^{\cdot} \quad (224)$$

With the exception of dibenzyl disulfide the relative reactivities of the disulfides towards the tributyltin radical and towards the polystyryl radical are quite similar in spite of the million-fold difference in absolute rate constants. An explanation for this similarity will have to await relative and absolute reactivity data for S_H2 reactions at disulfides by a wider variety of attacking radicals.

Homolytic cleavage of the S—S bond by phosphino radicals probably occurs in the reaction between alkyl alkylphosphinates and dialkyl-[253a] and diaryl[197,253b] disulfides.

[250] M. Pang and E. I. Becker, *J. Org. Chem.*, **29**, 1948 (1964).
[251] See footnote 163, Chapter VI.
[252] See footnote 164, Chapter VI.
[253] E. C. Stump, Jr., and C. D. Padgett, *Inorg. Chem.*, **3**, 610 (1964).
[253a] H. P. Benschop and D. H. J. M. Platenburg, *Chem. Comm.*, 1098 (1970).
[253b] L. P. Reiff and H. S. Aaron, *ibid.*, **92**, 5275 (1970).

$$(RO)MeP(=O)\cdot + PhSSPh \rightarrow (RO)MeP(=O)SPh + PhS\cdot \qquad (225)$$

$$(RO)MeP(=O)H + PhS\cdot \rightarrow PhSH + (RO)MeP(=O)\cdot \qquad (226)$$

U.V. irradiation of a benzene solution of alkyl methylphosphinate and diphenyl disulfide gave a product quantum yield of 18.6 indicating a relatively short chain length.[197] Reaction of optically active menthyl methylphosphinate with dimethyl disulfide gave a product in which the configuration at phosphorus was retained.[253a] Similarly, retention was observed in the cleavage of diphenyl disulfide by optically active isopropyl methylphosphinate.[253b] These results indicate that the phosphino radical is configurationally stable, at least as a transient intermediate.

Homolytic scission of S–S bonds extends to compounds such as thiosulfones,[254] for example,

$$PhCH_2\cdot + ArSSO_2Ar \longrightarrow PhCH_2SAr + PhCH_2SO_2Ar + \underset{\downarrow}{ArS\cdot} + \underset{\downarrow}{ArSO_2\cdot} \quad \text{dimers, etc.} \qquad (227)$$

The products do not necessarily imply that the benzyl radical can attack both the sulfide sulfur and the sulfone sulfur since the benzyl arylsulfone could be formed simply by a radical coupling reaction.

(d) SULFIDES

Homolytic substitution at sulfide sulfur is less frequently encountered than substitution at disulfides and polysulfides. It occurs most readily when the departing radical formed only a weak bond to sulfur. Thus, homolytic substitution tends to be faster with dibenzoyl sulfide or dibenzyl sulfide than with dibutyl sulfide.

Substitution by polysulfenyl or thiyl radicals requires rather high temperatures. For example,[255] diphenyl sulfone-^{35}S is reduced to diphenyl sulfide with a 75% loss of the original ^{35}S activity by heating with sulfur at 300°. The main reaction is not a reduction but is an S_H2 process with scission of the phenyl-sulfone bond being induced by polysulfenyl radicals. Because of the high temperatures required for homolytic substitutions by polysulfenyl and thiyl radicals other reactions such as hydrogen atom abstraction are also

[254] J. Degani, M. Tiecco, and A. Tundo, *Ann. Chim.* (*Rome*), **51,** 550 (1961).
[255] S. Ove and S. Kawamura, *Bull. Chem. Soc. Japan*, **36,** 163 (1963).

likely to take place. The complex product mixtures which may then result make the identification of an S_H2 reaction at the sulfur very difficult.

The homolytic substitution of sulfides by attacking carbon radicals has rarely been observed. Schmidt, Hochrainer, and Nikiforov[255a] have shown that 3-chlorobenzyl radicals (generated by the thermolysis or photolysis of the appropriate azo compound) attack dibenzyl sulfide to give benzyl-3-chlorobenzyl sulfide (13%) together with the three expected bibenzyl's, 3,3′-dichlorobibenzyl (76%), 3-chlorobibenzyl (9%) and bibenzyl (2%). These product yields indicate that the S_H2 reaction of 3-chlorobenzyl radicals with the sulfide is a facile process.

$$3\text{-}ClC_6H_4CH_2^{\cdot} + C_6H_5CH_2SCH_2C_6H_5 \longrightarrow 3\text{-}ClC_6H_4CH_2SCH_2C_6H_5 + C_6H_5CH_2^{\cdot} \quad (228)$$

Benzyl-3-chlorobenzyl sulfide was also formed by the reaction of benzyl radicals with bis(3-chlorobenzyl)sulfide. A tribenzyl sulfur species with nine electrons around sulfur must be an intermediate or transition state in these reactions. A triarylsulfur radical, $Ar_3S^{\cdot}$, which can break down reversibly to Ar_2S and $Ar^{\cdot}$, is believed to be formed during the photolysis of triarylsulfonium salts.[255b]

Kampmeir and Evans[256] have reported an S_Hi reaction in which phenyl substitutes for methyl in the photolysis of 2-iodo-2′-methylthiobiphenyl or in the free-radical reaction of this compound with tri-*n*-butyltin hydride.

$$\text{2-iodo-2'-(methylthio)biphenyl} \xrightarrow[Bu_3{}^nSn^{\cdot}]{h\nu} \text{2'-(methylthio)biphenyl-2-yl radical} \longrightarrow \text{dibenzothiophene} + CH_3^{\cdot} \quad (229)$$

The paucity of reports of S_H2 reactions at sulfide sulfur by carbon radicals is very surprising when it is realized that such reactions have apparently been

[255a] U. Schmidt, A. Hochrainer, and A. Nikiforov, *Tetrahedron Letters*, 3677 (1970).
[255b] J. W. Knapczyk and W. E. McEwen, *J. Org. Chem.*, **35**, 2539 (1970).
[256] J. A. Kampmeier and T. R. Evans, *J. Amer. Chem. Soc.*, **88**, 4096 (1966).

observed in compounds such as mercaptans[257] and sulfenyl halides[258] which might be expected to undergo terminal S_H2 reactions exclusively. The terminal reactions can certainly be important but the products indicate that nonterminal reactions at sulfur are also important. Thus, Greig and Thynne[257] found that in the reaction of $CH_3^{\cdot}$ radicals with CD_3SH in the gas phase at 139°–200°, CH_3SH was produced in addition to the expected CH_4 and CH_3D. The three reactions and their activation parameters and rate constants at 164° are listed below. The nonterminal substitution at sulfur and the terminal substitution at deuterium proceed at very similar rates.

	log A (M^{-1} sec^{-1})	E (kcal/mole)	$k_{164°}$ (M^{-1} sec^{-1})	
$CH_3^{\cdot} + CD_3SH \rightarrow CH_3SH + CD_3^{\cdot}$	7.73	7.6	8.5×10^3	(230)
$CH_3^{\cdot} + CD_3SH \rightarrow CH_4 + CD_3S^{\cdot}$	8.03	4.1	9.6×10^5	(231)
$CH_3^{\cdot} + CD_3SH \rightarrow CH_3D + {}^{\cdot}CD_2SH$	7.88	8.3	5.4×10^3	(232)

Free-radical chain reactions of alkanes with a number of compounds containing sulfur-chlorine bonds (e.g., sulfonyl chlorides and trichloromethanesulfenyl chloride) yield alkyl chlorides as the principal products.[259] However, the chain reaction with sulfur dichloride,[258] pentachlorobenzenesulfenyl chloride[260] and trifluoromethanesulfenyl chloride[259,261] can give both alkyl chlorides and sulfides.

$$R^{\cdot} + R'SCl \rightarrow RCl + R'S^{\cdot} \quad (233)$$

$$R^{\cdot} + R'SCl \rightarrow RSR' + Cl^{\cdot} \quad (234)$$

Thus, with CF_3SCl Harris reports[259] that primary alkyl radicals from toluene, *n*-butane, and isobutane give sulfides, $RSCF_3$, almost exclusively, only small amounts of the chloride, RCl, being formed.

$$CF_3SCl + CH_3Ph \xrightarrow{h\nu} \underset{70\%}{CF_3SCH_2Ph} + \underset{3\%}{ClCH_2Ph} + CF_3SSCF_3 + HCl \quad (235)$$

$$CF_3SCl + n\text{-}C_4H_{10} \xrightarrow{h\nu} \underset{13\%}{CF_3SC_4H_9\text{-}n} + \underset{46\%}{CF_3SCH(CH_3)C_2H_5} + \underset{12\%}{CH_3CHClCH_2CH_3} + \underset{1\%}{ClCH_2CH_2CH_2CH_3} + \underset{28\%}{CF_3SSCF_3} + HCl \quad (236)$$

[257] G. Greig and J. C. J. Thynne, *Trans. Faraday Soc.*, **62,** 379 (1966).
[258] E. Müller and E. W. Schmidt, *Chem. Ber.*, **96,** 3050 (1963).
[259] J. F. Harris, Jr., *J. Org. Chem.*, **31,** 931 (1966).
[260] N. Kharasch and Z. S. Ariyan, *Chem. Ind.* (*London*), 929 (1964).
[261] J. F. Harris, Jr., *J. Amer. Chem. Soc.*, **84,** 3148 (1962).

The secondary radicals from cyclohexane and *n*-butane form both sulfides and chlorides with the sulfide predominating.

$$CF_3SCl + C_6H_{12} \xrightarrow{h\nu} \underset{45\%}{CF_3SC_6H_{11}} + \underset{28\%}{C_6H_{11}Cl} + \underset{35\%}{CF_3SSCF_3} + HCl \quad (237)$$

The *t*-butyl radical from isobutane also gives both products but in this case the chloride predominates.

$$CF_3SCl + (CH_3)_3CH \xrightarrow{h\nu} \underset{24\%}{CF_3SC(CH_3)_3} + \underset{12\%}{CF_3SCH_2CH(CH_3)_2} + \underset{33\%}{(CH_3)_3CCl} + \underset{1\%}{(CH_3)_2CHCH_2Cl} + \underset{47\%}{CF_3SSCF_3} + HCl \quad (238)$$

These differences in product composition are presumably due to steric effects. The results suggest that alkyl radicals intrinsically prefer to attack the sulfur atom of CF_3SCl and the primary radicals do so almost exclusively. Secondary and tertiary alkyl radicals experience steric difficulties in attacking sulfur and they therefore tend to attack the more exposed chlorine atom. Steric factors may also account for the fact that cyclohexyl radicals attack the sulfur of Cl_3CSCl much less readily than they attack the sulfur of CF_3SCl. Cyclohexyl chloride is the only product from the former compound.[262,263]

The oxidation of alkyl radicals by metal ions in a high valency state can occur either by electron transfer to give a carbonium ion,

$$R^{\cdot} + M^{(n+1)+} \rightarrow R^{+} + M^{n+} \quad (239)$$

or by ligand transfer to give a substitution product directly.[264]

$$R + MX_m \rightarrow RX + MX_{(m-1)} \quad (240)$$

The relative importance of these two types of oxidation process depend not only on the metal and the ligands associated with it, but also on the solvent and on how easily the radical forms a carbonium ion. The majority of ligand transfers involve halogen atoms, for example, chlorine or bromine from Cu^{II} and Fe^{III} halides. However, thiocyanate, cyanide, and azide ligands can also be transferred to alkyl radicals from metal species such as Fe^{III}, Cr^{VI}, Pt^{IV}, and Pb^{IV}. The transfer of thiocyanate can be regarded as a non-terminal S_H2 reaction at sulfur which may involve fairly extensive charge separation in the

[262] V. Prey, E. Gutschik, and H. Berbalk, *Monatsch. Chem.*, **91**, 556 (1960).
[263] H. Kloosterziel, *Rec. Trav. Chim.*, **82**, 497, 508 (1963).
[264] For a general review of these classes of reactions see, J. K. Kochi, *Science*, **155**, 415 (1967).

transition state,[265] for example,

$$HO\overset{\overset{\large O}{\|}}{C}(CH_2)_4CH_2^{\cdot} + Fe(SCN)_6^{3-} \longrightarrow HO\overset{\overset{\large O}{\|}}{C}(CH_2)_4CH_2S{\equiv}CN \qquad (241)$$

Gritter and Carey[265a] found that a small yield (1%) of diphenyl sulfoxide was formed on heating diphenyl sulfide with di-*t*-butyl peroxide at 150°. This was taken as evidence for a reaction analogous to the *t*-butoxy radical-triphenyl phosphine reaction, and for the intermediacy of a sulfonium radical, that is,

$$Bu^tO^{\cdot} + PhSPh \longrightarrow \left[\begin{array}{c} Ph{-}\underset{|}{S}{-}Ph \\ Bu^t{-}O \end{array}\right]^{\cdot} \longrightarrow Ph{-}\underset{\underset{O}{\downarrow}}{S}{-}Ph + Bu^{t\cdot} \qquad (242)$$

Reaction of this sulfide with *t*-butyl hypochlorite gives much higher yields of diphenyl sulfoxide but the reaction does not involve free *t*-butoxy radicals.[265b] Alkyl sulfides and disulfides react with *t*-butoxy radicals at room temperature to form radicals in which a hydrogen has been removed from the carbon which is bonded to sulfur.[265c] Oxygen atom transfer from peroxy radicals to sulfides or disulfides appears to be a more facile process than transfer from alkoxy radicals[265d]

$$ROO^{\cdot} + R'SR'' \longrightarrow \left[\begin{array}{c} R'{-}\underset{|}{S}{-}R'' \\ ROO \end{array}\right]^{\cdot} \longrightarrow R'\underset{\underset{O}{\downarrow}}{S}R'' + RO^{\cdot} \qquad (243)$$

Although hydroxy radicals have not been reported to enter into S_H2 reactions with sulfide sulfur they do react with sulfoxides to displace an alkyl radical. Thus, methyl radicals have been directly detected by e.s.r. spectroscopy in the reaction of hydroxy radicals with dimethyl sulfoxide.[265e]

$$HO^{\cdot} + Me\underset{\underset{O}{\downarrow}}{S}Me \longrightarrow \left[\begin{array}{c} H \\ O \\ MeSMe \\ | \\ \dot{O} \end{array}\right] \longrightarrow Me\underset{\underset{O}{\downarrow}}{S}OH + Me^{\cdot} \qquad (244)$$

This reaction has been examined in more detail by Lagercrantz and Forshult[265f] who used nitroso compounds to trap the alkyl radicals produced

[265] H. E. De La Mare, J. Kochi, and F. Rust, *J. Amer. Chem. Soc.*, **85**, 1437 (1963)
[265a] R. J. Gritter and D. J. Carey, *J. Org. Chem.*, **29**, 1160 (1964).
[265b] C. Walling and M. J. Mintz, *ibid.*, **32**, 1286 (1967).
[265c] J. Q. Adams, *J. Amer. Chem. Soc.*, **92**, 4535 (1970).
[265d] See for example, K. R. Hargrave, *Proc. Roy. Soc.* (*London*), **A235**, 55 (1956); A. Rahman and A. Williams, *J. Chem. Soc.* (*B*), 1391 (1970).
[265e] W. T. Dixon, R. O. C. Norman, and A. L. Buley, *J. Chem. Soc.*, 3625 (1964).
[265f] C. Lagercrantz and S. Forshult, *Acta. Chem. Scand.*, **23**, 811 (1969).

and e.s.r. to identify the resultant nitroxide radicals.

$$R^{\cdot} + R'NO \longrightarrow R'N(\dot{O})R \qquad (245)$$

The hydroxy radicals were generated by the photolysis of hydrogen peroxide. Methyl, trideuteromethyl, ethyl, propyl, isopropyl, butyl, isobutyl, *s*-butyl, pentyl, hexyl, and benzyl radicals were formed from the corresponding symmetric sulfoxides. Analogously, the unsymmetric sulfoxides, methyl ethyl sulfoxide, methyl propyl sulfoxide, and methionine sulfoxide ($CH_3SOCH_2CH(NH_2)COOH$), gave both possible radicals. Tetrahydrothiophene sulfoxide gave a spectrum consistent with a spin adduct derived from cleavage of the alicyclic ring.

HO˙ + (tetrahydrothiophene S→O) ⟶ [ring S(HO)(O˙)] ⟶ ˙CH₂CH₂CH₂CH₂S(=O)OH (246)

No radicals could be detected in the reaction with diallyl sulfoxide, diphenyl sulfoxide, and bis(*p*-chlorophenyl) sulfoxide. Methyl phenyl sulfoxide gave only methyl radicals. Analogous S_H2 reactions at sulfur do not occur with sulfones though some sulfones gave radicals as a result of a hydrogen atom abstraction from the alkyl group by the hydroxyl radical.

Triphenyltin hydride reacts with alkyl sulfides and mercaptans to give organo-tin sulfides.[250] Rather high temperatures are required but the reactions are catalyzed by sources of free-radicals and presumably proceed by an S_H2 process, the driving force for the reaction coming from the very strong tin-sulfur bond. For example,[250]

$$Ph_3SnH + PhCH_2SCH_2Ph \xrightarrow[3\ hrs]{130^\circ} Ph_3SnSCH_2Ph + PhCH_3 \qquad (247)$$

(27% reaction uncat., 42% cat.)

$$Ph_3SnH + PhCH_2SH \xrightarrow[5\ hrs]{85^\circ} Ph_3SnSSnPh_3 + H_2S + PhCH_3 \qquad (248)$$

(0% reaction uncat., 60% cat.)

The photo-induced reaction of tributyltin hydride with benzyl mercaptan at 30° is much slower than the reaction with any disulfide.[198] The rate constant for reaction 249 was estimated to be $\sim 2 \times 10^4\ M^{-1}\ sec^{-1}$.

$$n\text{-}Bu_3Sn^{\cdot} + PhCH_2SH \rightarrow n\text{-}Bu_3SnSH + PhCH_2^{\cdot} \qquad (249)$$

Tributyltin hydride did not react with dibenzyl sulfide at 30° at a measurable rate.[198]

Tri-organotin hydrides induce the decomposition of aryl arylazo sulfides and sulfones.[142,266] Reaction with the sulfides is probably an S_H2 process at sulfur,

$$R_3Sn^{\cdot} + ArN{=}NSAr' \xrightarrow{50^\circ} R_3SnSAr' + N_2 + Ar^{\cdot} \tag{250}$$

However, a rate controlling electron transfer reaction is also possible.

$$R_3Sn^{\cdot} + ArN{=}NSAr' \rightarrow [R_3Sn^+ + Ar^{\cdot} + N_2 + {}^-SAr'] \rightarrow R_3SnSAr' + N_2 + Ar^{\cdot} \tag{251}$$

The products of reaction with the sulfone indicate an attack on oxygen or on the S=O bond.

$$R_3Sn^{\cdot} + ArN{=}N\overset{\overset{O}{\|}}{\underset{\underset{O}{\|}}{S}}Ar' \longrightarrow R_3SnO\overset{\overset{O}{\|}}{S}Ar' + N_2 + Ar^{\cdot} \tag{252}$$

The tin hydride can also add across the N=N bond to give

$$ArNHN(SnR_3)SO_2Ar'.$$

The azo-sulfides are thermally unstable at 50° and decompose slowly in dilute solution ($\leqslant 0.25\ M$).

$$ArN{=}NSAr' \rightarrow Ar^{\cdot} + N_2 + Ar'S^{\cdot} \tag{253}$$

In more concentrated solutions the rate of decomposition is accelerated, presumably because of an S_H2 process initiated by aryl or arylthiyl radicals. In dilute solutions the rate of decomposition of the sulfide is accelerated by aryl mercaptan and iodine as well as by the tin hydrides. It is possible that the decomposition is a chain reaction in these cases also. That is, the aryl radical reacts with the added material to yield a radical ($ArS^{\cdot}$ or $I^{\cdot}$) which then enters into an S_H2 reaction at the sulfur atom of the azo-sulfide and, thus, regenerates the aryl radical. The products establish this mechanism for the tin hydride reaction, but it has not been established for the mercaptan and iodine catalyzed decompositions.

The desulfuration of organic sulfides with Raney nickel catalyst usually involves the formation of free radicals.[267,268] This reaction may be regarded as a

[266] W. P. Neumann, H. Lind, and G. Alester, *Chem. Ber.*, **101,** 2845 (1968).

[267] W. A. Bonner and R. A. Grimm, in *The Chemistry of Organic Sulfur Compounds*, N. Kharasch and C. Y. Meyers, (Eds.) Pergamon Press, 1966, vol. 2, Ch. 2 and pp. 410–413 and references cited therein.

[268] For a possible exception see, A. C. Cope and J. E. Engelhart, *J. Org. Chem.*, **34,** 3199 (1969).

homolytic substitution by nickel accompanied, or followed, by scission of the nickel alkyl sulfide to give a second free-radical

$$RSR' + RaNi \longrightarrow \underset{Ni}{RS\cdot R'} \longrightarrow \underset{Ni\quad Ni}{R\cdot S\cdot R'} \longrightarrow R_2, RR', R'_2 \text{ etc.} \qquad (254)$$

Desulfuration by other metallic catalysts may proceed by a similar mechanism.

C. SELENIUM, TELLURIUM, AND POLONIUM

Although S_H2 reactions at the three heaviest atoms in Group VIA are expected to occur with great facility the rather sparse chemistry of these elements provides few examples of such processes.

The inherent similarity in the behavior of selenium to that of sulfur is illustrated by the existence of long-chain polymers in the liquid state of both elements. The occurrence of polymer is shown, for example, by the high viscosity of both liquids. There is, however, a profound difference in the manner of appearance of the polymer in the liquid state. As has already been described, sulfur melts (at 112.8° or 115.2° depending on its modification) to a liquid of relatively low viscosity consisting almost exclusively of S_8 rings. At ~160° the viscosity rises sharply due to the appearance of long-chain material. In contrast to crystalline sulfur, hexagonal (metallic) selenium consists of linear, long-chain, material so that it is not surprising for polymer to appear at the melting point (217°). Other selenium modifications which consist in whole or in part of Se_8 rings also yield polymeric material on melting and the viscosity of liquid selenium decreases uniformly as the temperature is increased. Therefore, in contrast to sulfur, no liquid region exists in which selenium is present exclusively as Se_8 rings and the "transition" temperature of selenium must lie below its melting point.

Although the equilibria of the various species which may be present in liquid selenium are not as well established as are those for sulfur,[269] Eisenberg, and Tobolsky[270] have utilized the theory they developed for sulfur to estimate thermodynamic quantities for two of the more probable equilibria, namely,

$$Se_8 \text{ (ring)} \rightleftharpoons \cdot Se_8\cdot \text{ (diradical)} \qquad (255)$$

$$\Delta H^\circ = 25{,}000 \text{ cal/mole} \qquad \Delta S^\circ = 23.0 \text{ cal/deg/mole}$$

[269] See for example, P. I. Sampath, *J. Chem. Phys.*, **45,** 3519 (1966); I. Chen, *ibid.*, **45,** 3536 (1966); G. B. Abdullaev, N. I. Ibragimov, Sh. V. Mamedov, and T. Ch. Dzhuvarly, *Phys. Status Solidi.*, **16,** K 113 (1966); M. Abkowitz, *J. Chem. Phys.*, **46,** 4537 (1967).
[270] A. Eisenberg and A. V. Tobolsky, *J. Polymer Sci.*, **46,** 19 (1960).

and,

$$\cdot Se_8^{\cdot} + Se_8\ (\text{ring}) \rightleftharpoons \cdot Se_{16}^{\cdot}\ (\text{diradical}) \qquad (256)$$

$$\Delta H^\circ = 2{,}270\ \text{cal/mole} \qquad \Delta S^\circ = 5.47\ \text{cal/deg/mole}$$

These thermodynamic quantities are similar to the values estimated for the analogous equilibria in sulfur.

Alkyl polyselenides undergo redistribution reactions reminiscent of the reactions of alkyl polysulfides. Thus,[181] dimethyldiselenide mixed with dichloro diselenide rapidly yields CH_3Se_4Cl. This compound then undergoes a sequence of condensation plus scrambling reactions until longer chains reorganize to give elemental selenium, leaving the terminal substituents (with a selenium atom still attached to methyl) in short chain molecules. In the range of equimolar proportions of the reagents the precipitated selenium consists of long spiralled chains, while for ratios of the starting reagents quite far from the equilibrium mixture the onset of precipitation is slower and the precipitate is Se_8 rings. Similar reactions occur with mixtures of dimethyl selenide and dichloro diselenide.[181] It seems likely that the growth of the selenium chains may involve a complex set of free-radical reactions in which Se–Se bonds are cleaved by attacking polyselenyl radicals.

$$RSe_n^{\cdot} + RSeSe_mR \rightleftharpoons RSe_{n+m}R + RSe^{\cdot} \qquad (257)$$

Selenium has been incorporated in polymerizing styrene by carrying out the reaction at 140° in the presence of diphenyl selenide.[271] Although an S_H2 reaction at selenium may be involved, this will be thermodynamically disfavored. It seems more likely that the polystyryl radical attacks one of the phenyl rings of the selenide.

$$M_n^{\cdot} + C_6H_5{-}Se{-}C_6H_5 \longrightarrow (H)(M_n)C_6H_5^{\cdot}{-}Se{-}C_6H_5 \qquad (258)$$

$$M_n^{\cdot} + (H)(M_n)C_6H_5^{\cdot}{-}Se{-}C_6H_5 \longrightarrow M_nH + M_n{-}C_6H_4{-}Se{-}C_6H_5 \qquad (259)$$

[271] E. E. Baroni, S. F. Kilin, T. N. Lebsadze, I. M. Rozman, and V. M. Shoniya, *At. Energy* (*USSR*), **17**, 497 (1964); *Chem. Abstr.*, **62**, 10604h (1965).

VIII S_H2 Reactions at Group VIIA Elements

Many organic halogen compounds readily undergo S_H2 reactions at the halogen atom. However, in most of their compounds the halogen is a terminal atom and so these reactions will not be discussed. There does not appear to be an example of an S_H2 reaction at a non-terminal halogen.

IX S_H2 Reactions at Transition Elements

Free radicals can interact with transition metal compounds in a number of different ways, and the diversity of behavior may be ascribed to the readily variable oxidation state of the transition metal.

Electron transfer[1] or ligand transfer[1] may occur as exemplified in equations 1 and 2, respectively.

$$CH_3\dot{C}H_2 + Cu^{2+}aq. \rightarrow CH_2{=}CH_2 + Cu^+ + H^+ \quad (1)$$

$$CH_3\dot{C}H_2 + CuCl_2 \rightarrow CH_3CH_2Cl + CuCl \quad (2)$$

Reaction 2 may be thought of as an S_H2 attack on (abstraction of) chlorine. The attacking radical may become permanently attached to the metal atom to form a metal–carbon σ-bond.[2–5]

$$R^{\cdot} + [Co(CN)_5]^{3-} \rightarrow [RCo(CN)_5]^{3-} \quad (3)$$

$$R^{\cdot} + Cr^{2+}aq. \rightarrow RCr^{2+}aq. \quad (4)$$

Attack of a radical on the transition metal to displace a second radical represents the S_H2 process 5 which is difficult to distinguish from electron transfer from an ionic ligand to the incoming radical, 6.

$$Y^{\cdot} + MX \rightarrow YM + X^{\cdot} \quad (5)$$

$$Y^{\cdot} + M^+X^- \rightarrow Y^-M^+ + X^{\cdot} \quad (6)$$

Minisci and Galli[6] have reported the addition of two azido groups across an olefinic double bond, using the decomposition of hydrogen peroxide by

[1] See footnote 264, Chapter VII.
[2] J. Kwiatek and J. K. Seyler, *J. Organometal. Chem.*, **3**, 421 (1965).
[3] J. Halpern and J. P. Maher, *J. Amer. Chem. Soc.*, **87**, 5361 (1965).
[4] C. E. Castro and W. C. Kray, Jr., *ibid.*, **85**, 2768 (1963).
[5] J. K. Kochi and D. D. Davis, *ibid.*, **86**, 5264 (1964).
[6] F. Minisci and R. Galli, *Tetrahedron Lett.*, 533 (1962); 357 (1963).

ferrous sulphate in the presence of sodium azide. The proposed mechanism is shown in equations 7–10.

$$Fe^{2+} + H_2O_2 \longrightarrow FeOH^{2+} + HO^{\cdot} \tag{7}$$

$$HO^{\cdot} + FeN_3^{2+} \longrightarrow FeOH^{2+} + N_3^{\cdot} \tag{8}$$

$$N_3^{\cdot} + \rangle C{=}C\langle \longrightarrow N_3{-}\overset{|}{\underset{|}{C}}{-}\overset{|}{\underset{|}{C}}\cdot \tag{9}$$

$$N_3{-}\overset{|}{\underset{|}{C}}{-}\overset{|}{\underset{|}{C}}\cdot + FeN_3^{2+} \longrightarrow N_3{-}\overset{|}{\underset{|}{C}}{-}\overset{|}{\underset{|}{C}}{-}N_3 + Fe^{2+} \tag{10}$$

Reaction 8 could be a homolytic displacement at iron, but an electron transfer process of the type shown in 6 is perhaps more likely.

Ferrocene reacts with benzoyl peroxide in benzene to give ferric benzoate in high yield.[7] The intermediacy of the ferricinium ion was established and it was proposed that whilst this ion was highly reactive towards free radicals ferrocene itself was rather inert.

$$(C_5H_5)_2Fe + PhCO_2^{\cdot} \longrightarrow (C_5H_5)_2Fe^+ + PhCO_2^- \tag{11}$$

$$(C_5H_5)_2Fe^+ \xrightarrow{PhCO_2^{\cdot}} (PhCO_2)_3Fe \tag{12}$$

The mechanism of reaction 12 was not discussed but it might involve attack of the benzoyl radical at iron to expel a cyclopentadienyl radical.

There are a large number of examples of compounds containing alkyl or aryl groups σ-bonded to transition metals,[8,9] and the reactions of these organometallic compounds with free radicals might be expected to provide examples of S_H2 processes occurring at the metal center. However, no such reaction has been studied mechanistically.

Transition metal organometallic compounds are usually subject to rapid autoxidation,[8,9] a reaction which may involve homolytic substitution by alkylperoxy and alkoxy radicals at the metal center. For example, many organotitanium compounds are very readily autoxidized to products in which

[7] A. L. J. Beckwith and R. J. Leydon, *ibid.*, 385 (1963).
[8] G. A. Razuvaev and V. N. Latyaeva, *Uspekhi Khimii*, **34**, 585 (1965).
[9] M. L. H. Green, *Organometallic Compounds*, vol. II, Methuen, London, 1968.

an oxygen atom has been inserted into the Ti–C bond.[10] By analogy with the mechanism of autoxidation of other organometallic compounds it seems reasonable that these reactions should involve free-radical chain processes with the initial products formed by an S_H2 reaction of a peroxy radical at titanium. Presumably these peroxides are then reduced by unreacted starting material, that is,

$$ROO^{\cdot} + RTiX_3 \rightarrow ROOTiX_3 + R^{\cdot} \quad (13)$$

$$ROOTiX_3 + RTiX_3 \rightarrow 2ROTiX_3 \quad (14)$$

Organotitanium compounds containing two cyclopentadienyl rings are less susceptible to autoxidation than the compounds containing one cyclopentadienyl ring or no such ring. Thus, for example, *bis*-(π-cyclopentadienyl)-dimethyltitanium and *bis*-(π-cyclopentadienyl)methyltitanium chloride are stable to oxygen at room temperature.[10] However, *bis*-(π-cyclopentadienyl)-diphenyltitanium, although stable to oxygen in the solid state, is autoxidized in solution at temperatures below its temperature of decomposition[11,12] (60–80°). The rate of autoxidation depends to a marked extent on the solvent in which the reaction is conducted.[11] In CCl_4 at 45–55° the products were *bis*-(π-cyclopentadienyl)titanium dichloride together with phosgene, chlorobenzene, biphenyl, and some phenol. In isopropanol the oxidation yields acetone (1.1 moles/mole of titanium compound), phenol (0.8 moles/mole) and traces of benzene and biphenyl. In this solvent autoxidation occurs readily even at 0°. *Bis*-(π-cyclopentadienyl)diphenyltitanium also reacts rapidly with benzoyl peroxide to yield *bis*-(π-cyclopentadienyl)titanium dibenzoate.[10] It is possible that this reaction involves homolytic substitution at titanium of benzoyloxy for phenyl radicals.

Cyclopentadienyltrimethyltitanium, tetramethyltitanium and tetraphenyltitanium react very rapidly with oxygen at room temperature and may ignite spontaneously.[10] Phenyltripropoxytitanium and phenyltributoxytitanium autoxidize readily to give benzene, biphenyl, and phenoxytitanium compounds which yield phenol on hydrolysis.[15]

$$PhTi(OR)_3 \xrightarrow{O_2} PhOTi(OR)_3 \xrightarrow{H_2O} PhOH \quad (15)$$

[10] See footnote 1, Chapter III.

[11] G. A. Razuvaev, V. N. Latyaeva, and L. I. Vyshinskaya, *Zhur. Obshch. Khim.*, **31,** 2667 (1961); *Trudy Khim. i Khim. Tekhnol*, **4,** 616 (1962); *Chem. Abstr.*, **58,** 2463h (1963).

[12] The thermal decomposition of bis-(π-cyclopentadienyl) diphenyltitanium has recently been shown to be a nonradical reaction,[13] rather than a radical process as had previously been supposed.[11,14]

[13] J. Dvorak, R. J. O'Brien, and W. Santo, *Chem. Comm.*, 411 (1970).

[14] G. A. Razuvaev, V. N. Latyaeva, L. I. Vyshinskaya, and G. A. Kilyakova, *Zhur. Obshch. Khim.*, **36,** 1491 (1966).

[15] D. F. Herman and W. K. Nelson, *J. Amer. Chem. Soc.*, **74,** 2693 (1952); **75,** 3877 (1953)

Alkyltitanium trichlorides are also readily autoxidized. Methyltitanium trichloride in hydrocarbon solutions absorbs exactly 0.5 mol oxygen to give methoxytitanium trichloride.[16] Hydrolysis of autoxidized ethyltitanium trichloride yields ethanol.[17]

The organometallic compounds of other transition elements are also frequently subject to rapid autoxidation. However, in no case has a free-radical chain mechanism been demonstrated. It is possible that some of these autoxidations are nonchain S_H1 reactions in which cleavage of the metal–carbon bond is rate controlling. It would be informative if workers who prepare complex organometallic compounds which are found to be unstable in air would, in the future, check whether the stability of their compounds is at all affected by the addition of galvinoxyl or other potent antioxidant. A relatively simple experiment would determine whether or not the new organometallic compound was oxidized by a free-radical chain process.

The transition metal–carbon bond is often readily cleaved by halogens to give the metal- and organic-halides. For example, diethylgold bromide reacts with bromine in carbon tetrachloride to give ethylgold dibromide.[18]

$$Et_2AuBr + Br_2 \rightarrow EtAuBr_2 + EtBr \qquad (16)$$

By analogy with the mechanism established for halogenodealkylation of organomercury compounds (see Chapter III), reaction 16 could involve displacement of an ethyl radical from gold by a bromine atom as a propagation step in a chain mechanism.

$$Br^{\cdot} + Et_2AuBr \rightarrow EtAuBr_2 + Et^{\cdot} \qquad (17)$$

$$Et^{\cdot} + Br_2 \rightarrow EtBr + Br^{\cdot} \qquad (18)$$

Gilman and Woods[19] noted that thiophenol cleaved the Au–C bond in trimethylgold while the more acidic phenol or trichloracetic acid did not.

$$Me_3Au + PhSH \rightarrow Me_2AuSPh(\text{dimeric}) + MeH \qquad (19)$$

This anomalous result was compared with the ready cleavage of organobismuth compounds by thiophenol—a free-radical chain reaction (see Chapter VI). It is possible that reaction 19 involves attack by the phenylthiyl radical at gold to displace a methyl radical.

$$PhS^{\cdot} + Me_3Au \rightarrow Me_2AuSPh + Me^{\cdot} \qquad (20)$$

[16] C. Beerman and H. Bestian, *Angew. Chem.*, **71,** 618 (1959).
[17] C. E. Bawn and J. Gladstone, *Proc. Chem. Soc.*, 227 (1959).
[18] A. Buraway and C. S. Gibson, *J. Chem. Soc.*, 860 (1934).
[19] H. Gilman and L. A. Woods, *J. Amer. Chem. Soc.*, **70,** 550 (1948).

X S_H2 Reactions at Rare Earth Elements

Little is known about the interaction of free radicals with compounds of the rare earth elements, and there are no authenticated examples of S_H2 reactions which occur at the metal center. The few organometallic compounds of these elements that are known[1] are air sensitive, as would be expected, and autoxidative cleavage of the metal–carbon bond clearly might involve homolytic substitution by oxygen-centered radicals.

To account for a quantum yield for formation of carbon dioxide of slightly greater than one in the photo-induced decarboxylation of pivalic acid by cerium (IV) salts, Sheldon and Kochi[2] proposed reaction 2 as a step in a possible chain reaction.

$$RCO_2^{\cdot} \rightarrow R^{\cdot} + CO_2 \tag{1}$$

$$R^{\cdot} + Ce^{IV}O_2CR \rightarrow RCe^{IV} + RCO_2^{\cdot} \tag{2}$$

Alternative explanations were also suggested, and even if 2 does occur it may proceed by electron transfer from the carboxylate ligand and not involve S_H2 attack on the metal.

[1] F. A. Hart, A. G. Massey, and M. Singh Saran, *J. Organometal. Chem.*, **21**, 147 (1970).
[2] R. A. Sheldon and J. K. Kochi, *J. Amer. Chem. Soc.*, **90**, 6688 (1968).

XI S_H2 Reactions at Group O Elements

Since the discovery in 1962 of the first chemical compounds of xenon stable at room temperature,[1] a large number of papers have been published describing the preparation and properties of compounds of the noble gases. The occurrence of a homolytic substitution process would require both attacking and displaced radicals to be capable of forming stable bonds to the noble gas atom, and it is not surprising that no examples of an S_H2 reaction at a Group O atom are known. In contrast, an S_H2 reaction at an atom bound to a Group O atom would seem more probable, for example,

$$Me^{\cdot} + XeF_6 \rightarrow MeF + [XeF_5^{\cdot}] \tag{1}$$

[1] N. Bartlett, *Proc. Chem. Soc.*, 218 (1962).

Author Index